Optimal Control

Weakly Coupled Systems
and Applications

AUTOMATION AND CONTROL ENGINEERING

A Series of Reference Books and Textbooks

Series Editors

FRANK L. LEWIS, Ph.D.,
FELLOW IEEE, FELLOW IFAC

Professor
Automation and Robotics Research Institute
The University of Texas at Arlington

SHUZHI SAM GE, Ph.D.,
FELLOW IEEE

Professor
Interactive Digital Media Institute
The National University of Singapore

Optimal Control

Weakly Coupled Systems and Applications

Zoran Gajić
Rutgers University
Piscataway, New Jersey, U.S.A.

Myo-Taeg Lim
Korea University
Seoul, South Korea

Dobrila Škatarić
Belgrade University
Belgrade, Serbia

Wu-Chung Su
National Chung-Hsing University
Taichung, Taiwan, Republic of China

Vojislav Kecman
Virginia Commonwealth University
Richmond, Virginia, U.S.A.

CRC Press
Taylor & Francis Group
Boca Raton   London   New York

CRC Press is an imprint of the
Taylor & Francis Group, an **informa** business

CRC Press
Taylor & Francis Group
6000 Broken Sound Parkway NW, Suite 300
Boca Raton, FL 33487-2742

© 2009 by Taylor & Francis Group, LLC
CRC Press is an imprint of Taylor & Francis Group, an Informa business

No claim to original U.S. Government works
Printed in the United States of America on acid-free paper
10 9 8 7 6 5 4 3 2 1

International Standard Book Number-13: 978-0-8493-7429-6 (Hardcover)

Visit the Taylor & Francis Web site at
http://www.taylorandfrancis.com

and the CRC Press Web site at
http://www.crcpress.com

Contents

PART I *Recursive Approach for Linear Weakly Coupled Control Systems*

PART II   Hamiltonian Approach for Linear Weakly Coupled Control Systems

PART III   Bilinear Weakly Coupled Control Systems

Preface

This book is intended for engineers, mathematicians, physicists, and computer scientists interested in control theory and its applications. It describes a special class of linear and bilinear control systems known as weakly coupled systems. These systems, characterized by the presence of weak coupling among subsystems, describe dynamics of many real physical systems such as chemical plants, power systems, aircraft, satellites, machines, cars, and computer/communication networks.

Weakly coupled control systems have become an extensive area of research since the end of the 1960s when the original papers of Professor Kokotović and his coworkers and graduate students were published. A relatively large number of journal papers on weakly coupled control systems were published from the 1970s through the 1990s. The approaches taken during the 1970s and 1980s were based on expansion methods (power series, asymptotic expansions, and Taylor series). These approaches were in most cases accurate only with an $O(\varepsilon^2)$ accuracy, where ε is a small, weak coupling parameter. Generating high-order expansions for these methods has been analytically cumbersome and numerically inefficient, especially for higher dimensional control systems. Moreover, for some applications it has been demonstrated in the control literature that $O(\varepsilon^2)$ accuracy is either not satisfactory or in some cases has not solved weakly coupled control problems.

The development of high-accuracy efficient techniques for weakly coupled control systems began at the end of the 1980s in the published papers of Professor Gajić and his graduate students and coworkers. The corresponding approach was recursive in nature and based on fixed-point iterations. In the early 1990s, the fixed-point recursive approach culminated in the so-called Hamiltonian approach for the exact decomposition of weakly coupled, linear-quadratic, deterministic and stochastic, optimal control, and filtering problems. In the new millennium, Professor Kecman developed the generalized Hamiltonian approach based on the eigenvector method. At the same time, Professor Mukaidani and his coworkers discovered a new approach for studying various formulations of optimal linear, weakly coupled control systems.

This book represents a comprehensive overview of the current state of knowledge of both the recursive approach and the Hamiltonian approach to weakly coupled linear and bilinear optimal control systems. It devises unique powerful methods whose core results are repeated and slightly modified over and over again, while the methods solve more and more challenging problems of linear and bilinear weakly coupled, optimal, continuous- and discrete-time systems. It should be pointed out that some related problems still remain unsolved, especially corresponding problems in the discrete-time domain, and the optimization problems over a finite horizon. Such problems are identified as open problems for future research.

The presentation is based on the research work of the authors and their coworkers. The book presents a unified theme about the exact decoupling of the corresponding optimal control problems and decoupling of the nonlinear algebraic

Riccati equation into independent, reduced-order, subsystem-based algebraic Riccati equations.

Each chapter is organized to represent an independent entity so that readers interested in a particular class of linear and bilinear weakly completed control systems can find complete information within a particular chapter. The book demonstrates theoretical results on many practical applications using examples from aerospace, chemical, electrical, and automotive industries. To that end, we apply theoretical results obtained from optimal control and filtering problems represented by real mathematical models of aircraft, power systems, chemical reactors, and so on.

The authors are thankful for support and contributions from their colleagues, Professors S. Bingulac, H. Mukaidani, D. Petkovski, B. Petrović, N. Prljaca, and X. Shen, and Drs. D. Aganović, I. Borno, Y.-J. Kim, M. Qureshi, and V. Radisavljević.

Zoran Gajić
Myo-Taeg Lim
Dobrila Skatarić
Wu-Chung Su
Vojislav Kecman

1 Introduction

This book is intended for engineers, mathematicians, physicists, and computer scientists interested in control theory and its applications. It studies special classes of linear and bilinear control dynamic systems known as weakly coupled systems. These systems, characterized by the presence of small parameters causing weak connections among subsystems, represent many real physical systems such as absorption columns, catalytic crackers, chemical plants, chemical reactors, helicopters, satellites, flexible beams, cold-rolling mills, power systems, electrical circuits, large space flexible structures, computer/communication networks, paper making machines, etc. The techniques presented show how to study independently, from the subsystem level perspective and with a high accuracy deterministic and stochastic, continuous- and discrete-time, optimal control and filtering problems for the considered class of systems.

Each chapter is organized to represent an independent entity so that readers interested in a particular class of weakly coupled control systems can find complete information within the particular chapter. The book demonstrates theoretical results on many practical applications using examples from aerospace, chemical, electrical, and automotive industries.

This book presents reduced-order (subsystem level) algorithms and techniques for optimal control of weakly coupled linear and bilinear dynamic systems composed, in general, of n subsystems. For the reason of simplicity, at many places we consider only two weakly coupled subsystems. The book is written in the spirit of parallel and distributed computations (Bertsekas and Tsitsiklis 1989, 1991) and parallel processing of information in terms of reduced-order controllers and filters (Gajić et al. 1990; Gajićc and Shen 1993; Aganović and Gajić 1995; Gajić and Lim 2001). It covers almost all important aspects of optimal control theory in the context of continuous and discrete, deterministic and stochastic weakly coupled linear systems, and major aspects of optimal control theory of bilinear weakly coupled systems.

The material considered in the book is mostly based on the authors' research accomplishments during the last 20 years, which resulted in many journal and conference papers and three monographs (Gajić et al. 1990; Gajić and Shen 1993; Aganović and Gajić 1995) on analysis and synthesis of optimal controllers and filters for weakly coupled control systems. Consequently, the material presented in this monograph in an integral part of all our previous publications. It also represents extensions, improvements, corrections, new ideas, and overviews of all our previous work on weakly coupled control systems.

The initial idea of weak coupling dealing with eigenvalues and eigenvectors of a weakly coupled system matrix can be found in the work of Milne (1965). The linear weakly coupled control systems were introduced to the control audience by Professor Petar Kokotović in 1969 (Kokotović et al. 1969; see also Kokotović 1972), and since then they have been studied in different setups by many well-respected control engineering researchers, for example (to name a few), Sundararajan and Cruz (1970), Haddad and Cruz (1970), Kokotović and Singh (1971), Medanić and Avramović (1975), Ishimatsu et al. (1975), Ozguner and Perkins (1977), Delacour et al. (1975), Ozguner and Perkins (1977), Delacour et al. (1978), Mahmoud (1978), Khalil and Kokotović (1978), Petkovski and Rakić (1979), Washburn and Mendel (1980), Kokotović (1981), Looze and Sandell (1982), Peponides and Kokotović (1983), Tzafestas and Anagnostou (1984), Sezer and Siljak (1986, 1991), Calvet and Title (1989), Kaszkurewicz et al. (1990), Siljak (1991), Srikant and Basar (1991, 1992a,b), Basar and Srikant (1991), Su and Gajić (1991, 1992), Al-Saggaf (1992), Aganović and Gajić (1993), Riedel (1993), Geray and Looze (1996), Finney and Heck (1996), Hoppensteadt and Izhikevich (1997), Derbel (1999), Lim and Gajić (1999), Gajić and Borno (2000), Mukaidani (2006a,b, 2007a–c), Kecman (2006), Huang et al. (2005), and Kim and Lim (2006, 2007).

Traditionally, solutions of the main equations of analysis and synthesis of *linear* optimal controllers and filters (Anderson and Moore 1990), Riccati-type (Lancaster and Rodman 1995) and Lyapunov-type equations (Gajić and Qureshi 1995) were obtained for weakly coupled systems in terms of Taylor series and power-series expansions with respect to a small weak coupling parameter ε. Approximate feedback control laws were derived by truncating expansions of the feedback coefficients of the optimal control law (Kokotović et al. 1969; Haddad and Cruz 1970; Ozguner and Perkins 1977; Delacour 1978; Petkovski and Rakić 1979). Such approximations have been shown to be near-optimal with performance that can made as close to the optimal performance as desired by including enough terms in the truncated expansions.

In this book, we will study linear weakly coupled control systems by using two new approaches developed by the authors during the last 20 years: the recursive approach (based on fixed point iterations) and the so-called Hamiltonian approach (based on block diagonalization of the Hamiltonian matrix of optimal control theory of linear systems). Consistently, the book is divided into three parts: Part I—Recursive approach for linear weakly coupled control systems, Part II—Hamiltonian approach for linear weakly coupled control systems, and Part III—Bilinear weakly coupled control systems.

The *recursive approach* to weakly coupled control systems (based on fixed point iterations) originated in the late 1980s and at the beginning of the 1990s in the papers by Gajić and his coworkers (Petrović and Gajić 1988; Harkara et al. 1989; Shen and Gajić 1990a–c; Shen 1990; and Qureshi 1992). It has been shown that the recursive methods are particularly useful when the coupling parameter ε is not extremely small and/or when any desired order of accuracy is required, namely, $O(\varepsilon^k)$,* where $k = 2, 3, 4, \dots$. In some applications a very good approximation is required, such as for a plant-filter augmented system (Shen and Gajić 1990a), where the accuracy of $O(\varepsilon^k)$, $k \geq 6$ was needed to stabilize considered real world

---

* $O(\varepsilon^k)$ stands for $C\varepsilon^k$, where C is a bounded constant and k is any arbitrary constant.

closed-loop electric power system. The recursive methods are particularly important for optimal output feedback control problems, where the solution of highly nonlinear algebraic equations is required. The effectiveness of the corresponding reduced-order algorithm and its advantages over the global full-order algorithm are demonstrated in Harkara et al. (1989) on a 12-plate chemical absorption column example. Obtained results strongly support the necessity for the existence of reduced-order recursive numerical techniques for solving corresponding nonlinear algebraic equations. In addition to the reduction in required computations, it can be easier to find a good initial guess and to handle the problem of nonuniqueness of the solution of corresponding nonlinear equations. The recursive approach to continuous and discrete, deterministic and stochastic, linear weakly coupled control systems was further advanced in the papers by Skataric et al. (1991, 1993), Skataric (1993), Hogan and Gajic (1994), Borno (1995), Borno and Gajic (1995), Gajic and Borno (2000), and Skataric (2005). The recursive approach to bilinear weakly coupled control systems was considered in Aganovic and Gajic (1995).

The linear weakly coupled system composed of two subsystems is defined by

$$\frac{dx_1(t)}{dt} = A_1 x_1(t) + \varepsilon A_2 x_2(t)$$
$$\frac{dx_2(t)}{dt} = \varepsilon A_3 x_1(t) + A_4 x_2(t) \tag{1.1}$$

where ε is a small weak coupling parameter and $x_1(t) \in R^{n_1}$ and $x_2(t) \in R^{n_2}$ are state space variables ($n_1 + n_2 = n$, n is the system order). Matrices A_i, $i = 1, 2, 3, 4$, are constant and $O(1)$. It is assumed that magnitudes of all the system eigenvalues are $O(1)$, that is $|\lambda_j| = O(1)$, $j = 1, 2, \ldots, n$, implying that matrices A_1 and A_4 are non-singular with $\det\{A_1\} = O(1)$ and $\det\{A_4\} = O(1)$. This is the standard assumption for weakly coupled linear systems, which also corresponds to the block diagonal dominance of the system matrix A (Chow and Kokotovic 1983). Hence, the main results presented in this book are valid under the following weak coupling assumption.

Assumption 1.1 Matrices A_i, $i = 1, 2, 3, 4$, are constant and $O(1)$. In addition, magnitudes of all system eigenvalues are $O(1)$, that is, $|\lambda_j| = O(1)$, $j = 1, 2, \ldots, n$, which implies that the matrices A_1, A_4 are nonsingular with $\det\{A_1\} = O(1)$ and $\det\{A_4\} = O(1)$.

This assumption in fact indicates block diagonal dominance of the system matrix. It states the condition which guarantees that weak connections among the subsystems will indeed imply weak dynamic coupling. Note that when this assumption is not satisfied, the system defined in Equation 1.1, in addition of weak coupling can also display multiple timescale phenomena (singular perturbations), as considered in Phillips and Kokotovic (1981), Delebeque and Quadrant (1981), and Chow (1982), for large-scale Markov chains and power systems. In the case when Assumption 1.1 is not satisfied, the slow coherency method (Chow 1982) can be used to form a reduced-order slow aggregate model that represents a long-term equivalent of the original system. Using the slow coherency method, the system (Equation 1.1) will be decoupled into three subsystems. The slow coherency method will not be covered in this book. However, in Chapter 3, we will present a class of systems that display both weak coupling and singular perturbations phenomena. The reader interested in coherency based decomposition methods is referred also to Kokotović et al. (1982).

The following simple example demonstrated importance of Assumption 1.1 for the definition of weakly coupled linear systems.

Example 1.1

Consider two "weakly" coupled linear systems. The first one satisfies the weak coupling Assumption 1.1, that is, both its eigenvalues are $O(1)$, and the second system has one eigenvalue of $O(\varepsilon)$ and two eigenvalues of $O(1)$

$$\frac{dx(t)}{dt} = \begin{bmatrix} -1 & 2\varepsilon \\ -1.5\varepsilon & -2 \end{bmatrix} x(t)$$

$$\frac{dz(t)}{dt} = \begin{bmatrix} -1 & 2\varepsilon & \varepsilon \\ 1.5\varepsilon & -2 & \varepsilon \\ \varepsilon & \varepsilon & -2\varepsilon \end{bmatrix} z(t)$$

The decoupled, reduced-order, state models of these systems can be obtained by neglecting $O(\varepsilon)$ terms, that is

$$\frac{dx(t)}{dt} = \begin{bmatrix} -1 & 0 \\ 0 & -1 \end{bmatrix} x(t)$$

$$\frac{dz(t)}{dt} = \begin{bmatrix} -1 & 0 & 0 \\ 0 & -1 & 0 \\ 0 & 0 & 0 \end{bmatrix} z(t)$$

Assuming that initial conditions for these two systems are given by $x(0) = [1\ 1]^T$ and $z(0) = [1\ 1\ 1]^T$, we have presented in Figures 1.1 and 1.2, respectively for the second- and third-order systems, the system state responses due to initial conditions (zero-input responses) for both the original and decoupled systems. The responses for the decoupled subsystems (obtained by setting $\varepsilon = 0$) are denoted by the dashed lines. It can be seen from Figure 1.1 that the response of the original and decoupled systems are close to each other, $O(\varepsilon)$ apart for all times, which is expected from a weak coupling system that satisfies Assumption 1.1. However, Figure 1.2 indicates, that the third-order system state space response for one of the state variables, corresponding to $O(\varepsilon)$ eigenvalue, is not close to the corresponding response of the decoupled subsystem. Even more, it can be seen from the same figure that this state variable is much slower than the remaining two state space variables indicating the presence of two timescales in this system.

For linear weakly coupled systems, the development of the decoupling transformation of Gajić and Shen (1989) is particularly important. With this nonsingular

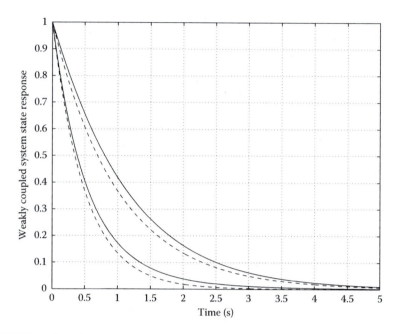

FIGURE 1.1 Zero-input response of the linear weakly coupled system (dashed lines denote the decoupled system response).

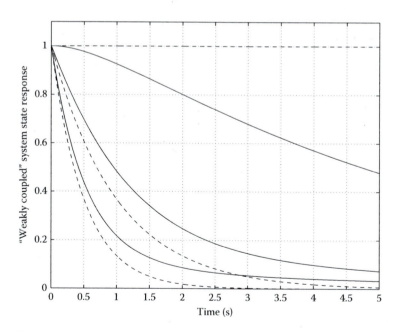

FIGURE 1.2 Zero-input response of a linear weakly coupled system (dashed lines denote the decoupled system response).

state transformation, the original system (Equation 1.1) is transformed into decoupled subsystems, that is, it is transformed into

$$\frac{d\eta_1(t)}{dt} = (A_1 - \varepsilon^2 LA_3)\eta_1(t)$$
$$\frac{d\eta_2(t)}{dt} = (A_4 + \varepsilon^2 A_2 L)\eta_2(t)$$

(1.2)

Matrix L is the unique solution of the nonlinear algebraic equation

$$A_1 L - LA_4 + A_2 - \varepsilon^2 LA_3 L = 0$$

(1.3)

Such a solution for L exists for sufficiently small values of the weak coupling parameter ε under the assumption that matrices A_1 and A_4 have no eigenvalues in common (Gajić and Shen 1989). Hence, another fundamental assumption used quite often in this book is related to the existence condition of the unique solutions for the decoupling Equation 1.3. That assumption is stated below.

Assumption 1.2 Matrices A_1 and A_4 have no common eigenvalues.

The original state variables can be recovered via the inverse transformation given by

$$\begin{bmatrix} x_1(t) \\ x_2(t) \end{bmatrix} = \begin{bmatrix} I & -\varepsilon L \\ \varepsilon H & I - \varepsilon^2 HL \end{bmatrix} \begin{bmatrix} \eta_1(t) \\ \eta_2(t) \end{bmatrix}$$

(1.4)

where matrix H satisfies the following algebraic equation

$$H(A_1 - \varepsilon^2 LA_3) - H(A_4 + \varepsilon^2 A_3 L) + A_3 = 0$$

(1.5)

The unique solution of Equation 1.5 exists under the same condition as the unique solution for Equation 1.3, that is, it is required that Assumption 1.2 be satisfied.

Another variant of this transformation that also decouples the L and H Equations 1.3 and 1.5, and hence achieves full parallelism in computations, will be presented in Chapter 5 (Qureshi 1992). The original transformation that decouples Equation 1.1 and produces Equation 1.2 can be obtained using the change of variables

$$\begin{bmatrix} \eta_1(t) \\ \eta_2(t) \end{bmatrix} = \begin{bmatrix} I & -\varepsilon L \\ \varepsilon H & I - \varepsilon^2 HL \end{bmatrix} \begin{bmatrix} x_1(t) \\ x_2(t) \end{bmatrix}$$

(1.6)

Generalization of the considered transformation (Equations 1.2 through 1.6) to more than two weakly coupled subsystems was studied by Gajić and Borno (2000) and in the recent paper by Prljaca and Gajić (2007). Lim and Gajić (1999) and Mukaidani (2005b, 2006b) considered linear-quadratic optimal control problems for weakly coupled systems composed of N subsystems.

The transformation of Gajić and Shen (1989) will be used throughout the book to simplify design of optimal controllers and filters for weakly coupled linear dynamic systems. This transformation is important, in general, for linear system theory and its applications, for example, to decouple equations coming from two inductively coupled electrical/electronic circuits and obtain equivalent reduced-order independent electrical circuits for which the effect of the inductive coupling is taken into consideration. Particularly, this transformation can be used to simplify the linear electric transformer equations, and for example, to evaluate the effective power inductively received in the process of wireless charging (Sawan et al. 2005) by the battery implemented in the human body. It can be also used in static problems, such as in statistics and stochastic processes to remove correlation among variables/ processes by mapping variables/processes into new coordinates and block diagonalizing corresponding correlation matrices. Such a decoupling that transforms dependence of variables to independence of variables plays an important role in statistics (de la Pena and Gine 1999).

The *Hamiltonian approach* to weakly coupled linear control systems was developed at the beginning of the 1990s in the papers by Su and Gajić (1991, 1992). The Hamiltonian approach is based on block diagonalization of the Hamiltonian matrix (Anderson and Moore 1990), corresponding to the algebraic Riccati equation of the optimal linear-quadratic control theory. The Hamiltonian matrix retains the weakly coupled form by interchanging state and costate variables so that it can be block diagonalized via the decoupling transformations introduced by Gajić and Shen (1989). The main idea is to obtain the solution of the full-order algebraic Riccati equation from two decoupled reduced-order subsystems, both leading to the nonsymmetric algebraic Riccati equations (Freiling 2002). It has been shown that the solutions of the reduced-order equations exist under stabilizability and detectability conditions imposed on subsystems (Su and Gajić 1992). The use of the nonsymmetric reduced-order Riccati equations can produce a lot of savings; that is, $O(n)$, in the size of computations required. Furthermore, the proposed method is very suitable for parallel computations since it allows complete parallelism. The Hamiltonian approach was further developed in a series of papers by the authors and used for various problem formulations of continuous and discrete, deterministic and stochastic linear weakly coupled systems (Gajić and Aganović 1995; Aganović et al. 1996; Gajić and Borno 2000; Kecman 2006). Particularly efficient is the "eigenvector method" of Kecman (2006). Several real physical weakly coupled system examples have been considered in Kecman (2006) in order to demonstrate the efficiency of the proposed eigenvector method.

The other classes of general linear optimal control problems can be studied by using the reduced-order algorithms presented in this book with the help of some standard control techniques such as the overlapping decomposition and prescribed degree of stability techniques. The overlapping decomposition methods of Siljak are very powerful tools in the system decomposition (Siljak 1991). The overlapping decomposition technique can influence weak coupling. The prescribed degree of stability requirement (Anderson and Moore 1990) imposed on the system in order to assure a prescribed stability margin can bring the system matrix into the block diagonally dominant form and make the system internally weakly coupled.

In addition, it will be also interesting to extend the ideas presented in this book to numerical linear algebra problems of weakly coupled systems. Some results in that direction have been already obtained by using the overlapping decomposition (Sezer and Siljak 1991). Even more, similar type of algorithms can be developed for solving (static) systems of nonlinear algebraic equations (Zecevic and Siljak 1994; Amano et al. 1996; Gacic et al. 1998). Another important future research topic is the development of asynchronous versions of the presented algorithms. The importance of asynchronous algorithms for block diagonally dominant systems (weakly coupled systems) is documented in Kaszkurewicz et al. (1990).

Note that the weakly coupled subsystems are also present in systems that display the so-called multimodeling structure. Those systems have slow and fast modes with the fast modes being mutually weakly coupled (Khalil and Kokotović 1978; Khalil 1980; Saksena and Cruz 1981a,b; Saksena and Basar 1982; Saksena et al. 1983; Dontchev and Veliov 1985; Gajić and Khalil 1986; Gajić 1988; Zhuang and Gajić 1990; Coumarbatch and Gajić 2000a,b; Mukaidani 2005).

The third part of this book presents *bilinear* optimal control problems for weakly coupled control systems. The stabilization problem of weakly coupled bilinear systems via state feedback was considered by Tzafestas and Anagnostou (1984). The study of the weakly coupled continuous-time optimal control problem for bilinear systems originated in Aganović (1993) and Aganović and Gajić (1993, 1995). Both open-loop and closed-loop optimal control problems of continuous-time bilinear systems were considered in Aganović and Gajić (1995). In this book, optimal control of weakly coupled bilinear systems is presented using the successive Galerkin approximation method. The corresponding numerical algorithm for optimal control of bilinear systems is derived. As a case study, the bilinear model of a paper making machine is presented. In addition, a robust H_∞ controller for continuous-time bilinear systems with parameter uncertainties is considered. The presentation of this chapter mostly follows the works of Kim and Lim (2006, 2007).

We hope that the recursive approach, based on reduced-order fixed-point iterations, can be extended to nonlinear weakly coupled control problems. The nonlinear weakly coupled systems were originally introduced to the control audience in Kokotović and Singh (1971). The study of weak coupling and system order reduction in nonlinear systems was considered in Peponides and Kokotović (1983). The nonlinear weakly coupled systems were studied in the context of differential games by Srikant and Basar (1991, 1992b). The stability problem of this kind of systems was considered in Martynyuk (1995).

It is interesting to point out that many dynamic nonlinear systems in physics are known to possess the weakly coupled form, for example, acoustic systems (Franzoni and Bliss 1998), temperature dissipation (Ritter and Figueiredo 2005), nonlinear oscillators (Aubry et al. 2001), and systems in photonics (Long et al. 1998). Dynamics of a nanomechanical resonator coupled to a single electron transistor (Averin and Likharev 1986) represents dynamics of a nonlinear weakly coupled system (Armor et al. 2004). Weakly coupled systems have been also studied in mathematics (e.g., Bhaya et al. 1991; Feingold and Varga 1962; Kaskurewics et al. 1990; Zecević and Siljak 1994; Nessyahu 1996; Thompson and Tisdell 1999;

Carrive et al. 2002), and in computer science (Jia and Leimkuhler 2003). The applications on nonlinear weakly coupled systems can be found even in medicine for a human scalp-recorded EEG (Sulimov 1998), and in ecology (Auger and Roussarie 1994). In addition, weakly coupled systems have been studied in economics (Simon and Ando 1963; Pierce 1974; Okuguchi 1978), management sciences (You 1998), and power system engineering (Medanić and Avramović 1975; Ilić-Spong et al. 1984; Crow and Ilić 1990; Ilić 2007) under the name of block diagonally dominant matrices and block diagonally dominant systems. Weak coupling linear structures also appear in nearly completely decomposable continuous- and discrete-time Markov chains (Philips and Kokotović 1981; Delebecque and Quadrant 1981; Aldhaheri and Khalil 1991; Stewart 1994; Hurie and Moresino 2007). Applications of weakly coupled systems to networking can be found in Jung et al. (2005). Several journal papers applied the weak coupling approach to linear models of power systems (Avramović and Medanić 1975; Shen and Gajić 1990; Momah and Shen 1991; Nuhanovic et al. 1998; Ilić 2007).

It is known that the linearized models of dynamical systems described by partial differential equations in the modal coordinates (Meirovich and Baruh 1983; Baruh and Choe 1990) consist of an infinite set of second-order internally decoupled differential equations. The coupling comes externally through the control input components. In practical applications, an infinite dimensional set of differential equations is approximated by a finite one of order $2n$. Using the techniques developed in this book we believe that corresponding control problems can be solved in terms of n parallel algorithms of order 2.

The system decomposition in this book is mostly presented for weakly coupled control systems composed of two subsystems. Corresponding parallel algorithms are solved by using two, three, or six processors working in parallel. However, under certain assumptions, the presented methods can be extended to weakly coupled systems with n subsystems. In those cases the original systems would be decomposed into n subsystems. The corresponding control algorithms could be solved by many parallel processors (Lim and Gajić 1999; Gajić and Borno 2000).

It is important to emphasize that weakly coupled linear systems are still an interesting research topic. Very recently, Professor Mukaidani, from Hiroshima University in Japan, in a series of papers, has been developing interesting and new approaches to weakly coupled linear control systems covering variety of research topics from linear quadratic-optimal controllers (including H_2 and H_∞ controllers) to Nash and zero-sum differential games (Mukaidani 2005b, 2006a,b, 2007a–c, 2008). His most recent paper (Mukaidani 2007d) is particularly interesting, in which a method is developed to study H_∞ control of strongly coupled systems using the weak coupling ideas.

The book contains several exercises, computer assignments, and formulations of the research problems to help the instructors who might be using this book as a graduate text on large-scale systems and/or parallel design of controllers. The required background for this book is a graduate level course on optimal control (Anderson and Moore 1990). For the related control theory concepts we refer the reader to the excellent books by Chen (1999) and Sontag (1998).

REFERENCES

Aganović, Z. and Z. Gajić, Optimal control of weakly coupled bilinear systems, *Automatica*, 29, 1591–1593, 1993.

Aganović, Z. and Z. Gajić, *Linear Optimal Control of Bilinear Systems—With Applications to Singular Perturbation and Weak Coupling*, Springer Verlag, New York, 1995.

Aganović, Z., Z. Gajić, and X. Shen, New method for optimal control and filtering of weakly coupled linear discrete stochastic systems, *Automatica*, 32, 83–88, 1996.

Al-Saggaf, U., Subsystem interconnection and near optimum control of discrete balanced systems, *IEEE Transactions on Automatic Control*, AC-37, 1026–1033, 1992.

Aldhaheri, R. and H. Khalil, Aggregation method for nearly completely decomposable Markov chains, *IEEE Transactions on Automatic Control*, 36, 178–187, 1991.

Amano, M., A. Zecevic, and D. Siljak, An improved block-parallel Newton method via epsilon decompositions for load-flow calculations, *IEEE Transactions on Power Systems*, 11, 1519–1525, 1996.

Anderson, B. and J. Moore, *Optimal Control: Linear Quadratic Method*, Prentice-Hall, Englewood Cliffs, NJ, 1990.

Armour, A., M. Blencowe, and Y. Zhang, Classical dynamics of a nanomechanical resonator coupled to a single electron transistor, *Physics Review B*, 69, Art. No. 125313, 2004.

Averin, D. and K. Likharev, Coulomb blockade of single-electron tunneling, and coherent oscillations in small tunnel junctions, *Journal of Low Temperature Physics*, 62, 345–373, 1986.

Aubry, S., G. Kopidakis, A. Morgante, and G. Tsironis, Analytical conditions for targeted energy transfer between nonlinear oscillators or discrete breathers, *Physica B—Condensed Matter*, 296, 222–236, 2001.

Auger, P. and R. Roussarie, Complex ecological models with simple dynamics—from individuals to populations, *ACTA Biotheoretica*, 42, 111–136, 1994.

Baruh, H. and K. Choe, Sensor placement in structural control, *AIAA J. Guidance, Dynamics and Control*, 13, 524–533, 1990.

Basar, T. and R. Srikant, Iterative Computation of Nash Equilibria in M-Player Games with Partial Weak Coupling, in *Differential Games—Developments in Modelling and Computation*, R. Hamalainen and H. Ehtamo (eds.), *Lecture Notes in Control and Information Sciences*, Vol. 156, 245–256, Springer Verlag, New York, 1991.

Bertsekas, D. and J. Tsitsiklis, *Parallel and Distributed Computation: Numerical Methods*, Prentice-Hall, Englewood Cliffs, NJ, 1989.

Bertsekas, D. and J. Tsitsiklis, Some aspects of parallel and distributed iterative algorithms— A survey, *Automatica*, 27, 3–21, 1991.

Bhaya, A., E. Kszkurewics, and F. Mota, Parallel block-iterative methods: For almost linear equations, *Linear Algebra and Its Applications*, 155, 487–508, 1991.

Borno, I., *Parallel Algorithms for Optimal Control of Linear Jump Parameter Systems and Markov Processes*, PhD dissertation, Rutgers University, NJ, 1995.

Borno, I. and Z. Gajić, Parallel algorithms for optimal control of weakly coupled and singularly perturbed jump linear systems, *Automatica*, 31, 85–988, 1995.

Calvet, J. and A. Titli, Overlapping and partitioning in block-iteration methods: Application in large scale system theory, *Automatica*, 25, 137–145, 1989.

Carrive, M., A. Miranville, A. Pietrus, and J. Rakotoson, Weakly coupled dynamical systems and applications, *Asymptotic Analysis*, 30, 161–185, 2002.

Chen, C., *Linear System Theory and Design*, Oxford University Press, New York, 1999.

Chow, J., *Time-Scale Modeling of Dynamic Networks with Applications to Power Systems*, Springer-Verlag, New York, 1982.

Coumarbatch, C. and Z. Gajić, Exact decomposition of the algebraic Riccati equation of deterministic multimodeling optimal control problems, *IEEE Transactions on Automatic Control*, 44, 790–794, 2000a.

Coumarbatch, C. and Z. Gajić, Parallel optimal Kalman filtering for stochastic systems in multimodeling form, *Transactions of ASME Journal Dynamic Systems, Measurement and Control*, 122, 542–550, 2000b.

Crow, M. and M. Ilić, The parallel implementation of the waveform relaxation method for transient stability simulation, *IEEE Transactions on Power Systems*, 5, 922–932, 1990.

Delacour, J., M. Darwish, and J. Fantin, Control strategies of large scale power systems, *International Journal of Control*, 27, 753–767, 1978.

Delebecque, F. and J. Quadrat, Optimal control of Markov-chains admitting strong and weak interactions, *Automatica*, 17, 281–296, 1981.

de la Pena, V. and E. Gine, *Decoupling: From Dependence to Independence*, Springer Verlag, New York, 1999.

Derbel, N., A new decoupling algorithm of weakly coupled systems, *System Analysis, Modeling and Simulation*, 35, 359–374, 1999.

Dontchev, A. and V. Veliov, Singular perturbations in linear control systems with weakly coupled stable and unstable fast subsystems, *Journal of Mathematical Analysis and Applications*, 110, 1–30, 1985.

Feingod, D. and M. Varga, Block diagonally dominant matrices and generalizations of the Greshgorin circle theorem, *Pacific Journal of Mathematics*, 12, 1241–1250, 1962.

Finney, J. and B. Heck, Matrix-scaling for large-scale system decomposition, *Automatica*, 32, 1177–1181, 1996.

Franzoni, L. and D. Bliss, A discussion of modal decoupling and an approximate closed-loop solution for weakly coupled systems with applications to acoustics, *Journal of the Acoustical Society of America*, 103, 1923–1932, 1998.

Freiling, G., A survey of nonsymmetric Riccati equations, *Linear Algebra and Its Applications*, 351–352, 243–270, 2002.

Gacic, N., A. Zecević, and D. Siljak, Coherency recognition using epsilon decomposition, *IEEE Transactions on Power Systems*, 13, 314–319, 1998.

Gajić, Z., The existence of a unique and bounded solution of the algebraic Riccati equation of the multimodel estimation and control problems, *Systems and Control Letters*, 10, 85–190, 1988.

Gajić, Z. and Z. Aganović, New filtering method for linear weakly coupled stochastic systems, *AIAA Journal Guidance, Control and Dynamics*, 18, 630–633, 1995.

Gajić, Z. and I. Borno, General transformation for block diagonalization of weakly coupled linear systems composed of N subsystems, *IEEE Transactions on Circuits and Systems—I: Fundamental Theory and Applications*, 47, 909–912, 2000.

Gajić, Z. and H. Khalil, Multimodel strategies under random disturbances and imperfect partial observations, *Automatica*, 22, 121–125, 1986.

Gajić, Z. and M. Lim, *Optimal Control of Singularly Perturbed Linear Systems and Applications: High Accuracy Techniques*, Marcel Dekker, New York, 2001.

Gajić, Z. and M. Qureshi, *Lyapunov Matrix Equation in Systems Stability and Control*, Academic Press, San Diego, CA, 1995.

Gajić, Z. and X. Shen, *Parallel Algorithms for Optimal Control of Large Scale Linear Systems*, Springer Verlag, London, 1993.

Gajić, Z. and D. Skatarić, Singularly perturbed weakly coupled linear control systems, *Proceedings European Control Conference*, Grenoble, France, 1607–1612, 1991.

Gajić, Z., D. Petkovski, and X. Shen, *Singularly Perturbed and Weakly Coupled Linear Control Systems—A Recursive Approach*, Springer-Verlag, New York, Lecture Notes in Control and Information Sciences, 140, 1990.

Geray, O. and D. Looze, Linear quadratic regulator loop shaping for high frequency compensation, *International Journal of Control*, 63, 1055–1068, 1996.

Haddad, A. and J. Cruz, ε-Coupling method for near-optimum design of large-scale linear systems, *Proceedings of IEE, Part D.*, 117, 223, 1970.

Harkara, N., D. Petkovski, and Z. Gajić, The recursive algorithm for the systems, *International Journal of Control*, 50, 1–11, 1989.

Haurie, A. and F. Moresino, Two-time scale controlled Markov chains: A decomposition and parallel processing approach, *IEEE Transactions on Automatic Control*, 52, 2325–2331, 2007.

Hoppensteadt, F. and E. Izhikevich, *Weakly Connected Neural Networks*, Springer Verlag, New York, 1997.

Huang, M., R. Malhame, and P. Caines, Nash equilibria for large-population linear stochastic systems of weakly coupled agents, in *Analysis, Control, and Optimization of Complex Systems*, E. Boukas and R. Malhame (eds.), Kluwer, 2005.

Ishimatsu, T., A. Mohri, and M. Takata, Optimization of weakly coupled systems by a two-level method, *International Journal of Control*, 22, 877–882, 1975.

Ilić, M., From hierarchical to open access electric power systems, *Proceedings of the IEEE*, 95, 1060–1084, 2007.

Ilić-Spong, M., M. Katz, M. Dai, and J. Zabusky, Block diagonal dominance for systems of nonlinear equations with applications to load flow calculations in power systems, *Mathematical Methods*, 5, 275–297, 1984.

Jia, Z. and B. Leimkuhler, A parallel multiple-time-scale reversible integrator for dynamic simulation, *Future Generation Computer Systems*, 19, 415–424, 2003.

Jung, W.-C., Y.-J. Kim, and M.-T. Lim, Design of an optimal controller for congestion in ATM networks, *Transactions of KIEE*, 5D, 359–365, 2005.

Kaszkurewicz, E., A. Bhaya, and D. Siljak, On the convergence of parallel asynchronous block-iterative computations, *Linear Algebra and Its Applications*, 131, 139–160, 1990.

Kim, Y.-J. and M.-T. Lim, Parallel robust H_∞ control for weakly coupled bilinear systems with parameter uncertainties using successive Galerkin approximation, *International Journal of Control, Automation, and Systems*, 4, 689–696, 2006.

Kim, Y.-J. and M.-T. Lim, Parallel optimal control for weakly coupled bilinear systems using successive Galerkin approximation, *Proceedings of IET—Control Theory and Applications*, 1, 909–914, 2007.

Kecman, V., Eigenvector approach for reduced-order optimal control problems of weakly coupled systems, *Dynamics of Continuous Discrete and Impulsive Systems*, 13, 569–588, 2006.

Khalil, H., Multi-model design of a Nash strategy, *Journal of Optimization Theory and Applications*, 31, 553–564, 1980.

Khalil, H. and P. Kokotović, Control strategies for decision makers using different models of the same system, *IEEE Transactions on Automatic Control*, AC–23, 289–298, 1978.

Kokotović, P., Feedback design of large linear systems, Chapter 4 in *Feedback Systems*, J. Cruz (Ed.), McGraw-Hill, New York, 1972.

Kokotović, P., Subsystems, time scales, and multimodeling, *Automatica*, 17, 789–795, 1981.

Kokotović, P. and G. Singh, Optimization of coupled non-linear systems, *International Journal of Control*, 14, 51–64, 1971.

Kokotović, P., B. Abramovic, J. Chow, and J. Winkelman, Coherency based decomposition and aggregation, *Automatica*, 18, 123–132, 1982.

Kokotović, P., W. Perkins, J. Cruz, and G. D'Ans, ε—coupling approach for near-optimum design of large scale linear systems, *Proceedings of IEE, Part D.*, 116, 889–892, 1969.

Lancaster, P. and L. Rodman, *Algebraic Riccati Equations*, Oxford University Press, 1995.

Lim, M. and Z. Gajić, Subsystem-level optimal control of weakly coupled linear stochastic systems composed of N subsystems, *Optimal Control Applications and Methods*, 20, 93–112, 1999.

Long, G., D. Ruan, W. Zhang, and S. Zhu, Splitting of one-phonon states in the weakly coupling model, *Chinese Physics Letters*, 15, 5–7, 1998.

Looze, D. and N. Sandell, Hierarchical control of weakly-coupled systems, *Automatica*, 18, 467–471, 1982.

Mahmoud, M., A quantitative comparison between two decentralized control approaches, *International Journal of Control*, 28, 261–275, 1978.

Martynyuk, V., Stability of nonlinear weakly coupled systems, *International Journal of Applied Mechanics*, 31, 312–316, 1995.

Medanić, J. and B. Avramović, Solution of load-flow problems in power stability by ε–coupling method, *Proceedings of IEE, Part D*, 122, 801–805, 1975.

Meirovich, L. and H. Baruh, On the problem of observation spillover in self-adjoint distributed parameter system, *Journal of Optimization Theory and Applications*, 39, 269–291, 1983.

Milne, R., The analysis of weakly coupled dynamic systems, *International Journal of Control*, 2, 171–199, 1965.

Momah, J. and X. Shen, Recursive approach to optimal control problem of multiarea electric energy system, *IEEE Proceedings Part D*, 138, 543–546, 1991.

Mukaidani, H., A new approach to robust guaranteed cost controller for uncertain multi-modeling systems, *Automatica*, 41, 1055–1062, 2005a.

Mukaidani, H., Numerical computation for H_2 state feedback control of large-scale systems, *Dynamics of Continuous, Discrete, and Impulsive Systems Series B: Applications and Algorithms*, 12, 281–296, 2005b.

Mukaidani, H., A numerical analysis of the Nash strategy for weakly coupled large-scale systems, *IEEE Transactions on Automatic Control*, 51, 1371–1377, 2006a.

Mukaidani, H., Optimal numerical strategy for Nash games of weakly coupled large-scale systems, *Dynamics of Continuous, Discrete, and Impulsive Systems*, 13, 249–268, 2006b.

Mukaidani, H., Numerical computation of sign indefinite linear quadratic differential games for weakly coupled linear large scale systems, *International Journal of Control*, 80, 75–86, 2007a.

Mukaidani, H., Newton method for solving cross-coupled sign-indefinite algebraic Riccati equations of weakly coupled large scale systems, *Applied Mathematics and Computation*, 188, 103–115, 2007b.

Mukaidani, H., Numerical computation for solving algebraic Riccati equation of weakly coupled systems, *Electrical Engineering of Japan*, 160, 39–48, 2007c.

Mukaidani, H., Numerical computation for H_∞ output feedback control for strongly coupled large-scale systems, *Applied Mathematics and Computations*, 197, 212–227, 2008.

Nessyahu, H., Convergence rate of approximate solutions to weakly coupled nonlinear systems, *Mathematics of Computation*, 65, 575–586, 1996.

Nuhanovic, A., M. Glavic, and N. Prljaca, Validation of a clustering algorithm for voltage stability analysis of Bosnian electric power system, *IEEE Proceedings Generation, Transmission and Distribution*, 145, 21–26, 1998.

Okuguchi, K., Matrices with dominant diagonal blocks and economic theory, *Journal of Mathematics and Economics*, 5, 43–52, 1978.

Ozguner, U. and W. Perkins, A series solution to the Nash strategies for large scale inter-connected systems, *Automatica*, 13, 313–315, 1979.

Peponides, G. and P. Kokotović, Weak connections, time scales, and aggregation of nonlinear systems, *IEEE Transactions on Automatic Control*, AC-28, 729–735, 1983.

Petkovski, D. and M. Rakić, On the calculation of optimum feedback gains for output constrained regulators, *IEEE Transactions on Automatic Control*, 23, 760, 1978.

Petrović, B. and Z. Gajić, The recursive solution of linear quadratic Nash games for weakly interconnected systems, *Journal of Optimization Theory and Applications*, 56, 463–477, 1988.

Phillips, R. and P. Kokotović, A singular perturbation approach to modeling and control of Markov chains, *IEEE Transactions on Automatic Control*, 26, 1087–1094, 1981.

Pierce, I., Matrices with dominating diagonal blocks, *Journal of Economic Theory*, 9, 159–170, 1974.

Prljaca, N. and Z. Gajić, A transformation for block diagonalization of weakly coupled linear systems composed of N subsystems, *WSEAS Transactions on Systems*, 6, 848–851, 2007.

Qureshi, M., *Parallel Algorithms for Discrete Singularly Perturbed and Weakly Coupled Filtering and Control Problems*, PhD dissertation, Rutgers University, NJ, 1992.

Reidel, K., Block diagonally dominant positive definite approximate filters and smoothers, *Automatica*, 29, 779–783, 1993.

Ritter, O. and W. Figueiredo, Useful work versus dissipation in weakly coupled systems at different temperatures, *Physica A—Statistical Mechanics and Its Applications*, 353, 101–113, 2005.

Saksena, V. and T. Basar, A multimodel approach to stochastic team problems, *Automatica*, 18, 713–720, 1982.

Saksena, V. and J. Cruz, A multimodel approach to stochastic Nash games, *Automatica*, 17, 295–305, 1981a.

Saksena, V. and J. Cruz, Nash strategies in decentralized control of multiparameter singularly perturbed large scale systems, *Large Scale Systems*, 2, 219–234, 1981b.

Saksena, V., J. Cruz, W. Perkins, and T. Basar, Information induced multimodel solution in multiple decision maker problems, *IEEE Transactions on Automatic Control*, AC-28, 716–728, 1983.

Sawan, M., Y. Hu, and J. Coulombe, Wireless smart implants dedicated to multichannel monitoring and microstimulation, *IEEE Circuits and Systems*, 5, 21–39, 2005.

Sezer, M. and D. Siljak, Nested ε—decomposition and clustering of complex systems, *Automatica*, 22, 321–331, 1986.

Sezer, M. and D. Siljak, Nested epsilon decomposition of linear systems: Weakly coupled and overlapping blocks, *SIAM Journal of Matrix Analysis and Applications*, 22, 521–533, 1991.

Siljak, D., *Decentralized Control of Complex Systems*, Academic Press, Cambridge, 1991.

Simon, H. and A. Ando, Aggregation of variables in dynamic systems, *Econometrica*, 29, 111–138, 1963.

Skatarić, D., *Parallel Algorithms for Reduced-Order Optimal Control of Quasi Singularly Perturbed and Weakly Coupled Systems*, PhD dissertation, University of Novi Sad, Novi Sad, 1993.

Skatarić, D., *Optimal Control of Quasi Singularly Perturbed and Weakly Coupled Systems*, Planeta Print, Belgrade, Serbia, 2005.

Skatarić, D., Z. Gajić, and D. Arnautovic, Reduced-order design of optimal controller for quasi-weakly coupled linear control systems, *Control—Theory and Advanced Technology*, 9, 481–490, 1993.

Skatarić, D., Z. Gajić, and D. Petkovski, Reduced-order solution for a class of linear quadratic optimal control problems, *Proceedings Allerton Conference on Communication, Control and Computing*, Urbana, IL, 440–447, 1991.

Shen, X., *Near-Optimum Reduced-Order Stochastic Control of Linear Discrete and Continuous Systems with Small parameters*, PhD dissertation, Rutgers University, NJ, 1990.

Shen, X. and Z. Gajić, Near-optimum steady state regulators for stochastic linear weakly coupled systems, *Automatica*, 26, 919–923, 1990a.

Shen, X. and Z. Gajić, Optimal reduced-order solution of the weakly coupled discrete Riccati equation, *IEEE Transactions on Automatic Control*, AC-35, 600–602, 1990b.

Shen, X. and Z. Gajić, Approximate parallel controllers for discrete weakly coupled linear stochastic systems, *Optimal Control Applications and Methods*, 11, 345–354, 1990c.

Sontag, E., *Mathematical Control Theory—Deterministic Finite Dimensional Systems*, Springer-Verlag, New York, 1998.

Srikant, R. and T. Basar, Iterative computation of noncooperative equilibria in nonzero-sum differential games with weakly coupled players, *Journal of Optimization Theory and Applications*, 71, 137–168, 1991.

Srikant, R. and T. Basar, Sequential decomposition and policy iteration schemes for M-player games with partial weak coupling, *Automatica*, 28, 95–105, 1992a.

Srikant, R. and T. Basar, Asymptotic solutions to weakly coupled stochastic teams with nonclassical information, *IEEE Transactions on Automatic Control*, AC-37, 163–173, 1992b.

Stewart, W., *Introduction to Numerical Solution of Markov Chains*, Princeton University Press, NJ, 1994.

Su, W. and Z. Gajić, Reduced-order solution to the finite time optimal control problems of linear weakly coupled systems, *IEEE Transactions Automatic Control*, AC-36, 498–501, 1991.

Su, W. and Z. Gajić, Decomposition method for solving weakly coupled algebraic Riccati equation, *AIAA Journal of Guidance, Dynamics and Control*, 15, 536–538, 1992.

Sulimov, A., Human scalp-recorded EEG may be a result of activity of weakly-coupled subsystems, *Neuroscience Letters*, 250, 72–74, 1998.

Sundararajan, N. and J. Cruz, ε-coupling method for near-optimum design of large-scale linear systems, *Proceedings of IEE, Part D*, 117, 223–224, 1970.

Thompson, H. and C. Tisdell, Nonlinear multipoint boundary value problems for weakly coupled systems, *Bulletin of the Australian Mathematical Society*, 60, 45–54, 1999.

Tzafestas, S. and K. Anagnostou, Stabilization of ε-coupled bilinear systems using state feedback, *International Journal of Systems Science*, 15, 639–646, 1984.

Washburn, H. and J. Mendel, Multistage estimation of dynamical and weakly coupled systems in continuous-time linear systems, *IEEE Transactions on Automatic Control*, AC-25, 71–76, 1980.

You, B.-W., On the development of lower order aggregated model for the linear large-scale model, *International Journal of Management Science*, 125–142, 1998.

Zecević, A. and D. Siljak, A block-parallel Newton method via overlapping decompositions, *SIAM Journal on Matrix Analysis and Applications*, 15, 824–844, 1994.

Zhuang, J. and Z. Gajić, Stochastic multimodel strategy with perfect measurements, *Control—Theory and Advanced Technology*, 7, 173–182, 1991.

Part I

Recursive Approach for Linear Weakly Coupled Control Systems

2 Linear Weakly Coupled Control Systems

2.1 INTRODUCTION

In this chapter, we study the main algebraic equations of the linear steady-state control theory: The Lyapunov and Riccati algebraic equations of weakly coupled systems are considered. We derive the corresponding recursive, reduced-order parallel algorithms for the solution of these equations in the most general case when the problem matrices are functions of a small weak coupling parameter. The numerical decomposition has been achieved, so that only low-order systems are involved in algebraic computations. The introduced recursive methods are of the fixed point type and can be implemented as synchronous parallel algorithms (Bertsekas and Tsitsiklis 1989, 1991).

Both continuous- and discrete-time versions of the algebraic Lyapunov and Riccati equations are studied. The partitioned expressions of the algebraic Riccati equation have very complicated forms in the discrete-time domain. We have overcome that problem by using the corresponding bilinear transformation, which is applicable under quite mild assumptions, so that the solution of the discrete algebraic Riccati equation of weakly coupled systems is obtained by using results for the corresponding continuous-time algebraic Riccati equation. It is shown that the recursive methods for weakly coupled linear systems converge with the rate of convergence of $O(\varepsilon^2)$.

Having obtained the approximate solutions of the algebraic Lyapunov and Riccati equations, the corresponding approximate linear-quadratic control problems are solved in terms of these solutions. Several real world examples are included in order to demonstrate the procedures: catalytic cracker and chemical plant.

2.2 WEAKLY COUPLED LINEAR CONTINUOUS SYSTEMS

Consider a linear dynamic system represented by a matrix differential equation

$$\dot{x}(t) = A(\varepsilon)x(t) + B(\varepsilon)u(t), \quad x(0) = x_0 \tag{2.1}$$

with a performance index

$$J(\varepsilon) = \frac{1}{2} \int_0^\infty \left[x^\mathrm{T} Q(\varepsilon)x + u^\mathrm{T} R(\varepsilon)u \right] \mathrm{d}t, \quad Q(\varepsilon) \geq 0, \quad R(\varepsilon) > 0 \tag{2.2}$$

which has to be minimized, where ε is a small weak coupling parameter and t represents time. $x(t) \in \Re^n$ and $u(t) \in \Re^m$ are state and control variables, respectively, with appropriate dimensions of the corresponding matrices. The optimal control $u(t)$ that minimizes Equation 2.2 along trajectories of Equation 2.1 is given by the well-known expression (Kwakernaak and Sivan 1972)

$$u(x(t)) = -R^{-1}(\varepsilon)B^{\mathrm{T}}(\varepsilon)P(\varepsilon)x(t) \tag{2.3}$$

where $P(\varepsilon)$ is the positive semidefinite stabilizing solution of the algebraic Riccati equation

$$P(\varepsilon)A(\varepsilon) + A^{\mathrm{T}}(\varepsilon)P(\varepsilon) + Q(\varepsilon) - P(\varepsilon)S(\varepsilon)P(\varepsilon) = 0$$
$$S(\varepsilon) = BR^{-1}B^{\mathrm{T}} \tag{2.4}$$

For $S(\varepsilon) = 0$, Equation 2.4 becomes the algebraic Lyapunov equation. In this section, we will also study a dual form of the algebraic Lyapunov equation that represents a variance equation of a linear system driven by white noise

$$\dot{x}(t) = A(\varepsilon)x(t) + G(\varepsilon)\omega(t) \tag{2.5}$$

where $\omega(t)$ is a zero-mean Gaussian stationary white noise process with a unity intensity matrix. The algebraic Lyapunov equation (Gajić and Qureshi 1995), corresponding to Equation 2.5, and representing the variance equation of $x(t)$, is given by

$$K(\varepsilon)A^{\mathrm{T}}(\varepsilon) + A(\varepsilon)K(\varepsilon) + G(\varepsilon)G^{\mathrm{T}}(\varepsilon) = 0 \tag{2.6}$$

The weakly coupled linear-quadratic control problem is defined by Equations 2.1 through 2.4, subject to the following partition of the problem matrices (Kokotović et al. 1969).

$$A(\varepsilon) = \begin{bmatrix} A_1(\varepsilon) & \varepsilon A_2(\varepsilon) \\ \varepsilon A_3(\varepsilon) & A_4(\varepsilon) \end{bmatrix}, \quad B(\varepsilon) = \begin{bmatrix} B_1(\varepsilon) & \varepsilon B_2(\varepsilon) \\ \varepsilon B_3(\varepsilon) & B_4(\varepsilon) \end{bmatrix}$$
$$Q(\varepsilon) = \begin{bmatrix} Q_1(\varepsilon) & \varepsilon Q_2(\varepsilon) \\ \varepsilon Q_2^{\mathrm{T}}(\varepsilon) & Q_3(\varepsilon) \end{bmatrix}, \quad R(\varepsilon) = \begin{bmatrix} R_1(\varepsilon) & 0 \\ 0 & R_2(\varepsilon) \end{bmatrix} \tag{2.7}$$

where ε is a small parameter. Dimensions of partitioned matrices are $A_1 \in \Re^{n_1 \times n_1}$, $A_4 \in \Re^{n_2 \times n_2}$, $R_1 \in \Re^{m_1 \times m_1}$, $R_2 \in \Re^{m_2 \times m_2}$, where $n = n_1 + n_2$, $m = m_1 + m_2$. It is assumed that magnitudes of all the system eigenvalues are $O(1)$, that is $|\lambda_j| = O(1)$, $j = 1, 2, \ldots, n$, implying that matrices $A_1(\varepsilon)$ and $A_4(\varepsilon)$ are nonsingular with $\det\{A_1(\varepsilon)\} = O(1)$ and $\det\{A_4(\varepsilon)\} = O(1)$. Hence, weakly coupled linear control systems are considered in this chapter under the following assumption (Chow and Kokotović 1983).

Assumption 2.1a (Weak Coupling Assumption). The magnitudes of all system eigenvalues are $O(1)$, $|\lambda_j| = O(1)$, $j = 1, 2, \ldots, n$, which implies $\det\{A_1(\varepsilon)\} = O(1)$ and $\det\{A_4(\varepsilon)\} = O(1)$.

This assumption in fact indicates block diagonal dominance of the system matrix. It states the condition which guarantees that weak connections among the subsystems will indeed imply weak dynamic coupling. Note that when this assumption is not satisfied, the system, in addition to weak coupling can also display multiple timescale phenomena (singular perturbations), as discussed in Phillips and Kokotović (1981), Delebecque and Quadrat (1981), and Chow (1982), for large-scale Markov chains and power systems.

The linear optimal control problem defined by Equations 2.1 and 2.2 subject to Equation 2.7 will be studied in terms of solutions of algebraic Lyapunov and Riccati equations. In this chapter, and in general in this book, we will require that all partitioned matrices involved in computations of such solutions are $O(1)$, which will provide that the solutions of the corresponding algebraic Lyapunov and Riccati equations also have the weakly coupled structure. Hence, the results presented in this chapter will be valid under the following assumption.

Assumption 2.1b All partitioned matrices defined in Equation 2.7, that is A_i, B_i, $i = 1, 2, 3, 4$, Q_i, $i = 1, 2, 3$, and R_j, $j = 1, 2$, are $O(1)$.

In Section 2.2.1, we will develop the recursive fixed point type parallel algorithms for solving the algebraic Lyapunov and Riccati equations of weakly coupled systems.

2.2.1 WEAKLY COUPLED ALGEBRAIC LYAPUNOV EQUATION

The algebraic Lyapunov equation of weakly coupled systems (regulator type) is given by

$$A^T(\varepsilon)P(\varepsilon) + P(\varepsilon)A(\varepsilon) + Q(\varepsilon) = 0 \tag{2.8}$$

Due to block dominant structure of matrices A and Q, the required solution P is properly scaled as follows (Assumptions 2.1a and b)

$$P(\varepsilon) = \begin{bmatrix} P_1(\varepsilon) & \varepsilon P_2(\varepsilon) \\ \varepsilon P_2^T(\varepsilon) & P_3(\varepsilon) \end{bmatrix} \tag{2.9}$$

Partitioned form of Equation 2.8 subject to Equation 2.7 produces

$$P_1 A_1 + A_1^T P_1 + Q_1 + \varepsilon^2 \left(P_2 A_3 + A_3^T P_2^T \right) = 0$$

$$P_1 A_2 + P_2 A_4 + A_1^T P_2 + A_3^T P_3 + Q_2 = 0 \tag{2.10}$$

$$P_3 A_4 + A_4^T P_3 + Q_3 + \varepsilon^2 \left(P_2^T A_2 + A_2^T P_2 \right) = 0$$

We define the $O(\varepsilon^2)$ approximation of Equation 2.10 as

$$
\begin{aligned}
\mathbf{P_1}A_1 + A_1^T\mathbf{P_1} + Q_1 &= 0 \\
\mathbf{P_2}A_4 + A_1^T\mathbf{P_2} &= -\mathbf{P_1}A_2 - A_3^T\mathbf{P_3} - Q_2 \\
\mathbf{P_3}A_4 + A_4^T\mathbf{P_3} + Q_3 &= 0
\end{aligned}
\tag{2.11}
$$

Note that we did not set $\varepsilon = 0$ in $A_i's$ and $Q_i's$, so that $\mathbf{P_i}$'s are functions of ε.

The unique solution of Equation 2.11 exists under the following assumption.

Assumption 2.2 Matrices $A_1(\varepsilon)$ and $A_4(\varepsilon)$ are stable.

Defining approximation errors as

$$
P_j = \mathbf{P_j} + \varepsilon^2 E_j, \quad j = 1, 2, 3
\tag{2.12}
$$

and subtracting Equation 2.11 from Equation 2.10 we obtain the following expression for the errors:

$$
\begin{aligned}
E_1A_1 + A_1^T E_1 + \mathbf{P_2}A_3 + A_3^T\mathbf{P_2^T} + \varepsilon^2\left(E_2A_3 + A_3^T E_2^T\right) &= 0 \\
E_2A_4 + A_1^T E_2 + E_1A_2 + A_3^T E_3 &= 0 \\
E_3A_4 + A_4^T E_3 + A_2^T\mathbf{P_2} + \mathbf{P_2^T}A_2 + \varepsilon^2\left(A_2^T E_2 + E_2^T A_2\right) &= 0
\end{aligned}
\tag{2.13}
$$

We propose the following algorithm, having reduced order and parallel structure, for solving Equation 2.13.

ALGORITHM 2.1

$$
\begin{aligned}
E_1^{(i+1)}A_1 + A_1^T E_1^{(i+1)} + P_2^{(i)}A_3 + A_3^T P_2^{(i)T} &= 0 \\
E_3^{(i+1)}A_4 + A_4^T E_3^{(i+1)} + A_2^T P_2^{(i)} + P_2^{(i)T}A_2 &= 0 \\
E_2^{(i+1)}A_4 + A_1^T E_2^{(i+1)} + E_1^{(i+1)}A_2 + A_3^T E_3^{(i+1)} &= 0, \quad i = 0, 1, 2, \ldots
\end{aligned}
\tag{2.14}
$$

with the starting point $E_2^{(0)} = 0$ and with

$$
P_j^{(i)} = \mathbf{P_j} + \varepsilon^2 E_j^{(i)}, \quad j = 1, 2, 3; \quad i = 0, 1, 2, \ldots
\tag{2.15}
$$

Using the stability property imposed in Assumption 2.2, it is easy to show that Equation 2.14 is a contraction mapping (Luenberger 1969)

$$
\left\| E_j^{(i)} - E_j \right\| = O(\varepsilon^i), \quad j = 1, 2, 3; \quad i = 1, 2, \ldots
\tag{2.16}
$$

Thus, the algorithm (Equation 2.14) is convergent. Using $E_j^{(\infty)}$, $j = 1, 2, 3$, in Equation 2.14 and comparing it to Equation 2.13, implies that the algorithm (Equation 2.14) converges to the unique solution of Equation 2.8. In summary, we have the following theorem.

THEOREM 2.1

Under stability assumptions imposed on matrices $A_1(\varepsilon)$ and $A_4(\varepsilon)$, the algorithm (Equation 2.14) converges to the exact solution E with the rate of convergence of $O(\varepsilon^2)$, and thus, the required solution P can be obtained with the accuracy of $O(\varepsilon^{2i})$ from Equation 2.17, that is

$$P_j = P_j^{(i)} + O(\varepsilon^{2i}), \quad j = 1, 2, 3; \quad i = 1, 2, 3, \ldots \tag{2.17}$$

2.2.2 WEAKLY COUPLED ALGEBRAIC RICCATI EQUATION

The algebraic Riccati equation (Equation 2.4), subject to the weakly coupled structure given in Equation 2.7, has the solution partitioned as in Equation 2.9. Substitution of Equations 2.7 and 2.9 in Equation 2.4 will produce the following partitioned equations

$$P_1 A_1 + A_1^T P_1 + Q_1 - P_1 S_1 P_1 + \varepsilon^2 \left(P_2 A_3 + A_3^T P_2^T \right)$$
$$- \varepsilon^2 \left[\left(P_1 S_{12} + P_2 Z^T \right) P_1 + \left(P_1 Z + P_2 \left(S_2 + \varepsilon^2 S_{21} \right) \right) P_2^T \right] = 0 \tag{2.18}$$

$$P_3 A_4 + A_4^T P_3 + Q_3 - P_3 S_2 P_3 + \varepsilon^2 \left(P_2^T A_2 + A_2^T P_2 \right)$$
$$- \varepsilon^2 \left[\left(P_3 S_{21} + P_2^T Z \right) P_3 + \left(P_3 Z^T + P_2^T \left(S_1 + \varepsilon^2 S_{12} \right) \right) P_2 \right] = 0 \tag{2.19}$$

$$P_1 A_2 + P_2 A_4 + A_1^T P_2 + A_3^T P_3 + Q_2 - P_1 S_1 P_2 - P_1 Z P_3 - P_2 S_2 P_3$$
$$- \varepsilon^2 \left[\left(P_1 S_{12} + P_2 Z^T \right) P_2 + P_2 S_{21} P_3 \right] = 0 \tag{2.20}$$

where

$$\begin{aligned} S_1 = B_1 R_1^{-1} B_1^T, \quad S_2 = B_4 R_2^{-1} B_4^T, \quad S_{12} = B_2 R_2^{-1} B_2^T \\ S_{21} = B_3 R_1^{-1} B_3^T, \quad Z = B_1 R_1^{-1} B_3^T + B_2 R_2^{-1} B_4^T \end{aligned} \tag{2.21}$$

The $O(\varepsilon^2)$ approximation of Equations 2.18 through 2.20 is defined as

$$\begin{aligned} \mathbf{P_1} A_1 + A_1^T \mathbf{P_1} - \mathbf{P_1} S_1 \mathbf{P_1} + Q_1 = 0 \\ \mathbf{P_3} A_4 + A_4^T \mathbf{P_3} - \mathbf{P_3} S_2 \mathbf{P_3} + Q_3 = 0 \end{aligned} \tag{2.22}$$

and

$$\mathbf{P_2} D_2 + D_1^T \mathbf{P_2} = -\left(\mathbf{P_1} A_2 + A_3^T \mathbf{P_3} + Q_2 - \mathbf{P_1} Z \mathbf{P_3} \right) \tag{2.23}$$

where

$$D_1(\varepsilon) = [A_1(\varepsilon) - S_1(\varepsilon)\mathbf{P}_1(\varepsilon)], \quad D_2(\varepsilon) = [A_4(\varepsilon) - S_2(\varepsilon)\mathbf{P}_3(\varepsilon)] \qquad (2.24)$$

The unique positive semidefinite stabilizing solution of Equation 2.22 exists under the following assumption.

Assumption 2.3 The triples $(A_1(\varepsilon),\ B_1(\varepsilon),\ \text{Chol}(Q_1(\varepsilon)))$ and $(A_4(\varepsilon),\ B_4(\varepsilon),\ \text{Chol}(Q_3(\varepsilon)))$ are stabilizable–detectable.*

Under Assumption 2.3 matrices $D_1(\varepsilon)$ and $D_2(\varepsilon)$ are stable so that the unique solution of Equation 2.23 exists also.

If the errors are defined as

$$P_j = \mathbf{P_j} + \varepsilon^2 E_j, \quad j = 1, 2, 3 \qquad (2.25)$$

then the exact solution will be of the form

$$P = \begin{bmatrix} \mathbf{P_1} + \varepsilon^2 E_1 & \varepsilon(\mathbf{P_2} + \varepsilon^2 E_2) \\ \varepsilon(\mathbf{P_2} + \varepsilon^2 E_2)^T & \mathbf{P_3} + \varepsilon^2 E_3 \end{bmatrix} \qquad (2.26)$$

Subtracting Equations 2.22 and 2.23 from the corresponding Equations 2.18 through 2.20 and using Equation 2.25 produces the following equations for the errors

$$E_1 D_1 + D_1^T E_1 = P_1 S_{12} P_1 + P_2 Z^T P_1 + P_1 Z P_2^T + P_2 S_2 P_2^T$$
$$- P_2 A_3 - A_3^T P_2^T + \varepsilon^2 \left(E_1 S_1 E_1 + P_2 S_{21} P_2^T \right) \qquad (2.27)$$

$$E_3 D_2 + D_2^T E_3 = P_3 S_{21} P_3 + P_2^T S_1 P_2 + P_3 Z^T P_2 + P_2^T Z\, P_3$$
$$- P_2^T A_2 - A_2^T P_2 + \varepsilon^2 (E_3 S_2 E_3 + P_2^T S_{12} P_2) \qquad (2.28)$$

$$D_1^T E_2 + E_2 D_2 = P_1 S_{12} P_2 + P_2 Z^T P_2 + P_2 S_{21} P_3$$
$$- E_1 D_{12} - D_{21}^T E_3 + \varepsilon^2 (E_1 S_1 E_2 + E_1 Z E_2 + E_2 S_2 E_3) \qquad (2.29)$$

where

$$D_{12} = A_2 - S_1 \mathbf{P_2} - Z\mathbf{P_3}, \quad D_{21} = A_3 - S_2 \mathbf{P_2^T} - Z^T \mathbf{P_1} \qquad (2.30)$$

It can be easily shown that the nonlinear Equations 2.27 through 2.29 have the form

$$E_1 D_1 + D_1^T E_1 = \text{const} + \varepsilon^2 f_1(E_1, E_2, \varepsilon^2)$$
$$E_3 D_2 + D_2^T E_3 = \text{const} + \varepsilon^2 f_3(E_2, E_3, \varepsilon^2) \qquad (2.31)$$
$$E_2 D_2 + D_1^T E_2 = \text{const} + \varepsilon^2 f_2(E_1, E_2, E_3, \varepsilon^2)$$

---

* It is common in the control literature to require that the triple $(A_1(\varepsilon),\ B_1(\varepsilon),\ \sqrt{(Q_1(\varepsilon))}\,)$ is stabilizable–detectable. Note that $\sqrt{Q_1(\varepsilon)}$ means $Q = M^2$. Notation used in Assumption 2.2 is more rigorous since the Cholesky factor of a positive semidefinite matrix is defined by $Q = C^T C$.

We can see that all cross-coupling terms and all nonlinear terms in Equations 2.27 through 2.29 are multiplied by ε^2, so that we propose the following reduced-order parallel algorithm for solving Equations 2.27 through 2.29.

ALGORITHM 2.2

$$E_1^{(i+1)}D_1 + D_1^T E_1^{(i+1)} = P_1^{(i)}S_{12}P_1^{(i)} + P_2^{(i)}Z^T P_1^{(i)} + P_1^{(i)}ZP_2^{(i)T}$$
$$- P_2^{(i)}A_3 - A_3^T P_2^{(i)T} + \varepsilon^2\left(E_1^{(i)}S_1 E_1^{(i)} + P_2^{(i)}S_{21}P_2^{(i)T}\right) + P_2^{(i)}S_2 P_2^{(i)T} \tag{2.32}$$

$$E_3^{(i+1)}D_2 + D_2^T E_3^{(i+1)} = P_3^{(i)}S_{21}P_3^{(i)} + P_2^{(i)T}ZP_3^{(i)} + P_3^{(i)}Z^T P_2^{(i)}$$
$$- P_2^{(i)T}A_2 - A_2^T P_2^{(i)} + \varepsilon^2\left(E_3^{(i)}S_2 E_3^{(i)} + P_2^{(i)T}S_{12}P_2^{(i)}\right) + P_2^{(i)T}S_1 P_2^{(i)} \tag{2.33}$$

$$D_1^T E_2^{(i+1)} + E_2^{(i+1)}D_2 = P_1^{(i+1)}S_{12}P_2^{(i)} + P_2^{(i)}Z^T P_2^{(i)}$$
$$- E_1^{(i+1)}D_{12} - D_{21}^T E_3^{(i+1)} + P_2^{(i)}S_{21}P_3^{(i)}$$
$$+ \varepsilon^2\left(E_1^{(i+1)}S_1 E_2^{(i)} + E_1^{(i+1)}ZE_2^{(i)} + E_2^{(i)}S_2 E_3^{(i+1)}\right) \tag{2.34}$$

with $E_1^{(0)} = 0$, $E_2^{(0)} = 0$, $E_3^{(0)} = 0$, where

$$P_j^{(i)} = \mathbf{P_j} + \varepsilon^2 E_j^{(i)}, \quad j = 1,2,3; \quad i = 1,2,3,\ldots \tag{2.35}$$

The following theorem indicates the features of the algorithm defined in Equations 2.32 through 2.35.

THEOREM 2.2

Under Assumption 2.3, the algorithm (Equations 2.32 through 2.35) converges to the exact solution of E with the rate of convergence of $O(\varepsilon^2)$, that is

$$\|E - E^{(i+1)}\| = O(\varepsilon^2)\|E - E^{(i)}\|, \quad i = 0,1,2,\ldots \tag{2.36}$$

or equivalently

$$\|E - E^{(i)}\| = O(\varepsilon^{2i}) \tag{2.37}$$

Proof The Jacobian of Equations 2.27 through 2.29, at some $\varepsilon = \varepsilon^*$, is given by

$$J(\varepsilon) = \begin{bmatrix} J_{11}(\varepsilon) & 0 & 0 \\ J_{21}(\varepsilon) & J_{22}(\varepsilon) & J_{23}(\varepsilon) \\ 0 & 0 & J_{23}(\varepsilon) \end{bmatrix} + \begin{bmatrix} O(\varepsilon^2) & O(\varepsilon^2) & 0 \\ O(\varepsilon^2) & O(\varepsilon^2) & O(\varepsilon^2) \\ 0 & O(\varepsilon^2) & O(\varepsilon^2) \end{bmatrix} \tag{2.38}$$

where

$$J_{11}(\varepsilon) = I_{n_1} \oplus D_1^T(\varepsilon) + D_1^T(\varepsilon) \oplus I_{n_1}$$
$$J_{22}(\varepsilon) = I_{n_2} \oplus D_2^T(\varepsilon) + D_1^T(\varepsilon) \oplus I_{n_1} \tag{2.39}$$
$$J_{33}(\varepsilon) = I_{n_2} \oplus D_2^T(\varepsilon) + D_2^T(\varepsilon) \oplus I_{n_2}$$

Since $D_1(\varepsilon)$ and $D_2(\varepsilon)$ are stable matrices (by Assumption 2.3), $J_{ii}(\varepsilon)$, $i = 1, 2, 3$, are nonsingular and hence the Jacobian will be nonsingular at $\varepsilon = \varepsilon^*$, assuming that ε is sufficiently small. Then, by the implicit function theorem (Ortega and Rheinboldt 2000), the existence of the unique bounded solution of Equations 2.27 through 2.79 is guaranteed.

In the next step, we have to prove convergence of the algorithm (Equations 2.32 through 2.35) and give an estimate of the rate of convergence. For $i = 0$, Equations 2.27 and 2.32 imply

$$\left(E_1 - E_1^{(1)} \right)D_1 + D_1^T\left(E_1 - E_1^{(1)} \right) = \varepsilon^2 f_1(E_1, E_2, \varepsilon^2) \tag{2.40}$$

Since $D_1(\varepsilon)$ is stable and E_1 and E_2 are bounded it follows that

$$\left\| E_1 - E_1^{(1)} \right\| = O(\varepsilon^2) \tag{2.41}$$

Similarly, from Equations 2.28 and 2.33 we have

$$\left(E_3 - E_3^{(1)} \right)D_2 + D_2^T\left(E_3 - E_3^{(1)} \right) = \varepsilon^2 f_3(E_2, E_3, \varepsilon^2) \tag{2.42}$$

and

$$\left\| E_3 - E_3^{(1)} \right\| = O(\varepsilon^2) \tag{2.43}$$

Using the same arguments in Equations 2.29 and 2.34 will produce

$$\left\| E_2 - E_2^{(1)} \right\| = O(\varepsilon^2) \tag{2.44}$$

Continuing the same procedure and by induction we conclude that

$$\left\| E_1 - E_1^{(i)} \right\| = O(\varepsilon^{2i})$$
$$\left\| E_2 - E_2^{(i)} \right\| = O(\varepsilon^{2i}) \tag{2.45}$$
$$\left\| E_3 - E_3^{(i)} \right\| = O(\varepsilon^{2i})$$

with $i = 1, 2, 3, \ldots$, which completes the proof of Theorem 2.2.   ∎

2.3 APPROXIMATE LINEAR REGULATOR FOR CONTINUOUS SYSTEMS

The positive semidefinite stabilizing solution of the algebraic Riccati equation (Equation 2.4), produces the answer to the optimal linear-quadratic steady-state control problem. Namely, the quadratic criterion (Equation 2.2) is minimized along trajectories of the linear dynamic system (Equation 2.1) by using the control input in the form of Equation 2.3. It is proved in Kokotović and Cruz (1969) that the near-optimal control given by

$$u^{(j)}\big(x^{(j)}(t)\big) = -R^{-1}B^{T}P^{(j)}x^{(j)}(t) = -F^{(j)}x^{(j)}(t) \tag{2.46}$$

where $P^{(j)}$ satisfies

$$P^{(j)} - P^{\text{opt}} = O(\varepsilon^{j}) \tag{2.47}$$

and

$$\dot{x}^{(j)}(t) = A(\varepsilon)x^{(j)}(t) + B(\varepsilon)u^{(j)}(t) \tag{2.48}$$

is near-optimal in the sense

$$J^{(j)} - J^{\text{opt}} = O(\varepsilon^{2j}) \tag{2.49}$$

The approximate performance $J^{(j)}$ can be obtained from the algebraic Lyapunov equation

$$\big(A - BF^{(j)}\big)^{T}K^{(j)} + K^{(j)}(A - BF^{(j)}) + Q + F^{(j)^{T}}RF^{(j)} = 0 \tag{2.50}$$

so that

$$J^{(j)} = \frac{1}{2}x^{T}(0)K^{(j)}x(0) \tag{2.51}$$

In Section 2.2, we have developed a very efficient technique for finding both $P^{(j)}$ and $K^{(j)}$. Hence, the presented algorithm represents a technique for finding the approximate solution of the required accuracy for the linear-quadratic optimal control problem of weakly coupled systems.

Exercise 2.1: Derive Algorithm 2.2 and write a MATLAB program (Hanselman and Littlefield 2001), for solving Equations 2.22 and 2.23 and Equations 2.32 through 2.35. Using the MATLAB® program find the fifth-order approximations of the optimal criterion and the optimal trajectories for a chemical plant given in Gomathi et al. (1980). Assume that the penalty matrices are identities. Plot the corresponding system trajectories for all five approximations and use a table to present the values for all approximations of the performance criterion.

2.4 WEAKLY COUPLED LINEAR DISCRETE SYSTEMS

The main goal in the theory of weakly coupled control systems is to obtain the required solution in terms of reduced-order problems, namely subsystems. In the case of the weakly coupled algebraic discrete Riccati equation, the inversion of the partitioned matrix $B^{T}PB + R$ will produce a lot of terms and make the corresponding approach computationally very involved, even though one is faced with the reduced-order numerical problems. To solve this problem, we have used the bilinear transformation of Kondo and Furuta (1986) to transform the discrete-time Riccati equation into the continuous-time algebraic Riccati equation such that the solutions of these algebraic Riccati equations are identical. The continuous-time algebraic Riccati equation can be solved in terms of the reduced-order problems very efficiently by using the recursive method of Section 2.3.2, which converges with the rate of convergence of $O(\varepsilon^{2})$. A model of a discrete chemical plant is considered as an illustrative example.

For the reason of completeness, we present first the results for the Lyapunov equation. Corresponding parallel reduced-order algorithm for solving discrete Lyapunov equation of weakly coupled systems is derived and demonstrated on a discrete catalytic cracker model.

As before, algorithms for both the Lyapunov and Riccati equations are implemented as synchronous ones. Their implementation as the asynchronous parallel algorithms is under investigation.

2.4.1 WEAKLY COUPLED DISCRETE ALGEBRAIC LYAPUNOV EQUATION

A discrete-time constant linear system with the zero-input

$$x(k + 1) = Ax(k) \qquad (2.52)$$

is asymptotically stable if and only if the solution of the algebraic discrete-time Lyapunov equation

$$A^{T}PA - P = -Q \qquad (2.53)$$

is positive definite for any positive definite symmetric matrix Q. The transpose of Equation 2.53 represents a variance equation of a linear stochastic system driven by zero-mean stationary Gaussian white noise $\omega(k)$ with the intensity matrix Q

$$x(k + 1) = Ax(k) + \omega(k) \qquad (2.54)$$

Consider the algebraic discrete Lyapunov Equation 2.53. In the case of a weakly coupled linear discrete system corresponding matrices are partitioned as

$$A = \begin{bmatrix} A_1 & \varepsilon A_2 \\ \varepsilon A_3 & A_4 \end{bmatrix}, \quad Q = \begin{bmatrix} Q_1 & \varepsilon O_2 \\ \varepsilon Q_2^{T} & Q_3 \end{bmatrix}, \quad P = \begin{bmatrix} P_1 & \varepsilon P_2 \\ \varepsilon P_2^{T} & P_3 \end{bmatrix} \qquad (2.55)$$

where A_i, $i = 1, 2, 3, 4$, and Q_j, $j = 1, 2, 3$, are assumed to be continuous functions of ε. Matrices P_1 and P_3 are of dimensions $n_1 \times n_1$ and $n_2 \times n_2$, respectively. The remaining matrices are of compatible dimensions. All partitioned matrices defined in Equation 2.55 are assumed to be $O(1)$, see Assumption 2.1b.

The partitioned form of Equation 2.53 subject to Equation 2.55 is

$$A_1^T P_1 A_1 - P_1 + Q_1 + \varepsilon^2 (A_1^T P_2 A_3 + A_3^T P_2^T A_1 + A_3^T P_3 A_3) = 0 \tag{2.56}$$

$$A_1^T P_1 A_2 - P_2 + Q_2 + A_1^T P_2 A_4 + A_3^T P_3 A_4 + \varepsilon^2 A_3^T P_2^T A_2 = 0 \tag{2.57}$$

$$A_4^T P_3 A_4 - P_3 + Q_3 + \varepsilon^2 (A_2^T P_1 A_2 + A_2^T P_2 A_4 + A_4^T P_2^T A_2) = 0 \tag{2.58}$$

Define $O(\varepsilon^2)$ perturbations of Equations 2.56 through 2.58 by

$$A_1^T \bar{P}_1 A_1 - \bar{P}_1 + Q_1 = 0 \tag{2.59}$$

$$A_1^T \bar{P}_1 A_2 + A_1^T \bar{P}_2 A_4 + A_3^T \bar{P}_3 A_4 - \bar{P}_2 + Q_2 = 0 \tag{2.60}$$

$$A_4^T \bar{P}_3 A_4 - \bar{P}_3 + Q_3 = 0 \tag{2.61}$$

Note that we did not set $\varepsilon = 0$ in $A_i's$ and $Q_j's$. Assume that the matrices A_1 and A_4 are stable (Assumption 2.2). Then, the unique solutions of Equations 2.59 through 2.61 exist.

Define errors as

$$\begin{aligned}
P_1 &= \bar{P}_1 + \varepsilon E_1 \\
P_2 &= \bar{P}_2 + \varepsilon E_2 \\
P_3 &= \bar{P}_3 + \varepsilon E_3
\end{aligned} \tag{2.62}$$

Subtracting Equations 2.59 through 2.61 from Equations 2.56 through 2.58, the following error equations are obtained:

$$\begin{aligned}
A_1^T E_1 A_1 - E_1 &= -A_1^T P_2 A_3 - A_3^T P_2^T A_1 - A_3^T P_3 A_3 \\
A_4^T E_3 A_4 - E_3 &= -A_2^T P_1 A_2 - A_2^T P_2 A_4 - A_4^T P_2^T A_2 \\
A_1^T E_2 A_4 - E_2 &= -A_1^T E_1 A_2 - A_3^T P_2^T A_2 - A_3^T E_3 A_4
\end{aligned} \tag{2.63}$$

The proposed parallel synchronous algorithm for the numerical solution of Equation 2.63 is as follows (Shen et al. 1991).

ALGORITHM 2.3

$$\begin{aligned}
A_1^T E_1^{(i+1)} A_1 - E_1^{(i+1)} &= -A_1^T P_2^{(i)} A_3 - A_3^T P_2^{(i)^T} A_1 - A_3^T P_3^{(i)} A_3 \\
A_4^T E_3^{(i+1)} A_4 - E_3^{(i+1)} &= -A_2^T P_1^{(i)} A_2 - A_2^T P_2^{(i)} A_4 - A_4^T P_2^{(i)^T} A_2 \\
A_1^T E_2^{(i+1)} A_4 - E_2^{(i+1)} &= -A_1^T E_1^{(i+1)} A_2 - A_3^T P_2^{(i)^T} A_2 - A_3^T E_3^{(i+1)} A_4
\end{aligned} \tag{2.64}$$

with starting points $E_1^{(0)} = E_2^{(0)} = E_3^{(0)} = 0$ and

$$P_j^{(i)} = \bar{P}_j + \varepsilon^2 E_j^{(i)}, \quad j = 1, 2, 3; \quad i = 0, 1, 2, \ldots \tag{2.65}$$

The convergence results of the presented algorithm can be summarized in the following theorem.

THEOREM 2.3

Under stability Assumption 2.2, the algorithm (Equations 2.64 through 2.65) converges to the exact solutions for E_j's with the rate of convergence of $O(\varepsilon^2)$.

The proof of this theorem is similar to the proof of the corresponding algorithm for continuous-time algebraic weakly coupled Lyapunov equation studied in Section 2.2.1. It uses the bilinear transformation from Power (1967) to transform the discrete-time Lyapunov equation into the continuous one and then follows the ideas of Section 2.2.1.

2.4.2   CASE STUDY: DISCRETE CATALYTIC CRACKER

A fifth-order model of a catalytic cracker (Kando et al. 1988) demonstrates the efficiency of the proposed method. The problem matrix A (after performing discretization with the sampling period $T = 1$) is given by

$$A_d = \begin{bmatrix} 0.011771 & 0.046903 & 0.096679 & 0.071586 & -0.019178 \\ 0.014096 & 0.056411 & 0.115070 & 0.085194 & -0.022806 \\ 0.066395 & 0.252260 & 0.580880 & 0.430570 & -0.11628 \\ 0.027557 & 0.104940 & 0.240400 & 0.178190 & -0.048104 \\ 0.000564 & 0.002644 & 0.003479 & 0.002561 & -0.000656 \end{bmatrix}$$

The small weak coupling parameter is $\varepsilon = 0.21$ and the state penalty matrix is chosen as $Q = I$.

The simulation results are presented in Table 2.1.

2.4.3   WEAKLY COUPLED DISCRETE ALGEBRAIC RICCATI EQUATION

The algebraic Riccati equation of weakly coupled linear discrete systems is given by

$$P = A^T P A + Q - A^T P B (B^T P B + R)^{-1} B^T P A, \quad R > 0, \quad Q \geq 0 \tag{2.66}$$

with

$$A = \begin{bmatrix} A_1 & \varepsilon A_2 \\ \varepsilon A_3 & A_4 \end{bmatrix}, \quad B = \begin{bmatrix} B_1 & \varepsilon B_2 \\ \varepsilon B_3 & B_4 \end{bmatrix}$$

$$Q = \begin{bmatrix} Q_1 & \varepsilon Q_2 \\ \varepsilon Q_2^T & Q_3 \end{bmatrix}, \quad R = \begin{bmatrix} R_1 & 0 \\ 0 & R_2 \end{bmatrix} \tag{2.67}$$

TABLE 2.1

Reduced-Order Solution of Discrete Weakly Coupled Algebraic Lyapunov Equation ($P^{(7)} = P_{\text{exact}}$)

i	$P_1^{(i)}$	$P_2^{(i)}$	$P_3^{(i)}$
0	1.0030 0.00135 0.00135 1.00540	0.54689 0.40537 −0.10944 2.08640 1.54650 −0.41752	1.93020 0.68954 −0.18620 0.068954 1.51110 −0.13802 −0.18620 −0.13802 1.03730
1	1.01390 0.05290 0.052897 1.20180	0.66593 0.49359 −0.13322 2.54040 1.88290 −0.50820	2.20320 0.89183 −0.24071 0.89183 1.66100 −0.17841 −0.24071 −0.17841 1.04820
2	1.01620 0.06184 0.06184 1.23600	0.69091 0.51209 −0.13821 2.63570 1.93550 −0.52722	2.26010 0.93400 −0.25208 0.93400 1.69230 −0.18683 −0.25208 −0.18683 1.05040
3	1.01670 0.06371 0.06371 1.24310	0.69604 0.515900 −0.13923 2.65520 1.96800 −0.53113	2.27170 0.94260 −0.25439 0.94260 1.69860 −0.18855 −0.25439 −0.18855 1.05090
4	1.01680 0.06409 0.06409 1.24450	0.69710 0.51668 −0.13944 2.65930 1.97100 −0.53193	2.27410 0.94437 −0.25487 0.94437 1.70000 −0.18891 −0.25487 −0.18891 1.05100
5	1.01680 0.06417 0.06417 1.24480	0.69731 0.51684 −0.13948 2.66010 1.97160 −0.53210	2.27460 0.94473 −0.25497 0.94473 1.70020 −0.18898 −0.25497 −0.18898 1.05100
6	1.01680 0.06418 0.06418 1.24490	0.69736 0.51687 −0.13949 2.66010 1.97170 −0.53213	2.27470 0.94481 −0.25499 0.94481 1.70030 −0.18899 −0.25499 −0.18899 1.05100
7	1.01680 0.06419 0.06419 1.24490	0.69737 0.51688 −0.13950 2.66030 1.97180 −0.53214	2.27470 0.94482 −0.25499 0.94482 1.70030 −0.18900 −0.25499 −0.18900 1.05100

and ε is a small weak coupling parameter. Due to block dominant structure of the problem matrices, and assuming that all submatrices defined in Equation 2.67 are $O(1)$ (Assumption 2.1b), the required solution P also has the weakly coupled form

$$P = \begin{bmatrix} P_1 & \varepsilon P_2 \\ \varepsilon P_2^{\mathrm{T}} & P_3 \end{bmatrix} \tag{2.68}$$

The bilinear transformation states that algebraic Riccati Equations 2.66 and

$$A_c^{\mathrm{T}} P_c + P_c A_c + Q_c - P_c S_c P_c = 0, \quad S_c = B_c R_c^{-1} B_c^{\mathrm{T}} \tag{2.69}$$

have the same solutions if the following relations hold

$$A_c = I - 2D^{-T}$$
$$S_c = 2(I + A)^{-1}S_dD^{-1}, \quad S_d = BR^{-1}B^T$$
$$Q_c = 2D^{-1}Q(I + A)^{-1}$$
$$D = (I + A)^T + Q(I + A)^{-1}S_d \cdot$$

(2.70)

assuming that $(I+A)^{-1}$ exists. It can be seen that for weakly coupled systems the matrix

$$(I + A)^{-1} = \begin{bmatrix} O(1) & O(\varepsilon) \\ O(\varepsilon) & O(1) \end{bmatrix}$$

(2.71)

is invertible for small values of e. It can be verified that the weakly coupled structure of the matrices defined in Equation 2.67 will produce the weakly coupled structure of the transformed continuous-time matrices defined in Equation 2.70. It follows from the fact that S_d from Equation 2.70 and Q from Equation 2.67 have the same weakly coupled structure as Equation 2.71, so does D in Equation 2.70. The inverse of D is also in the weakly coupled form as defined in Equation 2.71. From Equation 2.70, the weakly coupled structure of matrices A_c and Q_c follows directly since they are given in terms of the same and/or products of weakly coupled matrices.

Using the standard result from Stewart (1973), it follows that the method proposed in this section is applicable under the following assumption.

Assumption 2.4   The system matrix A has no eigenvalues located at -1.

It is important to point out that the eigenvalues located in the neighborhood of -1 will produce ill-conditioning with respect to matrix inversion and make the algorithm numerically unstable.

Let us introduce the following notation for the compatible partitions of the transformed weakly coupled matrices, that is

$$A_c = \begin{bmatrix} A_{11} & \varepsilon A_{12} \\ \varepsilon A_{21} & A_{22} \end{bmatrix}, \quad S_c = \begin{bmatrix} S_{11} & \varepsilon S_{12} \\ \varepsilon S_{12}^T & S_{22} \end{bmatrix}$$

(2.72)

$$P_c = \begin{bmatrix} P_1 & \varepsilon P_2 \\ \varepsilon P_2^T & P_3 \end{bmatrix}, \quad Q_c = \begin{bmatrix} Q_{11} & \varepsilon Q_{12} \\ \varepsilon Q_{12}^T & Q_{22} \end{bmatrix}$$

(2.73)

These partitions have to be performed by a computer only, in the process of calculations, and there is no need for the corresponding analytical expressions.

The solution of Equation 2.69 can be found in terms of the reduced-order problems by imposing standard stabilizability–detectability assumptions on the subsystems. The efficient recursive reduced-order algorithm for solving Equation 2.69 is obtained in Section 2.2.2. It will be briefly summarized here taking into account the specific features of the problem under consideration.

The $O(\varepsilon^2)$ approximation of Equation 2.69 subject to Equations 2.72 through 2.73 can be obtained from the following decoupled set of reduced-order algebraic equations

$$
\begin{aligned}
\mathbf{P}_1 A_{11} + A_{11}^{\mathrm{T}} \mathbf{P}_1 - \mathbf{P}_1 S_{11} \mathbf{P}_1 + Q_{11} = 0 \\
\mathbf{P}_3 A_{22} + A_{22}^{\mathrm{T}} \mathbf{P}_3 - \mathbf{P}_3 S_{22} \mathbf{P}_3 + Q_{22} = 0
\end{aligned}
\tag{2.74}
$$

and

$$
\mathbf{P}_2 \Delta_2 + \Delta_1^{\mathrm{T}} \mathbf{P}_2 = -(\mathbf{P}_1 A_{12} + A_{21}^{\mathrm{T}} \mathbf{P}_3 + Q_{12} - \mathbf{P}_1 S_{12} \mathbf{P}_3)
\tag{2.75}
$$

where

$$
\Delta_1(\varepsilon) = [A_{11}(\varepsilon) - S_{11}(\varepsilon)\mathbf{P}_1(\varepsilon)], \quad \Delta_2(\varepsilon) = [A_{22}(\varepsilon) = S_{22}(\varepsilon)\mathbf{P}_3(\varepsilon)]
\tag{2.76}
$$

The unique positive semidefinite stabilizing solutions of Equation 2.74 exist under the following assumption.

Assumption 2.5 The triples $(A_{ii}(\varepsilon), \mathrm{Chol}(S_{ii}(\varepsilon)), \mathrm{Chol}(Q_{ii}(\varepsilon)))$, $i = 1, 2$, are stabilizable–detectable.

Under Assumption 2.5 matrices $\Delta_1(\varepsilon)$ and $\Delta_2(\varepsilon)$ are stable so that the unique solution of Equation 2.75 also exists.

If the errors are defined as

$$
P_j = \mathbf{P}_j + \varepsilon^2 E_j, \quad j = 1, 2, 3
\tag{2.77}
$$

then the exact solution will be of the form

$$
P = \begin{bmatrix} \mathbf{P}_1 + \varepsilon^2 E_1 & \varepsilon(\mathbf{P}_2 + \varepsilon^2 E_2) \\ \varepsilon(\mathbf{P}_2 + \varepsilon^2 E_2)^{\mathrm{T}} & \mathbf{P}_3 + \varepsilon^2 E_3 \end{bmatrix}
\tag{2.78}
$$

The fixed point parallel reduced-order algorithm for the error terms, obtained by using results from Section 2.2.2, has the form (Shen and Gajić 1990).

ALGORITHM 2.4

$$
\begin{aligned}
E_1^{(i+1)} \Delta_1 + \Delta_1^{\mathrm{T}} E_1^{(i+1)} &= P_1^{(i)} S_{12} P_2^{(i)\mathrm{T}} + P_2^{(i)} S_{12}^{\mathrm{T}} P_1^{(i)} \\
&\quad + P_2^{(i)} S_{22} P_2^{(i)\mathrm{T}} - P_2^{(i)} A_{21} - A_{21}^{\mathrm{T}} P_2^{(i)\mathrm{T}} + \varepsilon^2 E_1^{(i)} S_{11} E_1^{(i)}
\end{aligned}
\tag{2.79}
$$

$$
\begin{aligned}
E_3^{(i+1)} \Delta_2 + \Delta_2^{\mathrm{T}} E_3^{(i+1)} &= P_2^{(i)\mathrm{T}} S_{11} P_2^{(i)} + P_3^{(i)} S_{12}^{\mathrm{T}} P_2^{(i)} \\
&\quad + P_2^{(i)\mathrm{T}} S_{12} P_3^{(i)} - P_2^{(i)\mathrm{T}} A_{12} - A_{12}^{\mathrm{T}} P_2^{(i)} + \varepsilon^2 E_3^{(i)} S_{22} E_3^{(i)}
\end{aligned}
\tag{2.80}
$$

$$\Delta_1^T E_2^{(i+1)} + E_2^{(i+1)} \Delta_2 + E_1^{(i+1)} \Delta_{12} + \Delta_{21}^T E_3^{(i+1)}$$
$$= P_2^{(i)} S_{12}^T P_2^{(i)} + \varepsilon^2 \left(E_1^{(i+1)} S_{11} E_2^{(i)} + E_1^{(i+1)} S_{12} E_3^{(i)} + E_2^{(i)} S_{22} E_3^{(i+1)} \right) \qquad (2.81)$$

with $E_1^{(0)} = 0$, $E_2^{(0)} = 0$, $E_3^{(0)} = 0$, where

$$P_j^{(i)} = \mathbf{P_j} + \varepsilon^2 E_j^{(i)}, \quad j = 1, 2, 3; \quad i = 1, 2, 3, \ldots \qquad (2.82)$$

and

$$\Delta_{12} = [A_{12} - S_{11} \mathbf{P_2} - S_{12} \mathbf{P_3}], \quad \Delta_{21} = \left[A_{21} - S_{22} \mathbf{P_2^T} - S_{12}^T \mathbf{P_1} \right] \qquad (2.83)$$

This algorithm satisfies all conditions given in Theorem 2.2, so that it converges to the exact solution of E, with the rate of convergence of $O(\varepsilon^2)$, that is

$$\|E - E^{(i+1)}\| = O(\varepsilon^2) \|E - E^{(i)}\|, \quad i = 0, 1, 2, \ldots \qquad (2.84)$$

or equivalently

$$\|E - E^{(i)}\| = O(\varepsilon^{2i}) \qquad (2.85)$$

In summary, the proposed parallel algorithm for the reduced-order solution of the weakly coupled discrete algebraic Riccati equation has the following form:

1. Transform Equation 2.66 into Equation 2.69 by using the bilinear transformation defined in Equation 2.70.
2. Solve Equation 2.69 by using the recursive reduced-order parallel algorithm defined by Equations 2.74 through 2.83.

2.5 APPROXIMATE LINEAR REGULATOR FOR DISCRETE SYSTEMS

The positive semidefinite stabilizing solution of the algebraic discrete Riccati Equation 2.66, produces the answer to the optimal linear-quadratic steady-state control problem. Namely, a quadratic criterion

$$J = \frac{1}{2} \sum_{k=0}^{\infty} \left(x^T(k) Q x(k) + u^T(k) R u(k) \right) \qquad (2.86)$$

is minimized along trajectories of a linear dynamic system

$$x(k+1) = Ax(k) + Bu(k) \qquad (2.87)$$

by using the control input of the form

$$u(x(k)) = -(R + B^T PB)^{-1} B^T PAx(k) \qquad (2.88)$$

where P is obtained from Equation 2.66 (Dorato and Levis 1971). It is derived in Litkouhi and Khalil (1984) that the near-optimal control given by

$$u^{(j)}(k) = -(R + B^T P^{(j)} B)^{-1} B^T P^{(j)} Ax(k) = -F^{(j)}x(k) \tag{2.89}$$

where $P^{(j)}$ satisfies

$$P^{(j)} - P^{opt} = O(\varepsilon^j) \tag{2.90}$$

is near-optimal in the sense

$$J_f^{(j)} - J_f^{opt} = O(\varepsilon^{2i}) \tag{2.91}$$

The approximate performance $J^{(j)}$ can be obtained from the discrete algebraic Lyapunov equation

$$K^{(j)} = \left(A - BF^{(j)}\right)^T K^{(j)} \left(A - BF^{(j)}\right) + Q + F^{(j)^T} RF^{(j)} \tag{2.92}$$

so that

$$J^{(j)} = \frac{1}{2} x^T(0) K^{(j)} x(0) \tag{2.93}$$

In Section 2.4 we have developed a very efficient technique for generating $P^{(j)}$ by using the recursive reduced-order schemes (Equations 2.79 through 2.83), such that each iteration improves the accuracy by an order of magnitude.

2.5.1   CASE STUDY: DISCRETE MODEL OF A CHEMICAL PLANT

A continuous-time real world physical example (a chemical plant model [Gomathi et al. 1980]) is used to demonstrate the efficiency of the proposed method. The discrete-time system matrices are obtained by performing discretization with the sampling rate $T=0.5$ of the corresponding continuous-time model. The discrete-time problem matrices are represented by the following data:

$$A = 10^{-2} \begin{bmatrix} 95.407 & 1.9643 & 0.3597 & 0.0673 & 0.0190 \\ 40.849 & 41.317 & 16.084 & 4.4679 & 1.1971 \\ 12.217 & 26.326 & 36.149 & 15.930 & 12.383 \\ 4.1118 & 12.858 & 27.209 & 21.442 & 40.976 \\ 0.1305 & 0.5808 & 1.8750 & 3.6162 & 94.280 \end{bmatrix}$$

$$B^T = 10^{-2} \begin{bmatrix} 0.0434 & 2.6606 & 3.7530 & 3.6076 & 0.4617 \\ -0.0122 & -1.0453 & -5.5100 & -6.6000 & -0.9148 \end{bmatrix}$$

$$Q = I_5, \quad R = I_2$$

The initial condition is assumed to be

$$x(0) = [1.5 \quad -1.0 \quad 0.5 \quad -0.5 \quad 2.0]$$

The small weak coupling parameter ε is built into the problem and can be roughly estimated from the strongest coupled matrix (matrix B). The strongest coupling is in the third row, where

$$\varepsilon = \frac{b_{31}}{b_{32}} = \frac{3.753}{5.510} = 0.68$$

Simulation results are obtained using the MATLAB package for computer-aided control system design. The components P_1, P_2, P_3 of the solution of the algebraic Riccati equation, obtained from Algorithm 2.4, is presented in Table 2.2.

TABLE 2.2
Reduced-Order Solution of the Discrete Weakly Coupled Algebraic Riccati Equation

J	$P_1^{(j)}$	$P_2^{(j)}$	$P_3^{(j)}$
0	20.9062 0.9202 0.9202 1.2382	1.8864 1.4363 18.5512 0.5259 0.3129 2.1851	1.2937 0.1971 1.2516 0.1971 1.1514 1.2887 1.2516 1.2887 21.0090
1	39.2212 2.5451 2.5453 1.5406	3.4274 2.3995 28.9280 0.7579 0.4431 3.3320	1.4754 0.2982 2.0621 0.2982 1.2067 1.7456 2.0621 1.7456 25.1918
2	50.6715 3.6522 3.6522 1.6830	4.2820 2.8654 32.9971 0.8642 0.5009 3.8322	1.5559 0.3424 2.4466 0.3424 1.2305 1.9460 2.4466 1.9460 26.7855
3	56.2107 4.2214 4.2214 1.7496	4.6836 3.0669 34.4667 0.9115 0.5253 4.0211	1.5914 0.3611 2.5976 0.3611 1.2400 2.0170 2.5976 2.0170 27.2170
4	58.6608 4.4804 4.4804 1.7791	4.8593 3.1514 35.0154 0.9317 0.5352 4.0904	1.6065 0.3687 2.6529 0.3687 1.2437 2.0421 2.6529 2.0421 27.3517
5	59.7068 4.5922 4.5922 1.7917	4.9338 3.1865 35.2273 0.9402 0.5393 4.1162	1.6128 0.3717 2.6727 0.3717 1.2451 2.0512 2.6727 2.0512 27.3995
6	60.1458 4.6393 4.6387 1.7969	4.9649 3.2009 35.3111 0.9436 0.5409 4.1260	1.6154 0.3730 2.6808 0.3730 1.2457 2.0567 2.6808 2.0567 27.4176
9	60.4348 4.6703 4.6703 1.8004	4.9853 3.2103 35.3637 0.9459 0.5420 4.1320	1.6171 0.3737 2.6853 0.3737 1.2461 2.0567 2.6853 2.0567 27.4286

TABLE 2.2 (continued)
Reduced-Order Solution of the Discrete Weakly Coupled Algebraic Riccati Equation

J	$P_1^{(j)}$	$P_2^{(j)}$	$P_3^{(j)}$
12	60.4549 4.6725	4.9867 3.2109 35.3672	1.6172 0.3738 2.6856
	4.6725 1.8006	0.9460 0.5420 4.1324	0.3718 1.2461 2.0569
			2.6856 2.0569 27.4293
16	60.4563 4.6727	4.9868 3.2109 35.3674	1.6172 0.3738 2.6856
	4.6727 1.8006	0.9460 0.5420 4.1324	0.3738 1.2461 2.0569
			2.6856 2.0569 27.4294
	$P_1 = P_1^{(15)}$	$P_2 = P_2^{(15)}$	$P_3 = P_3^{(15)}$

For this specific real world example the proposed algorithm perfectly matches the presented theory since convergence, with the accuracy of 10^{-4}, is achieved after nine iterations ($0.68^{18} \approx 10^{-4}$).

Note very dramatic changes in the element $P_1^{(j)}$ per iteration. Thus, in this example only higher order approximations produce satisfactory results. Corresponding differences between the optimal and approximate state trajectories for the corresponding components of the state vector are presented in Figures 2.1 through 2.5. The optimal and approximate control strategies are shown in Figures 2.6 and 2.7. In these figures, the solid lines represent the optimal quantities (state trajectories and controls); the point, dotted, dash-dotted, and dashed lines represent the approximate trajectories

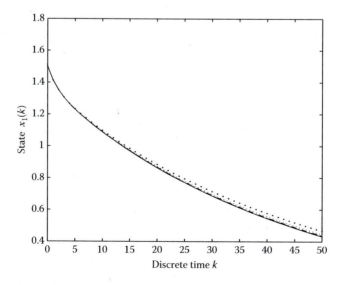

FIGURE 2.1 Approximate and optimal trajectories $x_1(k)$.

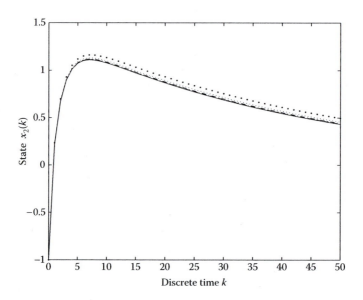

FIGURE 2.2　Approximate and optimal trajectories $x_2(k)$.

and controls, respectively, for $i = 1, 2, 3, 4$. The obtained figures justify the necessity for the existence of the higher order approximations for both the approximate control strategies and the approximate trajectories. We have also obtained the result for the approximate performance criterion in the form $|J - J^{(16)}| = 4.9454 \times 10^{-12}$.

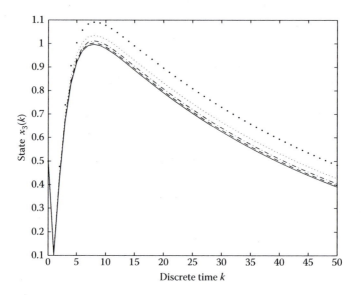

FIGURE 2.3　Approximate and optimal trajectories $x_3(k)$.

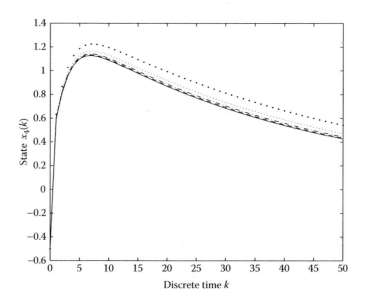

FIGURE 2.4 Approximate and optimal trajectories $x_4(k)$.

2.6 OUTPUT FEEDBACK CONTROL FOR LINEAR WEAKLY COUPLED SYSTEMS

The design of the optimal linear full state regulator requires measurements of all system states. In many practical applications, this is not feasible due to either the high cost of state measurements or the inaccessibility for measurement of some of the system states. The standard way to overcome these difficulties is to reconstruct

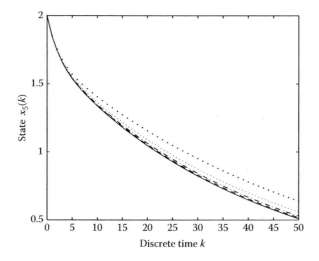

FIGURE 2.5 Approximate and optimal trajectories $x_5(k)$.

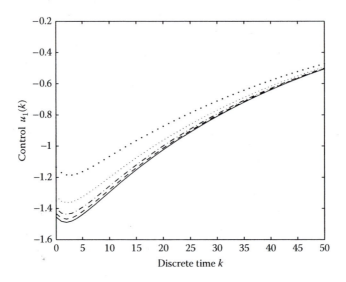

FIGURE 2.6   Approximate and optimal control strategies $u_1(k)$.

the full state vector from available measurements by using the Luenberger observer or, if measurements are noisy, by the Kalman filter. However, these state reconstruction methods will introduce an additional dynamical system. That is why, in the early 1970s, an increasing attention was given to the problem of designing output constrained regulators where a very limited member of state measurements were available for control implementation (Levine and Athans 1970; Levine et al. 1971;

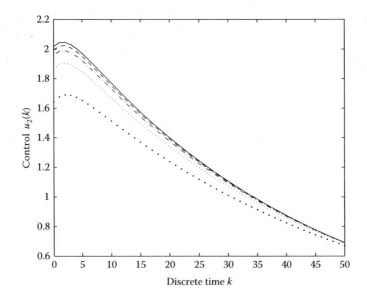

FIGURE 2.7   Approximate and optimal control strategies $u_2(k)$.

Mendel 1974; Petkovski and Rakić 1979). The optimal solution to this control problem has been obtained in terms of high-order nonlinear matrix algebraic equations. The convergence complexities of the algorithms suggested for the solution of these equations have hindered for quite a long time a wider application of this technique. The convergence problem was solved in Moerder and Calise (1985) and Toivonen (1985). Since then, the static output feedback control problem has become a very fruitful research area.

The output feedback control problem for weakly coupled linear systems has been studied in Petkovski and Rakić (1979) by using a series expansion approach. This approach is not recursive in application and it is numerically inefficient when a high order of accuracy is required or when the coupling parameter ε is not very small.

In this section, a recursive algorithm is developed for solving nonlinear algebraic equations comprising the solution of the optimal static output feedback control problem of linear weakly coupled systems. The numerical decomposition has been achieved so that only low-order systems are involved in algebraic computations. The effectiveness of the proposed reduced-order algorithm and its advantages over the global full-order algorithm are demonstrated on a 12-plate chemical absorption column. Obtained results strongly support necessity for reduced-order numerical techniques for solving corresponding nonlinear algebraic equations. In addition to reduction in required computations, it would be easier to find a good initial guess and to handle the problem of nonuniqueness of the solution of these nonlinear equations—they represent the necessary conditions only.

In this chapter, we have limited our attention to the deterministic continuous-time output feedback control problem. Stochastic output feedback makes no sense in the continuous-time domain since it is not rational to feedback the continuous-time white noise. In Chapter 6, the stochastic output feedback in the discrete-time domain is discussed.

Consider a linear system

$$\dot{x}(t) = Ax(t) + Bu(t), \quad x(t_0) = x_0 \tag{2.94}$$

$$y(t) = Cx(t) \tag{2.95}$$

where
  $x(t) \in \Re^n$ is the state vector
  $u(t) \in \Re^m$ is the control input
  $y(t) \in \Re^r$ is the measured output

In the following, A, B, and C are constant matrices of compatible dimensions; in general they are continuous functions of a small positive parameter ε. With Equations 2.94 and 2.95, consider the performance criterion

$$J = \frac{1}{2} \int\limits_0^\infty \left\{ x^T(t)Qx(t) + u^T(t)Ru(t) \right\} dt \tag{2.96}$$

with positive definite R and positive semidefinite Q, which has to be minimized. In addition, the control input $u(t)$ is constrained to

$$u(y(t)) = -Fy(t) \tag{2.97}$$

The optimal constant output feedback gain F is given by Levine and Athans (1970)

$$F = R^{-1}B^{T}KLC^{T}(CLC^{T})^{-1} \tag{2.98}$$

where matrices K and L satisfy high-order nonlinear coupled algebraic equations

$$(A - BFC)L + L(A - BFC)^{T} + x_0 x_0^{T} = 0 \tag{2.99}$$

$$(A - BFC)^{T}K + K(A - BFC) + Q + C^{T}F^{T}RFC = 0 \tag{2.100}$$

It is shown in Moerder and Calise (1985) that the algorithm proposed for the numerical solution of Equations 2.98 through 2.100 is defined by the following.

ALGORITHM 2.5

Choose $F^{(0)}$ such that $A - BF^{(0)}C$ is a stable matrix $\qquad$ (2.101)

$$\left(A - BF^{(i)}C\right)L^{(i+1)} + L^{(i+1)}\left(A - BF^{(i)}C\right)^{T} + x_0 x_0^{T} = 0 \tag{2.102}$$

$$\left(A - BF^{(i)}C\right)^{T}K^{(i+1)} + K^{(i+1)}\left(A - BF^{(i)}C\right) + Q$$
$$+ C^{T}F^{(i)^{T}}RF^{(i)}C = 0 \tag{2.103}$$

$$F^{(i+1)} = R^{-1}B^{T}K^{(i+1)}L^{(i+1)}C^{T}\left(CL^{(i+1)}C^{T}\right)^{-1} \tag{2.104}$$

with $i = 1, 2, \ldots$, converges to a local minimum under the nonrestrictive assumption. As a matter of fact, the updated value for F is defined in Moerder and Calise (1985) as

$$F_{N}^{(i+1)} = F^{(i)} + \alpha\left(F^{(i+1)} - F^{(i)}\right) \tag{2.105}$$

where $\alpha \in (0, 1)$ is chosen at each iteration to ensure that the minimum is not overshot, that is,

$$J_{i+1} = \mathrm{tr}\left\{K^{(i+1)}x_0 x_0^{T}\right\} < J_i = \mathrm{tr}\left\{K^{(i)}x_0 x_0^{T}\right\} \tag{2.106}$$

Note that in Toivonen and Makila (1987) the Newton method was derived for solving algebraic equations (Equations 2.98 through 2.100).

Exercise 2.2: Multiply Equation 2.100 from the left by CL and from the right by $L^{T}C^{T}$. Show that this procedure reduces Equation 2.100 to be standard algebraic Riccati equation. Comment on the procedure and obtained result.

Note that it has been customary in the control literature on the output feedback to assume that the initial conditions are uniformly distributed on the unit sphere, that is,

$$E\{x_0 x_0^{\mathrm{T}}\} = I_{(n_1+n_2)} \tag{2.107}$$

Consider now the weakly coupled linear system (Petkovski and Rakić 1979)

$$\dot{x}_1(t) = A_1 x_1(t) + \varepsilon A_2 x_2(t) + B_1 u_1(t) + \varepsilon B_2 u_2(t) \tag{2.108}$$

$$\dot{x}_2(t) = \varepsilon A_3 x_1(t) + A_4 x_2(t) + \varepsilon B_3 u_1(1) + B_4 u_2(t) \tag{2.109}$$

$$y(t) = \begin{bmatrix} C_1 & \varepsilon C_2 \\ \varepsilon C_3 & C_4 \end{bmatrix} \begin{bmatrix} x_1(t) \\ x_2(t) \end{bmatrix} = \begin{bmatrix} y_1(t) \\ y_2(t) \end{bmatrix} \tag{2.110}$$

where
  $x_1(t) \in \mathfrak{R}^{n_1}$ and $x_2(t) \in \mathfrak{R}^{n_2}$ are state space vectors
  $u_i(t) \in \mathfrak{R}^{m_i}$, $i = 1, 2$, are control inputs
  $y_i(t) \in \mathfrak{R}^{r}$, $i = 1, 2$ are measured outputs

In the following, A_i, B_i, and C_i, $i = 1, \ldots, 4$ are constant matrices of compatible dimensions; in general they are continuous functions of the small weak coupling parameter (Petrović and Gajić 1988).

With Equations 2.108 through 2.110, consider the performance criterion defined in Equation 2.96, with penalty matrices partitioned as

$$Q = \begin{bmatrix} Q_1 & \varepsilon Q_2 \\ \varepsilon Q_2^{\mathrm{T}} & Q_4 \end{bmatrix}, \quad R = \begin{bmatrix} R_1 & 0 \\ 0 & R_4 \end{bmatrix} \tag{2.111}$$

with positive definite R and positive semidefinite Q, which has to be minimized.

The control input $u(t)$ is constrained to be a direct feedback from the output $y(t)$ as given by formula 2.97 with the optimal feedback gain given by Equation 2.98, and matrices K and L satisfying high-order nonlinear coupled algebraic equations (Equations 2.99 and 2.100). The matrices A, B, and C are partitioned as

$$A = \begin{bmatrix} A_1 & \varepsilon A_2 \\ \varepsilon A_3 & A_4 \end{bmatrix}, \quad B = \begin{bmatrix} B_1 & \varepsilon B_2 \\ \varepsilon B_3 & B_4 \end{bmatrix}, \quad C = \begin{bmatrix} C_1 & \varepsilon C_2 \\ \varepsilon C_3 & C_4 \end{bmatrix} \tag{2.112}$$

Assuming that all partitioned matrices in Equations 2.111 through 2.112 are $O(1)$ (Assumption 2.1b), matrices K and L will be weakly coupled and partitioned as

$$K = \begin{bmatrix} K_1 & \varepsilon K_2 \\ \varepsilon K_2^{\mathrm{T}} & K_3 \end{bmatrix}, \quad L = \begin{bmatrix} L_1 & \varepsilon L_2 \\ \varepsilon L_2^{\mathrm{T}} & L_3 \end{bmatrix} \tag{2.113}$$

The algorithm given in Equations 2.101 through 2.104 can be applied for the numerical solution of the corresponding nonlinear matrix Equations 2.98 through 2.100. In order to simplify derivations, the following notation is introduced

$$[A - BFC] = \begin{bmatrix} D_1 & \varepsilon D_2 \\ \varepsilon D_3 & D_4 \end{bmatrix}$$

$$= \begin{bmatrix} A_1 - B_1 F_1 C_1 & \varepsilon(A_2 - B_1 F_2 C_4) \\ \varepsilon(A_3 - B_4 F_3 C_1) & A_4 - B_4 F_4 C_4 \end{bmatrix} \tag{2.114}$$

$$[Q + C^T F^T RFC] = \begin{bmatrix} q_1 & \varepsilon q_2 \\ \varepsilon q_3 & q_4 \end{bmatrix} \tag{2.115}$$

where

$$q_1 = Q_1 + C_1^T F_1^T R_1 F_1 C_1 + C_1^T F_3^T R_4 F_3 C_1$$

$$q_2 = Q_2 + C_1^T F_1^T R_1 F_2 C_4 + C_1^T F_3^T R_4 F_4 C_4$$

$$q_3 = Q_4 + C_4^T F_2^T R_1 F_2 C_4 + C_4^T F_4^T R_4 F_4 C_4$$

In addition, without loss of generality, it has been assumed that matrices B_2, B_3, C_2, and C_3 are zeros.

Partitioning Equations 2.102 and 2.103 compatible to Equations 2.112 and 2.113 and using Equations 2.114 and 2.115, we get the following set of equations

$$D_1^{(i)} L_1^{(i+1)} + L_1^{(i+1)} D_1^{(i)^T} + \varepsilon^2 \left(D_2^{(i)} L_2^{(i+1)^T} + L_2^{(i+1)} D_2^{(i)^T} \right) + I = 0 \tag{2.116}$$

$$L_2^{(i+1)} D_4^{(i)^T} + D_1^{(i)} L_2^{(i+1)} + L_1^{(i+1)} D_3^{(i)^T} + D_2^{(i)} L_3^{(i+1)} = 0 \tag{2.117}$$

$$L_3^{(i+1)} D_4^{(i)^T} + D_4^{(i)} L_3^{(i+1)} + \varepsilon^2 \left(D_3^{(i)} L_2^{(i+1)} + L_2^{(i+1)^T} D_3^{(i)^T} \right) + I = 0 \tag{2.118}$$

and

$$D_1^{(i)^T} K_1^{(i+1)} + K_1^{(i+1)} D_1^{(i)} + q_1^{(i)} + \varepsilon^2 \left(D_3^{(i)^T} K_2^{(i+1)^T} + K_2^{(i+1)} D_3^{(i)} \right) = 0 \tag{2.119}$$

$$K_2^{(i+1)} D_4^{(i)} + D_1^{(i)^T} K_2^{(i+1)} + D_3^{(i)^T} K_3^{(i+1)} + K_1^{(i+1)} D_2^{(i)} + q_2^{(i)} = 0 \tag{2.120}$$

$$K_3^{(i+1)} D_4^{(i)} + D_4^{(i)^T} K_3^{(i+1)} + q_3^{(i)} + \varepsilon^2 \left(D_2^{(i)^T} K_2^{(i+1)} + K_2^{(i+1)^T} D_2^{(i)} \right) = 0 \tag{2.121}$$

where

$$D_1^{(i)} = A_1 - B_1 F_1^{(i)} C_1, \quad D_2^{(i)} = A_2 - B_1 F_2^{(i)} C_4$$

$$D_3^{(i)} = A_3 - B_3 F_3^{(i)} C_1, \quad D_4^{(i)} = A_4 - B_4 F_4^{(i)} C_2$$

and

$$q_1^{(i)} = Q_1 + C_1^{\mathrm{T}} F_1^{(i)^{\mathrm{T}}} R_1 F_1^{(i)} C_1 + C_1^{\mathrm{T}} F_3^{(i)^{\mathrm{T}}} R_4 F_3^{(i)} C_1$$
$$q_2^{(i)} = Q_2 + C_1^{\mathrm{T}} F_1^{(i)^{\mathrm{T}}} R_1 F_2^{(i)} C_4 + C_1^{\mathrm{T}} F_3^{(i)^{\mathrm{T}}} R_4 F_4^{(i)} C_4$$
$$q_3^{(i)} = Q_4 + C_4^{\mathrm{T}} F_2^{(i)^{\mathrm{T}}} R_1 F_2^{(i)} C_4 + C_4^{\mathrm{T}} F_4^{(i)^{\mathrm{T}}} R_4 F_3^{(i)} C_4$$

with $i = 0, 1, 2, 3, \ldots$

Equations 2.116 through 2.118 and Equations 2.119 through 2.121 have been studied in Petkovski and Rakić (1979) by using a series expansion method. Their approach is not recursive in application and they numerically justify it for $O(\varepsilon)$ accuracy only. In this section, we develop a recursive scheme which will efficiently extend the main results of their work to any arbitrary order of accuracy, namely $O(\varepsilon^{2k})$, where k represents the number of required iterations of the proposed recursive scheme.

Note that all of matrices D_i, K_i, L_i, and q_i are functions of a small parameter ε. However, dependence on ε is suppressed in order to simplify notation.

Stability of the matrix

$$D^{(i)}(\varepsilon) = \begin{bmatrix} D_1^{(i)}(\varepsilon) & \varepsilon D_2^{(i)}(\varepsilon) \\ \varepsilon D_3^{(i)}(\varepsilon) & D_4^{(i)}(\varepsilon) \end{bmatrix}$$

is guaranteed by Moerder and Calise (1985). Due to block diagonal dominance of $D^{(i)}(\varepsilon)$, matrices $D_1^{(i)}(\varepsilon)$ and $D_4^{(i)}(\varepsilon)$ are stable for sufficiently small values of ε for $\forall i$.

In what follows Equations 2.116 through 2.121 will be numerically solved in terms of the reduced-order Lyapunov and Sylvester equations. Notice that Equations 2.116 through 2.121 represent standard Lyapunov equations of weakly coupled systems. An efficient recursive algorithm for their numerical solution, with $O(\varepsilon^2)$ rate of convergence is presented below.

The zeroth-order solutions of these equations are obtained by setting $\varepsilon = 0$ in Equations 2.116 through 2.121

$$D_1^{(i)} L_1^{(i+1)} + L_1^{(i+1)} D_1^{(i)^{\mathrm{T}}} + I = 0 \tag{2.122}$$

$$L_2^{(i+1)} D_4^{(i)^{\mathrm{T}}} + D_1^{(i)} L_2^{(i+1)} + L_1^{(i+1)} D_3^{(i)^{\mathrm{T}}} + D_2^{(i)} L_3^{(i+1)} = 0 \tag{2.123}$$

$$L_3^{(i+1)} D_4^{(i)^{\mathrm{T}}} + D_4^{(i)} L_3^{(i+1)} + I = 0 \tag{2.124}$$

and

$$D_1^{(i)^{\mathrm{T}}} K_1^{(i+1)} + K_1^{(i+1)} D_1^{(i)} + q_1^{(i)} = 0 \tag{2.125}$$

$$K_2^{(i+1)} D_4^{(i)} + D_1^{(i)^{\mathrm{T}}} K_2^{(i+1)} + D_3^{(i)^{\mathrm{T}}} K_3^{(i+1)} + K_1^{(i+1)} D_2^{(i)} + q_2^{(i)} = 0 \tag{2.126}$$

$$K_3^{(i+1)} D_4^{(i)} + D_4^{(i)^{\mathrm{T}}} K_3^{(i+1)} + q_3^{(i)} = 0 \tag{2.127}$$

It can be seen that the complete reduced-order decomposition is achieved in Equations 2.122 through 2.127, that is, one needs to solve four reduced-order Lyapunov and two reduced-order Sylvester equations.

The existence of the unique and bounded solutions of Equations 2.122 through 2.127 is guaranteed by the stability of $D^{(i)}(\varepsilon)$ (Moerder and Calise 1985). Due to stability of $D_1^{(i)}(\varepsilon)$ and $D_4^{(i)}(\varepsilon)$ the unique solutions of Equations 2.122 through 2.127 exist as well.

The zeroth-order solutions

$$\mathbf{L}^{(i+1)} = \begin{bmatrix} \mathbf{L}_1^{(i+1)} & \varepsilon \mathbf{L}_2^{(i+1)} \\ \varepsilon \mathbf{L}_2^{(i+1)^\mathsf{T}} & \mathbf{L}_3^{(i+1)} \end{bmatrix}, \quad \mathbf{K}^{(i+1)} = \begin{bmatrix} \mathbf{K}_1^{(i+1)} & \varepsilon \mathbf{K}_2^{(i+1)} \\ \varepsilon \mathbf{K}_2^{(i+1)^\mathsf{T}} & \mathbf{K}_3^{(i+1)} \end{bmatrix} \tag{2.128}$$

are $O(\varepsilon^2)$ close to the required ones $L^{(i+1)}$ and $K^{(i+1)}$ (Equations 2.116 through 2.121). We can relate them through the errors terms.

$$L^{(i+1)} - \mathbf{L}^{(i+1)} = \varepsilon^2 M = \varepsilon^2 \begin{bmatrix} M_1 & \varepsilon M_2 \\ \varepsilon M_2^\mathsf{T} & M_3 \end{bmatrix} \tag{2.129}$$

and

$$K^{(i+1)} - \mathbf{K}^{(i+1)} = \varepsilon^2 E = \varepsilon^2 \begin{bmatrix} E_1 & \varepsilon E_2 \\ \varepsilon E_2^\mathsf{T} & E_3 \end{bmatrix} \tag{2.130}$$

Clearly, the $O(\varepsilon^k)$ approximations for M and E will produce the $O(\varepsilon^{k+2})$ approximations of the required solutions. This is why we are interested in finding a convenient form for these error terms and the appropriate algorithm for their solution.

It can be shown that these error equations are given by

$$M_1 D_1^{(i)^\mathsf{T}} + D_1^{(i)} M_1 + D_2^{(i)} \left(\mathbf{L}_2^{(i+1)} + \varepsilon^2 M_2 \right)^\mathsf{T} + \left(\mathbf{L}_2^{(i+1)} + \varepsilon^2 M_2 \right) D_2^{(i)^\mathsf{T}} = 0 \tag{2.131}$$

$$M_2 D_4^{(i)^\mathsf{T}} + D_1^{(i)} M_2 + D_2^{(i)} M_3 + M_1 D_3^{(i)^\mathsf{T}} = 0 \tag{2.132}$$

$$M_3 D_4^{(i)^\mathsf{T}} + D_4^{(i)} M_3 + D_3^{(i)} \left(\mathbf{L}_2^{(i+1)} + \varepsilon^2 M_2 \right) + \left(\mathbf{L}_2^{(i+1)} + \varepsilon^2 M_2 \right)^\mathsf{T} D_3^{(i)^\mathsf{T}} = 0 \tag{2.133}$$

and

$$E_1 D_1^{(i)} + D_1^{(i)^\mathsf{T}} E_1 + D_3^{(i)^\mathsf{T}} \left(\mathbf{K}_2^{(i+1)} + \varepsilon^2 E_2 \right)^\mathsf{T} + \left(\mathbf{K}_2^{(i+1)} + \varepsilon^2 E_2 \right) D_3^{(i)} = 0 \tag{2.134}$$

$$E_2 D_4^{(i)} + D_1^{(i)^\mathsf{T}} E_2 + E_1 D_2^{(i)} + D_3^{(i)^\mathsf{T}} E_3 = 0 \tag{2.135}$$

$$E_3 D_4^{(i)} + D_4^{(i)^\mathsf{T}} E_3 + \left(\mathbf{K}_2^{(i+1)} + \varepsilon^2 E_2 \right)^\mathsf{T} D_2^{(i)} + D_2^{(i)^\mathsf{T}} \left(\mathbf{K}_2^{(i+1)} + \varepsilon^2 E_2 \right) = 0 \tag{2.136}$$

The weakly coupled and hierarchical structure of Equations 2.131 through 2.136 can be exploited by proposing the recursive scheme, which leads, after some algebra, to the six low-order completely decoupled recursive equations.

ALGORITHM 2.6

$$M_1^{(j+1)}D_1^{(i)^T} + D_1^{(i)}M_1^{(j+1)} + D_2^{(i)}\left(\mathbf{L}_2^{(i+1)} + \varepsilon^2 M_2^{(j)}\right)^T + \left(\mathbf{L}_2^{(i+1)} + \varepsilon^2 M_2^{(j)}\right)D_2^{(i)^T} = 0$$

(2.137)

$$M_2^{(j+1)}D_4^{(i)^T} + D_1^{(i)}M_2^{(j+1)} + D_2^{(i)}M_3^{(j+1)} + M_1^{(j+1)}D_3^{(i)^T} = 0 \tag{2.138}$$

$$M_3^{(j+1)}D_4^{(i)^T} + D_4^{(i)}M_3^{(j+1)} + D_3^{(i)}\left(\mathbf{L}_2^{(i+1)} + \varepsilon^2 M_2^{(j)}\right) + \left(\mathbf{L}_2^{(i+1)} + \varepsilon^2 M_2^{(j)}\right)^T D_3^{(i)^T} = 0$$

(2.139)

and

$$E_1^{(j+1)}D_1^{(i)} + D_1^{(i)^T}E_1^{(j+1)} + D_3^{(i)^T}\left(\mathbf{K}_2^{(i+1)} + \varepsilon^2 E_2^{(j)}\right)^T + \left(\mathbf{K}_2^{(i+1)} + \varepsilon^2 E_2^{(j)}\right)D_3^{(i)} = 0$$

(2.140)

$$E_2^{(j+1)}D_4^{(i)} + D_1^{(i)^T}E_2^{(j+1)} + E_1^{(j+1)}D_2^{(i)} + D_3^{(i)^T}E_3^{(j+1)} = 0 \tag{2.141}$$

$$E_3^{(j+1)}D_4^{(i)} + D_4^{(i)^T}E_3^{(j+1)} + \left(\mathbf{K}_2^{(i+1)} + \varepsilon^2 E_2^{(j)}\right)^T D_2^{(i)} + D_2^{(i)^T}\left(\mathbf{K}_2^{(i+1)} + \varepsilon^2 E_2^{(j)}\right) = 0$$

(2.142)

with $j = 1, 2, 3, \ldots$, and with initial conditions being chosen as $M_1^{(0)} = E_1^{(0)} = 0$, $M_2^{(0)} = E_2^{(0)} = 0$, and $M_3^{(0)} = E_3^{(0)} = 0$.

Observe the decoupled structure of Equations 2.137 through 2.142: the reduced-order Lyapunov equations (Equations 2.137, 2.139, 2.140, and 2.142) are solved first and then the Sylvester Equations 2.138 and 2.141 are solved.

The following theorem is proved in Harkara et al. (1989).

THEOREM 2.4

Algorithm (Equations 2.127 through 2.142) converges, for sufficiently small value of ε, to the solution of the error terms, and to the required solutions $L^{(i+1)}$ and $K^{(i+1)}$, with the rate of convergence of $O(\varepsilon^2)$.

Research Problem 2.1: Derive an asynchronous version of the parallel synchronous Algorithms 2.6. Establish a set of necessary and sufficient conditions for its convergence.

2.6.1 CASE STUDY: 12-PLATE ABSORPTION COLUMN

In order to illustrate the efficiency of the proposed algorithm for weakly coupled systems, the method is applied to the mathematical model of a 12-plate absorption column (Petkovski and Rakić 1979; Petkovski 1981). The system matrix is given by

$$A = \begin{bmatrix} A_1 & A_2 \\ A_3 & A_4 \end{bmatrix}$$

where

$$A_1 = A_4 = \begin{bmatrix} a_1 & a_2 & 0 & 0 & 0 & 0 \\ a_3 & a_1 & a_2 & 0 & 0 & 0 \\ 0 & a_3 & a_1 & a_2 & 0 & 0 \\ 0 & 0 & a_3 & a_1 & a_2 & 0 \\ 0 & 0 & 0 & a_3 & a_1 & a_2 \\ 0 & 0 & 0 & 0 & a_3 & a_1 \end{bmatrix}$$

with

$$a_1 = -1.73058, \quad a_2 = 0.634231, \quad a_3 = 0.538827$$

Here A_2 has all entries equal to zero except for $(A_2)_{6,1} = a_2$, and A_3 has all entries equal to zero except for $(A_3)_{1,6} = a_3$. The control matrix is partitioned as

$$[B_1 \ \varepsilon B_2] = \begin{bmatrix} b_1 & 0 \\ 0 & 0 \\ 0 & 0 \\ 0 & 0 \\ 0 & 0 \\ 0 & 0 \end{bmatrix}, \quad [\varepsilon B_3 \ B_4] = \begin{bmatrix} 0 & 0 \\ 0 & 0 \\ 0 & 0 \\ 0 & 0 \\ 0 & 0 \\ 0 & b_2 \end{bmatrix}$$

with

$$b_1 = 0.538827, \quad b_2 = 0.8809$$

and the input matrix is

$$C = \begin{bmatrix} 1 & 0 & 0 & 0 & 0 & 0 & 0 & 0 & 0 & 0 & 0 & 0 \\ 0 & 1 & 0 & 0 & 0 & 0 & 0 & 0 & 0 & 0 & 0 & 0 \\ 0 & 0 & 0 & 0 & 0 & 0 & 0 & 0 & 0 & 0 & 1 & 0 \\ 0 & 0 & 0 & 0 & 0 & 0 & 0 & 0 & 0 & 0 & 0 & 1 \end{bmatrix}$$

The initial conditions are

$$x_1^T(0) = [-0.036 \ -0.066 \ -0.092 \ -0.113 \ -0.132 \ -0.148]$$
$$x_2^T(0) = [-0.161 \ -0.173 \ -0.182 \ -0.190 \ -0.197 \ -0.203]$$

TABLE 2.3

Optimal and Approximate Criteria and Gains

I	$J_{opt}^{(i)}$	$J_{red}^{(i)}$	$\left\|F_{opt}^{(i)} - F_{opt}\right\|_{\infty}$	$\left\|F_{red}^{i} - F_{opt}^{(i)}\right\|_{\infty}$ $j = 6$
1	0.97305	0.97289	13.038	13.086
2	0.27731	0.27778	4.050	3.975
3	0.24112	0.24109	2.527	2.616
4	0.22316	0.22308	1.677	2.207
5	0.21596	0.21604	1.834	1.574
6	0.21355	0.21372	7.861	0.908
7	0.21286	0.21301	11.759	0.477
8	0.21277	0.21281	3.003	0.242
9	0.21274	0.21275	3.157	0.123
10	0.21274	0.21274	4.625	0.064
12	0.21273	0.21273	6.626	0.019
16	0.21273	0.21273	62.600	0.002
18	0.21273	0.21273	36.207	0.000
20	0.21273	0.21273	26.833	0.0000
22	a	0.21273	a	0.000

[a] The global algorithm fails to produce solution for $i > 21$.

The penalty matrices in the performance index are $Q = I_{12}$, $R = I_2$. The small coupling parameter ε is equal to 0.5.

In Table 2.3, we present results for the criterion and the gain error for the global algorithm of Moerder and Calise (1985), and the corresponding quantities for the proposed reduced-order recursive algorithm. The parameter α is chosen as $\alpha = 0.5$.

In order to facilitate finding the solution to the problem under study by using the global algorithm and to avoid problems of system instability, smaller values of α were used. The entries in Table 2.4 show the results obtained by using $\alpha = 0.05$ and $\alpha = 0.1$. The global algorithm fails to produce a unique value for the solution even though convergence to the optimal value of the criterion is achieved at $i = 116$ and $i = 55$ for $\alpha = 0.05$ and $\alpha = 0.1$, respectively.

The initial value for the gain $F^{(0)}$ is obtained by using the method proposed in Petkovski and Rakić (1978). The global algorithm takes 11 iterations to achieve the accuracy of up to 5 decimal digits, where $J_{opt} = 0.21273$.

It is important to note that the nonuniqueness of the solution of Equations 2.98 through 2.100 is shown by the entries in the fourth column of Table 2.3 which were obtained by using the global algorithm. It is seen that there are several possible solutions to the optimal control problem even though convergencies to the optimal value of the criterion is achieved at $i = 11$. Furthermore, for $i \geq 22$, with $\alpha = 0.5$ the global algorithm fails to produce the solution so that it cannot converge to the unique value of the gain. From the entries in the fifth column of Table 2.3, it is clear that by using the reduced-order algorithm proposed, the difficulty of nonuniqueness of the

TABLE 2.4

Nonuniqueness of the Global Algorithm

$\varepsilon = 0.5$	$\left\|F_{opt}^{(i)} - F_{opt}\right\|_\infty$	$\left\|F_{opt}^{(i)} - F_{opt}\right\|_\infty$
I	$\alpha = 0.05$	$\alpha = 0.1$
1	13.028	13.028
10	3.175	1.587
20	1.626	1.107
30	1.058	2.071
40	1.025	2.666
50	1.249	0.608
60	1.811	1.110
70	2.527	1.753
80	2.631	2.0119
90	1.632	2.069
100	0.826	2.009
110	0.714	1.903
120	0.949	1.783
130	1.253	1.666
140	1.510	1.555
150	1.692	1.456
160	1.804	1.366
170	1.864	1.283
180	1.889	1.210
190	1.888	1.144
200	1.867	1.082

solution to the optimal output control problem is resolved since the reduced-order algorithm produces a unique value of the feedback gain F. In addition, there are no problems with system instability when the reduced-order algorithm is used.

This example clearly shows the superiority of the reduced-order algorithm over the global algorithm.

2.7 NOTES AND COMMENTS

The presented algorithms are applicable to large-scale systems already in the weakly coupled form. It will be interesting to develop parallel algorithms for the optimal control of general large-scale systems (Siljak 1978), brought in the above forms by using the overlapping decomposition technique of Siljak (1991) and the optimal linear-quadratic control technique with the prescribed degree of stability (Anderson and Moore 1990). It will be also interesting to extend these results to the linear algebra problems of weakly coupled systems. Some results in that direction have been already obtained by using the overlapping decomposition (Sezer and Siljak 1991). Even more, similar type of algorithms can be developed for solving nonlinear algebraic equations (Zecević and Siljak 1994; Amano et al. 1996; Gacic et al. 1998).

Another important future research topic is the development of asynchronous versions of the presented algorithms. The importance of asynchronous algorithms for block diagonally dominant systems (weakly coupled systems) is documented in Kaszkurewicz et al. (1990).

Very recently, Professor Mukaidani has developed new interesting techniques for recursively finding solutions to optimal control problems of weakly coupled systems (Mukaidani 2006a,b, 2007a–c; see also Sagara et al. 2007). In addition to the linear-quadratic regulator problem and the corresponding Riccati equation of weakly coupled systems, studied in Mukaidani (2007c), the Nash games of weakly coupled systems were considered in Mukaidani (2006a,b), and the zero-sum differential games of weakly coupled systems were researched in Mukaidani (2007a). With his coworkers (Sagara et al. 2007), Professor Mukaidani has considered the stochastic H_∞ optimal control of weakly coupled systems with state-dependence noise. Another interesting approach to decoupling weakly coupled systems can be found in Derbel (1999). The same author considers the optimal control of weakly coupled systems in Derbel (1995, 1998, 2001).

REFERENCES

Amano, M., A. Zecević, and D. Siljak, An improved block-parallel Newton method via epsilon decompositions for load-flow calculations, *IEEE Transactions on Power Systems*, 11, 1519–1525, 1996.

Anderson, B. and J. Moore, *Optimal Control: Linear Quadratic Methods*, Prentice-Hall, Englewood Cliffs, NJ, 1990.

Bertsekas, D. and J. Tsitsiklis, *Parallel and Distributed Computation: Numerical Methods*, Prentice-Hall, Englewood Cliffs, NJ, 1989.

Bertsekas, D. and J. Tsitsiklis, Some aspects of parallel and distributed iterative algorithms— A survey, *Automatica*, 27, 3–21, 1991.

Chow, J., *Time-Scale Modeling of Dynamic Networks with Applications to Power Systems*, Springer Verlag, New York, 1982.

Chow, J. and P. Kokotović, Sparsity and time scales, *Proceedings of the American Control Conference*, San Francisco, CA, 656–661, 1983.

Delebecque, F. and J. Quadrat, Optimal control of Markov-chains admitting strong and weak interactions, *Automatica*, 17, 281–296, 1981.

Derbel, N., Optimal control of linear weakly coupled systems, *Proceedings of the Conference on Decision and Control*, New Orleans, LA, 643–648, 1995.

Derbel, N., A simple assignment of the weakly coupled system optimal control, *Proceedings of the IFAC Conference Large Scale Systems: Theory and Applications*, Patras, Greece, 1018–1023, 1998.

Derbel, N., A new decoupling algorithm of weakly coupled systems, *System Analysis, Modeling and Simulation*, 35, 359–374, 1999.

Derbel, N., How to solve Lyapunov iterative equations, *Computers and Electrical Engineering*, 27, 459–474, 2001.

Dorato, P. and A. Levis, Optimal linear regulators: The discrete time case, *IEEE Transactions on Automatic Control*, AC-16, 613–620, 1971.

Gacic, N., A. Zecević, and D. Siljak, Coherency recognition using epsilon decomposition, *IEEE Transactions on Power Systems*, 13, 314–319, 1998.

Gajić, Z. and M. Qureshi, *Lyapunov Matrix Equation in Systems Stability and Control*, Academic Press, San Diego, CA, 1995.

Gomathi, K., S. Prabhu, and M. Pai, A suboptimal controller for minimum sensitivity of closed-loop eigenvalues to parameter variations, *IEEE Transactions on Automatic Control*, AC-25, 587–588, 1980.

Harkara, N., D. Petkovski, and Z. Gajić, The recursive algorithm for the optimal static output feedback control problem of linear weakly coupled systems, *International Journal of Control*, 50, 1–11, 1989.

Hanselman, D. and B. Littlefield, *Mastering MATLAB 6*, Prentice-Hall, Upper Saddle River, NJ, 2001.

Kando, H., T. Iwazumi, and H. Ukai, Singular perturbation modeling of large-scale systems with multi-time scale property, *International Journal of Control*, 48, 2361–2387, 1988.

Kaszkurewicz, E., A. Bhaya, and D. Siljak, On the convergence of parallel asynchronous block-iterative computations, *Linear Algebra and Its Applications*, 131, 139–160, 1990.

Kokotović, P. and J. Cruz, An approximation theorem for linear optimal regulators, *Journal of Mathematical Analysis and Applications*, 27, 249–252, 1969.

Kokotović, P., W. Perkins, J. Cruz, and G. D'Ans, ε—coupling approach for near-optimum design of large scale linear systems, *Proceedings of IEE, Part D*, 116, 889–892, 1969.

Kondo, R. and K. Furuta, On the bilinear transformation of Riccati equations, *IEEE Transactions Automatic Control*, AC-31, 50–54, 1986.

Kwakernaak, H. and R. Sivan, *Linear Optimal Control Systems*, Wiley-Interscience, New York, 1972.

Levine, W. and M. Athans, On the determination of the optimal constant output feedback gains for linear multivariable systems, *IEEE Transactions on Automatic Control*, AC-15, 44–48, 1970.

Levine, W., T. Johnson, and M. Athans, Optimal limited state variable feedback controllers for linear systems, *IEEE Transactions Automatic Control*, AC-16, 785–793, 1971.

Litkouhi, B. and H. Khalil, Infinite-time regulators for singularly perturbed difference equations, *International Journal of Control*, 39, 587–598, 1984.

Luenberger, D. *Optimization by Vector Space Methods*, Wiley, New York, 1969.

Mendel, J., A concise derivation of optimal constant limited feedback gains, *IEEE Transactions on Automatic Control*, AC-19, 447–448, 1974.

Moerder, D. and A. Calise, Convergence of a numerical algorithm for calculating optimal output feedback gains, *IEEE Transactions Automatic Control*, AC-30, 900–903, 1985.

Mukaidani, H., A numerical analysis of the Nash strategy for weakly coupled large-scale systems, *IEEE Transactions on Automatic Control*, 51, 1371–1377, 2006a.

Mukaidani, H., Optimal numerical strategy for Nash games of weakly coupled large-scale systems, *Dynamics of Continuous, Discrete, and Impulsive Systems*, 13, 249–268, 2006b.

Mukaidani, H., Numerical computation of sign indefinite linear quadratic differential games for weakly coupled linear large systems, *International Journal of Control*, 80, 75–86, 2007a.

Mukaidani, H., Newton method for solving cross-coupled sign-indefinite algebraic Riccati equations of weakly coupled large scale systems, *Applied Mathematics and Computation*, 188, 103–115, 2007b.

Mukaidani, H., Numerical computation for solving algebraic Riccati equation of weakly coupled systems, *Electrical Engineering of Japan*, 160, 39–48, 2007c.

Ortega, J. and W. Rheinboldt, *Iterative Solution of Nonlinear Equations in Several Variables*, SIAM Publishers, Philadelphia, PA, 2000.

Petrović, B. and Z. Gajić, The recursive solution of linear quadratic Nash games for weakly interconnected systems, *Journal of Optimization Theory and Applications*, 56, 463–477, 1988.

Petkovski, D., Design of decentralized proportional-plus-integral controllers for multivariable systems, *Computer and Chemical Engineering*, 5, 51–56, 1981.

Petkovski, D. and M. Rakić, On the calculation of optimum feedback gains for output constrained regulators, *IEEE Transactions on Automatic Control*, 23, 760, 1978.

Petkovski, D. and M. Rakić, A series solution of feedback gains for output constrained regulators, *International Journal of Control*, 30, 661–669, 1979.

Phillips, R. and P. Kokotović, A singular perturbation approach to modeling and control of Markov chains, *IEEE Transactions on Automatic Control*, 26, 1087–1094, 1981.

Power, H., Equivalence of Lyapunov matrix equations for continuous and discrete systems, *Electronic Letters*, 3, 83, 1967.

Sagara, M., H. Mukaidani, and T. Yamamoto, Stochastic H_∞ control problem with state-dependent noise for weakly coupled systems, *IEEJ Transactions of Electrical and Electronic Engineering*, 127, 571–578, 2007.

Sezer, M. and D. Siljak, Nested epsilon decomposition of linear systems: Weakly coupled and overlapping blocks, *SIAM Journal of Matrix Analysis and Applications*, 12, 521–533, 1991.

Siljak, D., *Large-Scale Dynamic Systems: Stability and Structure*, North Holland, New York, 1978.

Siljak, D., *Decentralized Control of Complex Systems*, Academic Press, Cambridge, 1991.

Shen, X. and Z. Gajić, Optimal reduced-order solution of the weakly coupled discrete Riccati equation, *IEEE Transactions Automatic Control*, AC-35, 600–602, 1990.

Shen, X., Z. Gajić, and D. Petkovski, Parallel reduced-order algorithms for Lyapunov equations of large scale linear systems, *Proceedings of IMACS Symposium MCTS*, Lille, France, 697–702, 1991.

Stewart, G., *Introduction to Matrix Computation*, Academic Press, New York, 1973.

Toivonen, H., A globally convergent algorithm for the optimal constant output feedback problem, *International Journal of Control*, 41, 1589–1599, 1985.

Toivonen, H. and P. Makila, Newton's method for solving parametric linear quadratic control problems, *International Journal of Control*, 46, 897–911, 1987.

Zecević, A. and D. Siljak, A block-parallel Newton method via overlapping decompositions, *SIAM Journal on Matrix Analysis and Applications*, 15, 824–844, 1994.

3 Quasi-Weakly Coupled Linear Control Systems

Several structures of linear-quadratic control problems containing small parameters can be studied efficiently by using the methodology similar to the one presented in Chapter 2. We call these structures quasi-weakly coupled systems. Namely, the quasi-weakly coupled linear-quadratic optimal control problem is very closely related to the corresponding standard weakly coupled control problems. However, these similarities are not obvious, and very often, in many applications, quasi-weakly coupled structures produce simple parallel reduced-order algorithms converging under milder conditions than the standard weakly coupled linear-quadratic control algorithm.

3.1 OPTIMAL CONTROLLER FOR QUASI-WEAKLY COUPLED LINEAR SYSTEMS

In this section, we consider a special class of linear systems having block diagonally dominant system matrix and with the control input influencing only one of the subsystems. The optimal reduced-order controllers are designed through the recursive reduced-order algorithm which converges quickly to the required optimal solution. Many real world systems (such as power systems, chemical reactors, flexible structures, and, in general, systems with only few actuators) possess the control structure studied in this section. We call these structures quasi (nearly) weakly coupled since they contain the diagonally block dominant system matrix (like the standard weakly coupled systems), but they have only one decision maker (weakly coupled systems require at least two decision makers).

Consider a linear dynamical system composed of two subsystems in the form

$$\begin{bmatrix} \dot{x}_1(t) \\ \dot{x}_2(t) \end{bmatrix} = \begin{bmatrix} A_1 & \varepsilon A_2 \\ \varepsilon A_3 & A_4 \end{bmatrix} \begin{bmatrix} x_1(t) \\ x_2(t) \end{bmatrix} + \begin{bmatrix} B_1 \\ 0 \end{bmatrix} u(t) \tag{3.1}$$

where
$x_i(t) \in R^{n_i}$, $i = 1, 2$, are state vectors
$u(t) \in \Re^m$ is a control input vector
ε is a small parameter

This is a special class of linear dynamical systems represented, in general, by

$$\dot{x}(t) = Ax(t) + Bu(t) \tag{3.2}$$

with

$$x(t) = \begin{bmatrix} x_1(t) \\ x_2(t) \end{bmatrix}, \quad A = \begin{bmatrix} A_1 & \varepsilon A_2 \\ \varepsilon A_3 & A_4 \end{bmatrix}, \quad B = \begin{bmatrix} B_1 \\ 0 \end{bmatrix} \tag{3.3}$$

The results presented in this chapter are valid under the standard weak coupling assumption (Chow and Kokotović 1983).

Assumption 3.1a   The magnitudes of all the system eigenvalues are $O(1)$, $|\lambda_j| = O(1)$, $j = 1, 2, \ldots, n$, which implies $\det\{A_1(\varepsilon)\} = O(1)$ and $\det\{A_4(\varepsilon)\} = O(1)$.

As indicated in Section 2.2, this assumption in fact indicates block diagonal dominance of the system matrix. It states the condition which guarantees that weak connections among the subsystems will indeed imply weak dynamic coupling. Note that when this assumption is not satisfied, in addition of weak coupling, the system can also display the multiple timescale phenomenon known as singular perturbations.

A quadratic type functional to be minimized is associated with Equation 3.1 in the form

$$J = \frac{1}{2} \int_0^\infty \left(x^T(t)Qx(t) + u^T(t)Ru(t) \right) dt, \quad Q \geq 0, \quad R > 0 \tag{3.4}$$

For the purpose of this section we assume that the structure of the matrix Q is consistent with the system matrix A, that is

$$Q = \begin{bmatrix} Q_1 & \varepsilon Q_2 \\ \varepsilon Q_2^T & Q_3 \end{bmatrix} \tag{3.5}$$

All problem matrices defined in Equations 3.1 through 3.5 are constant and of appropriate dimensions.

The structure defined in Equation 3.1 corresponds to the weakly interconnected subsystems (Kokotović et al. 1969) with the control input influencing only one of them. This structure has not been studied in the literature from the order reduction point of view. The purpose of this section is not to derive new theoretical concepts. Instead, its main goal is to show that certain classes of linear optimal control problems can be studied by using the developed reduced-order recursive theory for the weakly coupled linear control systems.

The optimal problem of minimizing Equation 3.4 along trajectories of Equation 3.1 has the very well-known solution given by

$$u_{opt}(x(t)) = -F_{opt}x(t) = -R^{-1}B^T Px(t) \tag{3.6}$$

where P is the positive semidefinite stabilizing solution of the algebraic Riccati equation

$$A^\mathrm{T}P + PA + Q - PSP = 0, \quad S = BR^{-1}B^\mathrm{T} \tag{3.7}$$

To obtain the weakly coupled structure for the solution of the algebraic equation, we need the following assumption to be imposed to the problem matrices.

Assumption 3.1b Matrices A_i, $i = 1, 2, 3, 4$, B_1, Q_j, $j = 1, 2, 3$, and R are constant and $O(1)$.

Under Assumption 3.1b, the solution of Equation 3.7 has the weakly coupled form and it is given by

$$P = \begin{bmatrix} P_1 & \varepsilon P_2 \\ \varepsilon P_2^\mathrm{T} & P_3 \end{bmatrix} \tag{3.8}$$

Partitioning Equation 3.7 compatible to Equations 3.3, 3.5, and 3.8, we get three matrix algebraic equations

$$P_1 A_1 + A_1^\mathrm{T} P_1 + Q_1 - P_1 S_1 P_1 + \varepsilon^2 \left(P_2 A_3 + A_3^\mathrm{T} P_2 \right) = 0 \tag{3.9}$$

$$P_2 A_4 + A_1^\mathrm{T} P_2 + Q_2 - P_1 S_1 P_2 + A_3^\mathrm{T} P_3 + P_1 A_2 = 0 \tag{3.10}$$

$$P_3 A_4 + A_4^\mathrm{T} P_3 + Q_3 + \varepsilon^2 \left(P_2^\mathrm{T} A_2 + A_2^\mathrm{T} P_2 - P_2^\mathrm{T} S_1 P_2 \right) = 0 \tag{3.11}$$

where $S_1 = B_1 R^{-1} B_1^\mathrm{T}$.

Since ε is a small parameter, we can define $O(\varepsilon^2)$ approximation of Equations 3.9 through 3.11 as follows:

$$P_1^{(0)} A_1 + A_1^\mathrm{T} P_1^{(0)} + Q_1 - P_1^{(0)} S_1 P_1^{(0)} = 0 \tag{3.12}$$

$$P_2^{(0)} A_4 + \left(A_1 - S_1 P_1^{(0)} \right)^\mathrm{T} P_2^{(0)} + Q_2 + A_3^\mathrm{T} P_3^{(0)} + P_1^{(0)} A_2 = 0 \tag{3.13}$$

$$P_3^{(0)} A_4 + A_4^\mathrm{T} P_3^{(0)} + Q_3 = 0 \tag{3.14}$$

so that the sought solution Equation 3.8 satisfies

$$P^{(0)} = \begin{bmatrix} P_1^{(0)} & \varepsilon P_2^{(0)} \\ \varepsilon P_2^{(0)\mathrm{T}} & P_3^{(0)} \end{bmatrix} = P + O(\varepsilon^2) \tag{3.15}$$

On the contrary to the standard weakly coupled systems (Kokotović et al. 1969), where the zeroth-approximation is given in terms of two reduced-order Riccati equations, for the quasi-weakly coupled systems we need to solve only one reduced-order Riccati Equation 3.12.

The unique positive semidefinite stabilizing solution $P^{(0)}$, obtained from Equations 3.12 through 3.14, exists under the following assumption.

Assumption 3.2 The triple $(A_1, B_1, \mathrm{Chol}(Q_1))$ is stabilizable–detectable and the matrix A_4 is stable.

Defining the approximation errors as

$$P_j = P_j^{(0)} + \varepsilon^2 E_j, \quad j = 1, 2, 3 \tag{3.16}$$

and using Equation 3.16 in Equations 3.9 through 3.11 and Equations 3.12 through 3.14, we get the following error equations:

$$E_1 D_1 + D_1^T E_1 = -P_2 A_3 - A_3^T P_2^T + \varepsilon^2 E_1 S_1 E_1 \tag{3.17}$$

$$E_2 A_4 + D_1^T E_2 = \varepsilon^2 E_1 S_1 E_2 - E_1 A_2 - A_3^T E_3 - E_1 S_1 P_1^{(0)} \tag{3.18}$$

$$E_3 A_4 + A_4^T E_3 = P_2^T A_2 + A_2^T P_2 - P_2^T S_1 P_2 \tag{3.19}$$

where $D_1 = A_1 - S_1 P_1^{(0)}$, is a stable matrix (Gajić et al. 1990).

Let us propose the following reduced-order parallel algorithm for solving Equations 3.17 through 3.19.

ALGORITHM 3.1

$$E_1^{(i+1)} D_1 + D_1^T E_1^{(i+1)} = -P_2^{(i)} A_3 - A_3^T P_2^{(i)^T} + \varepsilon^2 E_1^{(i)} S_1 E_1^{(i)} \tag{3.20}$$

$$E_3^{(i+1)} A_4 + A_4^T E_3^{(i+1)} = P_2^{(i)^T} A_2 + A_2^T P_2^{(i)} - P_2^{(i)^T} S_1 P_2^{(i)} \tag{3.21}$$

$$E_2^{(i+1)} A_4 + D_1^T E_2^{(i+1)} = \varepsilon^2 E_1^{(i+1)} S_1 E_2^{(i)} - E_1^{(i+1)} D_{12} - A_3^T E_3^{(i+1)} \tag{3.22}$$

where

$$P_j^{(i)} = P_j^{(0)} + \varepsilon^2 E_j^{(i)}, \quad j = 1, 2, 3, \quad i = 0, 1, 2, 3, \ldots \tag{3.23}$$

and $D_{12} = A_2 - S_1 P_2^{(0)}$ with initial conditions

$$E_1^{(0)} = 0, \quad E_2^{(0)} = 0, \quad E_3^{(0)} = 0 \tag{3.24}$$

The following theorem indicates the features of the proposed algorithm (Equations 3.20 through 3.24).

THEOREM 3.1

Under conditions stated in Assumption 3.2, the algorithm (Equations 3.20 through 3.24) converges to the exact solution of the error term with the rate of convergence of $O(\varepsilon^2)$, that is

$$\left\| E_j - E_j^{(i+1)} \right\| = O(\varepsilon^2) \left\| E_j - E_j^{(i)} \right\| \tag{3.25}$$

or equivalently

$$\left\| E_j - E_j^{(i)} \right\| = O\big(\varepsilon^{2(i+1)}\big), \quad i = 0, 1, 2, \ldots \tag{3.26}$$

The proof of this theorem follows ideas reported in Gajić et al. (1990), and thus, is omitted. In the first step of the proof, the nonsingularity of the Jacobian of Equations 3.17 through 3.19 at $\varepsilon = 0$ has to be established. In the second step, the error estimates given by Equations 3.25 and 3.26 are obtained from Equations 3.17 through 3.19 and Equations 3.20 through 3.22. We will justify results stated in Theorem 3.1 on several real control system examples (Sections 3.1.1 through 3.1.3).

Notice that from Equations 3.25 and 3.26 we have

$$\left\| P - P^{(i)} \right\| = O\big(\varepsilon^{2(i+1)}\big) \tag{3.27}$$

where

$$P^{(i)} = \begin{bmatrix} P_1^{(0)} + \varepsilon^2 E_1^{(i)} & \varepsilon\big(P_2 + \varepsilon^2 E_2^{(i)}\big) \\ \varepsilon\big(P_2 + \varepsilon^2 E_2^{(i)}\big)^{\mathrm{T}} & P_3^{(0)} + \varepsilon^2 E_3^{(i)} \end{bmatrix} \tag{3.28}$$

The approximate optimal gain $F^{(i)}$ is defined by

$$F^{(i)} = -R^{-1} B^{\mathrm{T}} P^{(i)} \tag{3.29}$$

and the approximate criterion is obtained from

$$J_{\mathrm{app}}^{(i)} = \mathrm{tr}\big\{ V^{(i)} \big\} \tag{3.30}$$

where $V^{(i)}$ satisfies

$$\big(A - SP^{(i)}\big)^{\mathrm{T}} V^{(i)} + V^{(i)}\big(A - SP^{(i)}\big) + Q + P^{(i)} SP^{(i)} = 0 \tag{3.31}$$

Using the criterion approximation theorem (Kokotović and Cruz 1969), Equation 3.27 implies

$$\left| J_{\mathrm{opt}} - J_{\mathrm{app}}^{(i)} \right| = O\big(\varepsilon^{4(i+1)}\big), \quad i = 0, 1, 2, \ldots \tag{3.32}$$

In some applications, like power systems, the open-loop system matrix A is stable and the elements in the matrix B_1 are all of $O(\varepsilon)$. In such cases, the presented algorithm can be even more simplified under the following assumption.

Assumption 3.3 The stability of the matrix A implies stability of the partitioned matrices A_1 and A_4.

The zeroth-order approximations in Equations 3.12 and 3.13 are now defined by (note that Equation 3.14 remains unchanged)

$$P_1^{(0)}A_1 + A_1^{\mathrm{T}}P_1^{(0)} + Q_1 = 0 \tag{3.33}$$

$$P_2^{(0)}A_4 + A_1^{\mathrm{T}}P_2^{(0)} + Q_2 + A_3^{\mathrm{T}}P_3^{(0)} + P_1^{(0)}A_2 = 0 \tag{3.34}$$

Introducing the notation

$$B_1 = \varepsilon B_{1p}, \quad S_{1p} = B_{1p}R^{-1}B_{1p}^{\mathrm{T}} \tag{3.35}$$

the modified algorithm (Equations 3.20 through 3.22) gets the following form.

ALGORITHM 3.2

$$E_1^{(i+1)}A_1 + A_1^{\mathrm{T}}E_1^{(i+1)} = -P_2^{(i)}A_3 - A_3^{\mathrm{T}}P_2^{(i)^{\mathrm{T}}} + P_1^{(i)}S_{1p}P_1^{(i)} \tag{3.36}$$

$$E_3^{(i+1)}A_4 + A_4^{\mathrm{T}}E_3^{(i+1)} = -P_2^{(i)^{\mathrm{T}}}A_2 - A_2^{\mathrm{T}}P_2^{(i)} + \varepsilon^2 P_2^{(i)^{\mathrm{T}}}S_{1p}P_2^{(i)} \tag{3.37}$$

$$E_2^{(i+1)}A_4 + A_1^{\mathrm{T}}E_2^{(i+1)} = P_1^{(i+1)}S_{1p}P_1^{(i+1)} - E_1^{(i+1)}A_2 - A_3^{\mathrm{T}}E_3^{(i+1)} \tag{3.38}$$

Thus, the complete solution, in this case, is obtained in terms of algebraic Lyapunov equations only.

In Chapter 2, we have shown how to generate the solution of the algebraic Riccati equation in terms of algebraic equations corresponding to reduced-order subsystems. Having obtained this solution, Equation 3.28, allows us to construct an approximation to the optimal control as follows

$$u^{(i)}(x^{(i)}(t)) = F^{(i)}x^{(i)}(t)$$

where $F^{(i)}$ is given by Equation 3.29 and $x^{(i)}(t)$ satisfies

$$\dot{x}^{(i)}(t) = \left(A - BF^{(i)}\right)x^{(i)}(t)$$

Using Equations 3.27 and 3.29, it follows that the control law $u^{(i)}(t)$ and the approximate trajectories $x^{(i)}(t)$ are suboptimal in the sense

$$x^{(i)}(t) = x_{\mathrm{opt}}(t) + O\left(\varepsilon^{2(i+1)}\right), \quad u^{(i)}\left(x^{(i)}(t)\right) = u_{\mathrm{opt}}(x(t)) + O\left(\varepsilon^{2(i+1)}\right)$$

The approximate feedback control $u^{(i)}(x^{(i)}(t))$ applied to the system produces the approximate performance index (Equation 3.30) with its property established in Equation 3.32.

In the following sections, we consider three real physical system control problems: chemical reactor, F-4 fighter aircraft, and multimachine power system, and demonstrate the near-optimality of the presented algorithms with respect to the performance criterion.

3.1.1 CHEMICAL REACTOR

The model of a chemical reactor has been studied in Patnaik et al. (1980). The system and input matrices are given by

$$A = \begin{bmatrix} -4.019 & 5.12 & 0 & 0 & -2.082 & 0 & 0 & 0 & 0.87 \\ -0.346 & 0.986 & 0 & 0 & -2.34 & 0 & 0 & 0 & 0.97 \\ -7.909 & 15.407 & -4.069 & 0 & -6.45 & 0 & 0 & 0 & 2.68 \\ -21.816 & 35.606 & -0.339 & -3.87 & -17.8 & 0 & 0 & 0 & 7.39 \\ -60.196 & 98.188 & -7.907 & 0.34 & -53.008 & 0 & 0 & 0 & 20.4 \\ 0 & 0 & 0 & 0 & 94 & -147.2 & 0 & 53.2 & 0 \\ 0 & 0 & 0 & 0 & 0 & 94 & -147.2 & 0 & 0 \\ 0 & 0 & 0 & 0 & 0 & 12.8 & 0 & -31.6 & 0 \\ 0 & 0 & 0 & 0 & 12.8 & 0 & 0 & 18.8 & -31.6 \end{bmatrix}$$

$$B^{\mathrm{T}} = \begin{bmatrix} 0.010 & 0.003 & 0.009 & 0.024 & 0.068 & 0 & 0 & 0 & 0 \\ -0.011 & -0.021 & -0.059 & -0.162 & -0.445 & 0 & 0 & 0 & 0 \\ -0.151 & 0 & 0 & 0 & 0 & 0 & 0 & 0 & 0 \end{bmatrix}$$

Weighting matrices Q and R are chosen as identities.

This control system problem can be decoupled with $n_1 = 5$ and $n_2 = 4$, where the first five state variables comprise the first subsystem. Using the formula for an estimate of a small coupling parameter ε suggested by Shen and Gajić (1990), we have obtained $\varepsilon = 0.47 = 94/200.4$.

Simulation results are presented in Table 3.1. Obtained results reveal that the accuracy of $O(10^{-10})$ is obtained after only seven iterations despite a relatively big value of the coupling parameter ε. This is consistent with the results given in Theorem 3.2 and formula 3.32 since $(0.47)^{32} = 0.32146 \times 10^{-10}$. Thus, the

TABLE 3.1
Errors in the Performance Criterion per Iteration

i	$J_{\mathrm{app}} - J_{\mathrm{opt}}$
0	0.13910×10^{-2}
1	0.15714×10^{-3}
2	0.14805×10^{-4}
3	0.13045×10^{-5}
4	0.10936×10^{-6}
5	0.81286×10^{-8}
6	0.39972×10^{-9}
7	0.33651×10^{-10}

presented method is very efficient even in the case when ε is not "small enough"— the standard assumption for all small parameter theories. Even more, by using the presented method the accuracy of an arbitrary order is easily achieved.

3.1.2 F-4 FIGHTER AIRCRAFT

An F-4 fighter aircraft (the actuator case) is considered in Harvey and Stein (1978). This model is described by the following system and control input matrices:

$$B^T = \begin{bmatrix} 0 & 0 & 0 & 0 & 20 & 0 \\ 0 & 0 & 0 & 0 & 0 & 10 \end{bmatrix}$$

$$A = \begin{bmatrix} -0.746 & 0.387 & -12.9 & 0 & 0.952 & 6.05 \\ 0.024 & -0.174 & 4.31 & 0 & -1.76 & -0.416 \\ 0.006 & -0.999 & -0.0578 & 0.0369 & 0.0092 & -0.0012 \\ 1 & 0 & 0 & 0 & 0 & 0 \\ 0 & 0 & 0 & 0 & -10 & 0 \\ 0 & 0 & 0 & 0 & 0 & -5 \end{bmatrix}$$

Weighting matrices Q and R are chosen as $Q = I_6$, $R = I_2$.

Even though the aircraft is not inherently weakly coupled system, we will show that the presented algorithm can be applied to the reduced-order controller design of this aircraft with a prescribed degree of stability. The system is decomposed with $n_1 = 4$ and $n_2 = 2$, where the first four state variables comprise the first subsystem. The eigenvalues of matrix A are given by -0.006, -0.765, -0.103 $\pm j2.093$, -5, -10. In order to have the weakly coupled structure for matrix A we need that $\det\{A_1\} = O(1)$ and $\det\{A_4\} = O(1)$ (Chow and Kokotović 1983). However, in this example $\det\{A_1(0)\} = 0.021278$. This can be facilitated by choosing the performance criterion in the form (which assures a prescribed degree of stability)

$$J = \frac{1}{2} \int_0^\infty e^{2\alpha t} \left[x^T(t)Qx(t) + u^T(t)Ru(t) \right] dt \tag{3.39}$$

The consequence of this is that the actual system matrix that we are working with is $A + \alpha I$, Anderson and Moore (1990). For this modified system matrix and for $\alpha = -10$ the strongest coupling is in the first row, so that an estimate of ε according to Shen and Gajić (1990) is given by $7.002/24.033 = 0.291349$. Since $A + \alpha I$ is a stable matrix the required conditions from Assumption 3.2 are satisfied. Simulation results for the performance criterion are presented in Table 3.2.

Note that $(0.291349)^{24} = 0.1399 \times 10^{-12}$ so that this example perfectly matches the results established in Theorem 3.1 and formula Equation 3.32.

TABLE 3.2

Errors in the Performance Criterion per Iteration

i	$J_{app} - J_{opt}$
0	0.73118×10^{-3}
1	0.97281×10^{-5}
2	0.16248×10^{-6}
3	0.25505×10^{-8}
4	0.39449×10^{-10}
5	0.34106×10^{-12}

It is important to emphasize that since $A + \alpha I$ is diagonally dominant for α large enough, and thus, weakly coupled, the presented method is more general and applicable to systems which are not inherently weakly coupled.

The importance of the higher order approximations for weakly coupled systems is demonstrated in Shen and Gajić (1990), where the $O(\varepsilon^6)$ accuracy was required in order to stabilize the closed-loop system.

3.1.3 CASE STUDY: MULTIMACHINE POWER SYSTEM

The nearly weakly coupled structure studied in this section can be found in power systems. The efficiency of the proposed reduced-order algorithm (Equations 3.20 through 3.24) is demonstrated on the design example for the decentralized multi-variable excitation controllers in a multimachine power system.

We consider a complex multimachine power system, composed of N synchronous machine-regulator units and connected to the network which includes transformers, lines, and load. In these studies, it is customary to treat the synchronous generators in plant as one equivalent machine and to use the assumption that turbine torques are constant and that changes in these torques are slow in comparison to other phenomena of significance in the voltage regulation problem. Furthermore, the electromagnetic transient processes in armature windings of the machines and the elements of the network are usually neglected as well as transient processes in the damping winding are of less significance in the problem under consideration.

The linearized model obtained under these assumptions is used in this section. Each of the synchronous machines is described by Park's equations with a field circuit in the direct axis. The synchronous generators are assumed to be equipped with first-order exciters. The network is represented by constant admittances and reduced by eliminating nongenerator basis. Loads are represented by constant admittances and are included in the network admittance matrix.

The linearized equations of the considered multimachine power system are written in the state space form as

$$\dot{x}(t) = Ax(t) + \sum_{i=1}^{N} B_i u_i(t), \quad i = 1, 2, 3, \ldots, N \tag{3.40}$$

where the state vector $x^T(t) = \left[x_1^T(t) \ x_2^T(t), \ldots, x_N^T(t) \right]$ has components $x_i^T(t) = \left[\delta_{iN}(t) \ \omega_i(t) \ \Psi_{fi}(t) \ E_{fdi}(t) \right]$, $i = 1, 2, \ldots, N$, with $\delta_{iN}(t)$ represents the load angle with respect to the angle of the reference machine, $\omega_i(t)$ is the rotor angular velocity, $\Psi_{fi}(t)$ is the field flux linkage, and $E_{fdi}(t)$ is the exciter state variable of the ith machine. All variables represent small deviations from the operating point.

In this section, we study the real example that represents the portion of the Serbian grid in isolated operation composed of two hydropower plants (Arnautovic 1988; Arnautovic and Medanić 1990). Each machine is equipped with the fast exciter whose parameters and operating points are given in Appendix 3.1.

Matrices A and B of the corresponding linearized model are given by

$$A = \begin{bmatrix} 0 & -314.159 & 314.159 & 0 & 0 & 0 & 0 \\ 0.003 & -0.131 & -0.012 & -0.141 & -0.006 & 0 & 0 \\ -0.271 & -0.352 & -2.763 & -0.182 & -0.371 & 0 & 0 \\ 0.005 & -0.290 & -0.008 & -0.373 & 0.005 & 314.159 & 0 \\ -0.290 & -0.127 & -0.724 & 0.025 & -1.261 & 0 & 314.159 \\ 0 & 0 & 0 & 0 & 0 & -33.333 & 0 \\ 0 & 0 & 0 & 0 & 0 & 0 & -33.333 \end{bmatrix}$$

$$B^T = \begin{bmatrix} 0 & 0 & 0 & 0 & 0 & 0.062 & 0 \\ 0 & 0 & 0 & 0 & 0 & 0 & 0.201 \end{bmatrix}$$

The weighted matrices Q and R are chosen as $Q = I_7$ and $R = I_2$.

It can be seen that the matrix A has the weakly coupled structure. By interchanging rows in matrices A and B, we can obtain the nearly weakly coupled structure defined in Equation 3.33. The eigenvalues of the matrix A are given by $-0.048, -0.549, -0.822, -1.555 \pm j9.164, -33.33, -33.33$. It can be shown that both conditions of assumption 3.2 are satisfied for the power system example. Due to the special structure of the matrix B, the feedback control law will affect only slightly some of the very small eigenvalues so that the system will remain almost marginally stable under feedback control. In addition, in order to have the weakly coupled structure for matrix A we need that the following conditions be satisfied $\det\{A_1\} = O(1)$ and $\det\{A_4\} = O(1)$ (Chow and Kokotović 1983). The above problems can be facilitated by choosing the performance criterion in the following form:

$$J = \frac{1}{2} \int_0^\infty e^{2\alpha t} \left[x^T(t)Qx(t) + u^T(t)Ru(t) \right] dt$$

In order to improve the numerical behavior of the proposed algorithm, it is advisable to balance the elements in the matrix A (some of them are very large) by introducing

TABLE 3.3

Approximate Values

for the Performance Index

i	J_{app}	$J_{opt} - J_{app}$
0	122.144	26.198
1	96.285	0.339
2	96.008	0.062
3	95.954	0.008
4	95.948	1.5×10^{-3}
5	95.946	2.0×10^{-4}
6	95.946	8.0×10^{-5}
Optimal	95.946	

simple scaling in the form $\bar{x}_i(t) = k x_i(t)$, with $k = 0.1$, $i = 5, 6$, and $k = 0.3$ for $i = 1, 2$. Parameters α and ε are chosen as $\alpha = 1$ and $\varepsilon = 0.4$.

Simulation results are obtained by using the MATLAB$^{\circledR}$ package for computer-aided control system design. Obtained results are presented in Table 3.3. It can be seen that the convergence to the optimal performance criterion is pretty rapid, namely $O(\varepsilon^2)$ per iteration, which is consistent with the main result of Theorem 3.1.

3.2 REDUCED-ORDER CONTROLLER FOR A CLASS OF WEAKLY COUPLED SYSTEMS

In this section, the reduced-order solution is obtained for a class of linear-quadratic optimal control problems having weakly interconnected system matrix, strongly connected control matrix, and with a special structure for the state penalty matrix. An example is included to demonstrate effectiveness of the proposed reduced-order algorithm. The presented method is very well suited for parallel implementation.

Consider a linear dynamical system given by

$$\dot{x}(t) = Ax(t) + Bu(t) \qquad (3.41)$$

with

$$x(t) = \begin{bmatrix} x_1(t) \\ x_2(t) \end{bmatrix}, \quad A = \begin{bmatrix} A_1 & \varepsilon A_2 \\ \varepsilon A_3 & A_4 \end{bmatrix}, \quad B = \begin{bmatrix} B_1 \\ B_2 \end{bmatrix} \qquad (3.42)$$

where

$x_i(t) \in \Re^{n_i}$, $i = 1, 2$, are state vectors
$u(t) \in \Re^m$ is the control input
ε is a small parameter

A quadratic type functional to be minimized is associated with Equation 3.41 in the form

$$J = \frac{1}{2} \int_0^\infty \left[x^T(t)Qx(t) + u^T(t)Ru(t) \right] dt, \quad Q \ge 0, \quad R > 0 \tag{3.43}$$

All matrices defined in Equations 3.41 through 3.43 are constant and of appropriate dimensions.

The system matrix A, defined in Equation 3.42, has the structure of the weakly coupled systems (Kokotović et al. 1969; Gajić et al. 1990). However, due to strongly coupled control matrix B this system does not belong to the class of weakly coupled linear control systems. In this section, we will show that despite strong coupling coming from the input matrix, the order-reduction can be achieved (like in the case of purely weakly coupled systems) by using the specific structure for the state penalty matrix.

To the best of knowledge, our the problem order-reduction through the choice of the state penalty matrix Q has not been studied in the control literature. Thus, the engineering relevance of this section is to study the linear-quadratic optimal control problem of Equations 3.41 and 3.42, in the spirit of parallel and distributed reduced-order algorithms (Bertsekas and Tsitsiklis 1991), under the following assumption.

Assumption 3.4   The state penalty matrix Q has the structure

$$Q = \begin{bmatrix} Q_1 & \varepsilon Q_2 \\ \varepsilon Q_2 & \varepsilon Q_3 \end{bmatrix} \tag{3.44}$$

This choice of the matrix Q is quite common in engineering practice since the control engineers hardly penalize all state variables by weighting factors of the same magnitude, especially for large-scale systems.

It is assumed in this section that in addition to the weak coupling assumption (Assumption 3.1a), the problem matrices satisfy the following additional assumption.

Assumption 3.5   Matrices A_i, $i = 1, 2, 3, 4$, B_1, B_2, Q_j, $j = 1, 2, 3$, and R are constant and $O(1)$.

The optimal problem of minimizing Equation 3.43 along trajectories of Equation 3.41 has the very well-known solution given by

$$u_{opt}(x(t)) = -F_{opt}x(t) = -R^{-1}B^T Px(t) \tag{3.45}$$

where P is the positive semidefinite stabilizing solution of the algebraic Riccati equation

$$A^T P + PA + Q - PSP = 0, \quad S = BR^{-1}B^T \tag{3.46}$$

It can be shown that under Assumption 3.5, the nature of the solution of Equation 3.46 subject to the partition of the problem matrices as defined in Equations 3.42 through 3.44 is given by

$$P = \begin{bmatrix} P_1 & \varepsilon P_2 \\ \varepsilon P_2^{\mathrm{T}} & \varepsilon P_3 \end{bmatrix} \tag{3.47}$$

In the following, we will derive the reduced-order algorithm for finding P. Partitioning Equation 3.46 compatible to Equations 3.42, 3.44, and 3.47, we get three algebraic equations:

$$P_1 A_1 + A_1^{\mathrm{T}} P_1 + Q_1 - P_1 S_1 P_1 + \varepsilon^2 \left(P_2 A_3 + A_3^{\mathrm{T}} P_2^{\mathrm{T}} \right)$$
$$- \varepsilon \left(P_1 S_2 P_2^{\mathrm{T}} + P_2 S_2^{\mathrm{T}} P_1 + \varepsilon P_2 S_3 P_2^{\mathrm{T}} \right) = 0 \tag{3.48}$$

$$P_1 A_2 + P_2 A_4 + A_1^{\mathrm{T}} P_2 - P_1 S_1 P_2 - P_1 S_2 P_3 + Q_2$$
$$+ \varepsilon \left(A_3^{\mathrm{T}} P_3 - P_2 S_2^{\mathrm{T}} P_2 - P_2 S_3 P_3 \right) = 0 \tag{3.49}$$

$$P_3 A_4 + A_4^{\mathrm{T}} P_3 + Q_3$$
$$+ \varepsilon \left(P_2^{\mathrm{T}} A_2 + A_2^{\mathrm{T}} P_2 - P_2^{\mathrm{T}} S_1 P_2 - P_2^{\mathrm{T}} S_2 P_3 - P_3 S_2^{\mathrm{T}} P_2 - P_3 S_3 P_3 \right) = 0 \tag{3.50}$$

where

$$S_1 = B_1 R^{-1} B_1^{\mathrm{T}}, \quad S_2 = B_1 R^{-1} B_2^{\mathrm{T}}, \quad S_3 = B_2 R^{-1} B_2^{\mathrm{T}} \tag{3.51}$$

Since ε is a small parameter, we can define an $O(\varepsilon)$ approximation of Equations 3.48 through 3.50 as follows:

$$P_1^{(0)} A_1 + A_1^{\mathrm{T}} P_1^{(0)} + Q_1 - P_1^{(0)} S_1 P_1^{(0)} = 0 \tag{3.52}$$

$$P_1^{(0)} A_2 + P_2^{(0)} A_4 + A_1^{\mathrm{T}} P_2^{(0)} - P_1^{(0)} S_1 P_2^{(0)} - P_1^{(0)} S_2 P_3^{(0)} + Q_2 = 0 \tag{3.53}$$

$$P_3^{(0)} A_4 + A_4^{\mathrm{T}} P_3^{(0)} + Q_3 = 0 \tag{3.54}$$

The unique positive semidefinite stabilizing solution for $P^{(0)}$, obtained from Equations 3.52 through 3.54, and defined by

$$P^{(0)} = \begin{bmatrix} P_1^{(0)} & \varepsilon P_2^{(0)} \\ \varepsilon P_2^{(0)\mathrm{T}} & \varepsilon P_3^{(0)} \end{bmatrix} \tag{3.55}$$

exists under the following assumption.

Assumption 3.6 The triple $(A_1, B_1, \mathrm{Chol}(Q_1))$ is stabilizable–detectable and the matrix A_4 is stable.

Since all solutions obtained from Equations 3.52 through 3.54 are $O(1)$, it can be concluded that our staring assumption (Equation 3.47) about the nature of the solution of Equation 3.46 is correct.

From Equations 3.52 through 3.54 we have obtained the first-order approximation of the required solution in terms of the completely decomposed reduced-order algebraic equations.

In the next step, we will derive the reduced-order parallel algorithm, based on the fixed point iterations, for obtaining the solution of P up to any arbitrary degree of accuracy.

Defining the approximation errors as

$$P_j = P_j^{(0)} + \varepsilon E_j, \quad j = 1, 2, 3 \tag{3.56}$$

and using Equation 3.55 in Equations 3.48 through 3.50 and Equations 3.52 through 3.54, we get the following error equations:

$$E_1 D_1 + D_1^T E_1 = \varepsilon E_1 S_1 E_1 - \varepsilon \left(P_2 A_3 + A_3^T P_2^T \right)$$
$$+ P_1 S_2 P_2^T + P_2 S_2^T P_1 + \varepsilon P_2 S_3 P_2^T \tag{3.57}$$

$$E_2 A_4 + D_1^T E_2 = \varepsilon (E_1 S_1 E_2 + E_1 S_2 E_3) - E_1 D_2$$
$$+ P_1^{(0)} S_2 E_3 - A_3^T P_3 + P_2 S_2^T P_2 + P_2 S_3 P_3 \tag{3.58}$$

$$E_3 A_4 + A_4^T E_3 = P_2^T S_1 P_2 + P_2^T S_2 P_3$$
$$+ P_3 S_2^T P_2 + P_3 S_3 P_3 - P_2^T A_2 - A_2^T P_2 \tag{3.59}$$

where

$$D_1 = A_1 - S_1 P_1^{(0)}, \quad D_2 = A_2 - S_1 P_2^{(0)} - S_2 P_3^{(0)} \tag{3.60}$$

Let us propose the following reduced-order parallel algorithm for solving Equations 3.57 through 3.60.

ALGORITHM 3.3

$$E_1^{(i+1)} D_1 + D_1^T E_1^{(i+1)} = \varepsilon E_1^{(i)} S_1 E_1^{(i)} - \varepsilon \left(P_2^{(i)} A_3 + A_3^T P_2^{(i)^T} \right)$$
$$+ P_1^{(i)} S_2 P_2^{(i)^T} + P_2^{(i)} S_2^T P_1^{(i)} + \varepsilon P_2^{(i)} S_3 P_2^{(i)^T} \tag{3.61}$$

$$E_2^{(i+1)} A_4 + D_1^T E_2^{(i+1)} = \varepsilon \left(E_1^{(i+1)} S_1 E_2^{(i)} + E_1^{(i+1)} S_2 E_3^{(i+1)} \right) - E_1^{(i+1)} D_2 + P_1^{(0)} S_2 E_3^{(i+1)}$$
$$- A_3^T P_3^{(i+1)} + P_2^{(i)} S_2^T P_2^{(i)} + P_2^{(i)} S_3 P_3^{(i+1)} \tag{3.62}$$

$$E_3^{(i+1)} A_4 + A_4^T E_3^{(i+1)} = P_2^{(i)^T} S_1 P_2^{(i)} + P_2^{(i)^T} S_2 P_3^{(i)} + P_3^{(i)} S_2^T P_2^{(i)}$$
$$+ P_3^{(i)} S_3 P_3^{(i)} - P_2^{(i)^T} A_2 - A_2^T P_2^{(i)} \tag{3.63}$$

with initial conditions

$$E_1^{(0)} = 0, \quad E_2^{(0)} = 0, \quad E_3^{(0)} = 0 \tag{3.64}$$

The following theorem indicates the features of the proposed algorithm (Equations 3.61 through 3.64).

THEOREM 3.2

Under conditions stated in Assumptions 3.3 and 3.4, the algorithm (Equations 3.61 through 3.64) converges to the exact solution of the error term, and thus to the required solution P, with the rate of convergence of O(ε), that is

$$\left\| E_j - E_j^{(i+1)} \right\| = O(\varepsilon) \left\| E_j - E_j^{(i)} \right\| \tag{3.65}$$

or equivalently

$$\left\| E_j - E_j^{(i)} \right\| = O\left(\varepsilon^{(i+1)}\right), \quad i = 0, 1, 2, \ldots, \quad j = 1, 2, 3 \tag{3.66}$$

Proof As a starting point, we need to show the existence of a bounded solution of Equations 3.57 through 3.59 in the neighborhood of $\varepsilon = 0$. By the implicit function theorem (Ortega and Rheinboldt 1970), it is enough to show that the corresponding Jacobian is nonsingular at $\varepsilon = 0$. The Jacobian at $\varepsilon = 0$ is given by

$$J(\varepsilon)_{|\varepsilon=0} = \begin{bmatrix} J_{11} & 0 & 0 \\ J_{21} & J_{22} & J_{23} \\ 0 & 0 & J_{33} \end{bmatrix} \tag{3.67}$$

with

$$\begin{aligned}
J_{11} &= I_{n_1} \otimes D_1^{\mathsf{T}} + D_1^{\mathsf{T}} \otimes I_{n_1} \\
J_{22} &= I_{n_1} \otimes D_1^{\mathsf{T}} + A_4^{\mathsf{T}} \otimes I_{n_2} \\
J_{33} &= I_{n_2} \otimes A_4^{\mathsf{T}} + A_4^{\mathsf{T}} \otimes I_{n_2}
\end{aligned} \tag{3.68}$$

where $\otimes$ stands for the Kronecker product representation. For the Jacobian to be nonsingular, the block diagonal elements J_{ii}, $i = 1, 2, 3$, have to be nonsingular. The matrix D_1 is a closed-loop matrix, and thus stable by the well-known property of the algebraic Riccati equation and by Assumption 3.5. By the same assumption, the matrix A_4 is stable, so that by the property of the Kronecker product (Lancaster and Tismenetsky 2000), matrices J_{ii} are nonsingular. Thus, for ε small enough, the Jacobian is nonsingular.

In the next step, we have to show the convergence of the algorithm (Equations 3.61 through 3.64) and give an estimate of the rate of convergence. For $i = 0$, Equations 3.57 and 3.61 imply

$$\left(E_1 - E_1^{(0)}\right)D_1 + D_1^{\mathsf{T}}\left(E_1 - E_1^{(0)}\right) = \varepsilon f_1(E_1, E_2, \varepsilon) \tag{3.69}$$

Since D_1 is stable and E_1 and E_2 are bounded it follows that

$$\left\| E_1 - E_1^{(0)} \right\| = O(\varepsilon) \tag{3.70}$$

Similarly from Equations 3.59 and 3.63 we have

$$\left(E_3 - E_3^{(0)} \right) A_4 + A_4^{\mathrm{T}} \left(E_3 - E_3^{(0)} \right) = \varepsilon f_3(E_2, E_3, \varepsilon) \tag{3.71}$$

so that

$$\left\| E_3 - E_3^{(0)} \right\| = O(\varepsilon) \tag{3.72}$$

The use of the same arguments in Equations 3.58 and 3.62 produces

$$\left\| E_2 - E_2^{(0)} \right\| = O(\varepsilon) \tag{3.73}$$

Continuing the same procedure and by induction, we conclude that

$$\left\| E_1 - E_1^{(i)} \right\| = O\left(\varepsilon^{(i+1)} \right) \tag{3.74}$$

$$\left\| E_2 - E_2^{(i)} \right\| = O\left(\varepsilon^{(i+1)} \right) \tag{3.75}$$

$$\left\| E_3 - E_3^{(i)} \right\| = O\left(\varepsilon^{(i+1)} \right) \tag{3.76}$$

with $i = 1, 2, \ldots$, which completes the proof of Theorem 3.2. ∎

It is obvious that the proposed algorithm (Equations 3.61 through 3.64) can be implemented as a synchronous one (Bertsekas and Tsitsiklis 1991). The study is underway to prove the convergence of the corresponding asynchronous algorithm.

3.2.1 NUMERICAL EXAMPLE

The following fourth-order linear-quadratic control problem example demonstrates the efficiency of the proposed method. Problem matrices are taken from Shien and Tsay (1982).

$$A = \begin{bmatrix} -0.75 & 0.28125 & 0.15 & 0.31875 \\ -0.25 & -1.15625 & -0.35 & -0.24375 \\ -0.75 & 0.28125 & -3.25 & -0.28125 \\ -0.25 & -1.15625 & 1.25 & -1.84375 \end{bmatrix}, \quad B = \begin{bmatrix} 1 & 0 \\ 0 & 1 \\ 1 & 0 \\ 0 & 1 \end{bmatrix}$$

$$y = Cx = \begin{bmatrix} 1.5 & 1.9375 & 0.5 & 0.0625 \\ -0.25 & 0.71875 & 0.25 & 0.28125 \end{bmatrix} x$$

$$Q = C^{\mathrm{T}}C, \quad R = I_2, \quad \varepsilon = 0.1$$

TABLE 3.4

Errors in the Criterion Approximation per Iteration

i	$J^{(i)} - J_{\text{opt}}$
1	0.16855×10^{-2}
2	0.66638×10^{-4}
3	0.28597×10^{-5}
4	0.22424×10^{-6}
5	0.14966×10^{-7}
6	0.10897×10^{-8}
7	0.47606×10^{-10}
8	0.33538×10^{-11}
9	0.33538×10^{-12}

Simulation results are presented in Table 3.4. The optimal value for the criterion is $J_{\text{opt}} = 1.8222$. From Table 3.4 we can notice very good numerical behavior of the proposed algorithm consistent with the statement of Theorem 9.5.

Other examples of weakly coupled systems having strong coupling through the input matrix are binary distillation column considered in Bhattacharyya et al. (1983) and L-1011 fighter aircraft presented in Beale and Shafai (1989). These control systems can be numerically decomposed and solved in terms of the reduced-order problems by choosing the state penalty matrix according to Assumption 3.5 as presented in Sections 3.2.1 and 3.3.2.

The presented method is applicable to almost any linear control system with a prescribed degree of stability (Anderson and Moore 1990), since in that case we are working with $A + \alpha I$ which is block diagonally dominant for α large enough. In some cases, the overlapping idea of Siljak (1991) can be used to achieve the desired structure.

3.2.2 CASE STUDY 1: L-1011 FIGHTER AIRCRAFT

A mathematical model of L-1011 fighter aircraft can be found in Beale and Shafai (1989). The problem matrices are given by

$$A = \begin{bmatrix} 0 & 1 & 0 & 0 \\ 0 & -1.89 & 0.39 & -5.53 \\ 0 & -0.034 & -2.98 & 2.43 \\ 0.034 & -0.0011 & -0.99 & -0.21 \end{bmatrix}, \quad B = \begin{bmatrix} 0 & 0 \\ 0.36 & -1.6 \\ -0.95 & -0.032 \\ 0.03 & 0 \end{bmatrix}$$

$$Q = \begin{bmatrix} 2.313 & 2.727 & 0.688 & 0.023 \\ 2.727 & 4.271 & 1.148 & 0.323 \\ 0.688 & 1.148 & 0.313 & 0.102 \\ 0.023 & 0.323 & 0.102 & 0.083 \end{bmatrix}, \quad R = I_2$$

TABLE 3.5

Difference between Approximate and Optimal Criteria

i	$J^{(i)} - J_{opt}$
1	$0.1114 \times 10^{+1}$
2	0.8048×10^{-1}
3	0.1529×10^{-1}
4	0.3569×10^{-2}
5	0.4790×10^{-3}
6	0.6193×10^{-4}
7	0.1513×10^{-4}
8	0.2542×10^{-5}
9	0.2832×10^{-6}
10	0.5678×10^{-7}
11	0.1066×10^{-7}
12	0.1183×10^{-8}

This control system is decomposed into two subsystems, each of order two. Small parameter ε is chosen as $\varepsilon = 0.3$. The optimal performance index is $J_{opt} = 7.239$. Simulation results for the performance criterion are presented in Table 3.5.

It can be seen that the obtained numerical results are consistent with the established analytical relationship.

3.2.3 CASE STUDY 2: DISTILLATION COLUMN

Mathematical model of a binary distillation column with condenser, reboiler, and nine plates is given in Bhattacharyya et al. (1983).

$$
A = \begin{bmatrix}
-0.991 & 0.529 & 0 & 0 & 0 & 0 & 0 & 0 \\
0.522 & -1.051 & 0.596 & 0 & 0 & 0 & 0 & 0 \\
0 & 0.522 & -1.118 & 0.596 & 0 & 0 & 0 & 0 \\
0 & 0 & 0.522 & -1.548 & 0.718 & 0 & 0 & 0 \\
0 & 0 & 0 & 0.922 & -1.640 & 0.799 & 0 & 0 \\
0 & 0 & 0 & 0 & 0.922 & -1.721 & 0.901 & 0 \\
0 & 0 & 0 & 0 & 0 & 0.922 & -1.823 & 1.021 \\
0 & 0 & 0 & 0 & 0 & 0 & 0.922 & -1.943
\end{bmatrix}
$$

$$
B^T = 10^{-3} \begin{bmatrix}
3.84 & 4 & 37.6 & 3.08 & 2.36 & 2.88 & 3.08 & 3 \\
-2.88 & -3.04 & -2.80 & -2.32 & -3.32 & -3.82 & -4.12 & -3.96
\end{bmatrix}
$$

$$Q = \begin{bmatrix} 1 & 0 & 0 & 0 & 0.5 & 0 & 0 & 0.1 \\ 0 & 1 & 0 & 0 & 0.1 & 0 & 0 & 0 \\ 0 & 0 & 1 & 0 & 0 & 0.5 & 0 & 0 \\ 0 & 0 & 0 & 1 & 0 & 0 & 0 & 0 \\ 0.5 & 0.1 & 0 & 0 & 0.1 & 0 & 0 & 0 \\ 0 & 0 & 0.5 & 0 & 0 & 0.1 & 0 & 0 \\ 0 & 0 & 0 & 0 & 0 & 0 & 0.1 & 0 \\ 0.1 & 0 & 0 & 0 & 0 & 0 & 0 & 0.1 \end{bmatrix}$$

$R = I_2, \quad \varepsilon = 0.2$

The optimal value of the performance criterion is $J_{opt} = 6.1656$. Obtained simulation results are presented in Table 3.6. The results are consistent with the statement of the corresponding theorem and show the corresponding rate of convergence.

Research Problem 3.1: Power system load frequency control problem is modeled by a weakly coupled structure of the form (Kawashima, 1992, 1995)

TABLE 3.6

Difference between Approximate and Optimal Criteria

i	$J_{app}^{(i)} - J_{opt}^{(i)}$
0	0.0186
1	0.0060
2	0.0020
3	7.2365×10^{-4}
4	2.6123×10^{-4}
5	9.5219×10^{-5}
6	3.4870×10^{-5}
7	1.2799×10^{-5}
8	4.7030×10^{-6}
9	1.7291×10^{-6}
10	6.3585×10^{-7}
11	2.3386×10^{-7}
12	8.6016×10^{-8}
13	3.1639×10^{-8}
14	1.1638×10^{-8}
15	4.2807×10^{-9}
16	1.5746×10^{-9}
17	5.7919×10^{-10}
18	2.1305×10^{-10}
19	7.8363×10^{-11}
20	2.8817×10^{-11}

$$A = \begin{bmatrix} A_1 & \varepsilon A_2 \\ A_3 & A_4 \end{bmatrix}, \quad B = \begin{bmatrix} B_1 & 0 \\ 0 & B_2 \end{bmatrix}$$

Using the methodology presented in Sections 3.1 and 3.2, derive the recursive reduced-order algorithms for solving the corresponding linear-quadratic optimal control problem. For the state penalty matrix consider either the structure given in Equation 3.5 or Equation 3.44.

3.3 NOTES

Results presented in this chapter are mostly based on the work of Skatarić et al. (1991, 1993) and Skatarić (1993, 2005). The study of optimal control and decoupling of the quasi-weakly coupled systems is not complete. There are many other classes of the linear-quadratic optimal control problems with small parameters that can be decomposed into the reduced-order subproblems. The presented results, obtained in the continuous-time domain, can serve as a guideline. Their extension to the discrete-time domain is also an interesting research area.

APPENDIX 3.1

The operating points for these two machines are given as follows. Machine number 1: $P_1 = 170$ MW, $Q_1 = 82$ MVAr, $V_1 = 15.75$ kV. Machine number 2: $P_2 = 24.5$ MW, $Q_2 = -6$ MVAr, $V_2 = 6.5$ kV, $V_2 = 6.5 / -4.5°$. Machine and exciter data are presented in Table 3.7.

TABLE 3.7
Synchronous Machine and Fast Exciter Data

Unit No.	No. 1	No. 2
Synchronous Machine		
Rated MVA	190	28
Rated kV	15.75	6.3
Xd (p.u.)	1.245	1
Xq (p.u.)	0.925	0.7
Xd' (p.u.)	0.373	0.42
Xad (p.u.)	1.145	0.85
Tdo' (s)	6.5	1.65
ra (p.u.)	0.00285	0.0107
Ta (S)	11.06	2.45
D (p.u.)	1	1
Exciter		
Te (s)	0.03	0.03
Ke (p.u.)	0.00185	0.00604

REFERENCES

Anderson, B. and J. Moore, *Optimal Control—Linear quadratic Methods*, Prentice-Hall, Englewood Cliffs, NJ, 1990.

Arnautovic, D., *Multivariable Voltage Regulator Synthesis in Multimachine Power Systems by Projective Control Method*, Doctoral Dissertation, University of Belgrade, Belgrade, 1988.

Arnautovic, D. and J. Medanić, The sequential design of different multivariable excitation controllers in multimachine power systems, *Electric Power Systems Research*, 18, 37–46, 1990.

Beale, S. and B. Shafai, Robust control system design with a proportional-integral observer, *International Journal of Control*, 50, 97–111, 1989.

Bertsekas, D. and J. Tsitsiklis, Some aspects of parallel and distributed iterative algorithms— A survey, *Automatica*, 27, 3–21, 1991.

Bhattacharyya, S., A. Del Nero Gomes, and J. Howze, The structure of robust disturbance rejection control, *IEEE Transactions of Automatic Control*, AC-28, 874–881, 1983.

Chow, J. and P. Kokotović, Sparsity and time scales, *Proceedings of American Control Conference*, San Francisco, CA, 656–661, 1983.

Gajić, Z., D. Petkovski, and X. Shen, *Singularly Perturbed and Weakly Coupled Linear Control Systems—A Recursive Approach*, Springer-Verlag, New York, Lecture Notes in Control and Information Sciences, 140, 1990.

Harvey, H. and G. Stein, Quadratic weights for asymptotic regulator properties, *IEEE Transactions of Automatic Control*, AC-23, 378–387, 1978.

Kokotović, P. and J. Cruz, An approximation theorem for linear optimal regulators, *Journal of Mathematical Analysis and Applications*, 27, 249–252, 1969.

Kokotović, P., W. Perkins, J. Cruz, and G. D'Ans, ε—coupling approach for near-optimum design of large scale linear systems, *Proceedings of IEE, Part D*, 116, 889–892, 1969.

Kawashima, S., Integrator decoupling applied to power system load frequency control, *Proceedings of Intelligent Vehicle Symposium*, 330–335, 1992.

Kawashima, S., Integrator weakly decoupling method applied to power control, *Proceedings of Energy Management and Power Delivery Conference*, 613–619, 1995.

Lancaster, P. and M. Tismenetsky, *The Theory of Matrices*, Academic Press, Orlando, FL, 1985.

Ortega, J. and W. Rheinboldt, *Iterative Solution of Nonlinear Equations In Several Variables*, Academic Press, New York, 1970.

Patnaik, P., N. Viswanadham, and I. Sarma, Computer control algorithms for a tubular ammonia reactor, *IEEE Transactions on Automatic Control*, AC-25, 642–651, 1980.

Siljak, D., *Decentralized Control of Complex Systems*, Academic Press, Cambridge, 1991.

Skatarić, D., *Quasi Singularly Perturbed and Weakly Coupled Linear Control Systems*, Doctoral Dissertation, University of Novi Sad, Novi Sad, 1993.

Skatarić, D., *Optimal Control of Quasi Singularly Perturbed and Weakly Coupled Systems*, Planeta Print, Belgrade, Serbia, 2005.

Skatarić, D., Z. Gajić, and D. Arnautovic, Optimal reduced-order controllers for nearly weakly coupled linear systems, *Control—Theory and Advanced Technology*, 9, 481–490, 1993.

Skatarić, D., Z. Gajić, and D. Petkovski, Reduced-order solution for a class of linear quadratic optimal control problems, *Proceedings of Allerton Conference on Communication, Control and Computing*, Urbana, IL, 440–447, 1991.

Shen, X. and Z. Gajić, Near-optimum steady state regulators for stochastic linear weakly coupled systems, *Automatica*, 26, 919–923, 1990.

Shien, L. and Y. Tsay, Transformation of a class of multivariable control systems to block companion forms, *IEEE Transactions on Automatic Control*, AC-27, 199–202, 1982.

4 Weakly Coupled Singularly Perturbed Systems

4.1 INTRODUCTION

In mathematical models of many real physical systems small parameters appear. Two large classes of small parameter problems have been studied extensively in the context of control theory: (1) singularly perturbed systems (Kokotović et al. 1986; Gajić and Lim 2001) and (2) weakly coupled systems. Motivated by mathematical models of the real physical systems, we have found that many of them have both singularly perturbed and weakly coupled structures. Even more, the structure of many systems with slow–fast phenomena and weak coupling cannot be put either in the standard singularly perturbed or standard weakly coupled forms. In this chapter, we study systems that are at the same time both singularly perturbed and weakly coupled.

A special class of singularly perturbed weakly coupled linear systems has been studied within the concept of multimodeling (Khalil and Kokotović 1978; Khalil 1980; Saksena and Cruz 1981a,b; Saksena and Basar 1982; Saksena et al. 1983; Gajić and Khalil 1986; Gajić 1988; Zhuang and Gajić 1991; Coumarbatch and Gajić 2000a,b; Mukaidani 2001, 2005) where the weak coupling is allowed between fast variables only. In this chapter, we will study the effect of weak coupling between slow and fast variables. The obtained solution will be given in terms of a ratio of two small parameters. Let ε_1 and ε_2 represent small positive weak coupling and small positive singular perturbation parameters, respectively, then one can study any of the following cases:

$$
\begin{aligned}
&1.\ 0 < m \le \frac{\varepsilon_1}{\varepsilon_2} \le M < \infty \\[2mm]
&2.\ \frac{\varepsilon_1}{\varepsilon_2} \to 0 \\[2mm]
&3.\ \frac{\varepsilon_2}{\varepsilon_1} \to 0
\end{aligned}
\tag{4.1}
$$

In the first structure, which is the subject of this chapter, the system is both singularly perturbed and weakly coupled. In the second structure, it is predominantly weakly coupled, and in the third one it is predominantly singularly perturbed, so that they can be studied by using corresponding techniques developed for singularly perturbed and weakly coupled control systems. Note that pure singularly perturbed systems involving many small parameters of the same magnitude have been studied under the name of "multiparameter singular perturbations," (e.g., Khalil and Kokotović 1979a,b; Mukaidani et al. 2002a,b, 2003; Mukaidani 2003).

The approach taken in this chapter is in the spirit of the reduced-order fixed point iterations (Gajić and Shen 1993; Gajić and Lim 2001), and parallel synchronous algorithms (Bertsekas and Tsitsiklis 1989, 1991).

The study of this chapter reveals one very important feature of this kind of systems displaying slow–fast phenomena. Namely, the stabilizability–detectability condition is imposed directly on subsystems, in contrary to pure singularly perturbed systems where this condition has to be imposed on the slow subsystem matrices, which depend on the given problem matrices in quite a complicated manner.

The obtained results for singularly perturbed weakly coupled linear systems are extended in Section 4.4 to the so-called quasi singularly perturbed weakly coupled linear systems (Skatarić 1993). Several real world control problems are solved in order to demonstrate the efficiency of the proposed synchronous reduced-order parallel algorithms.

4.2　WEAKLY COUPLED SINGULARLY PERTURBED LINEAR CONTROL SYSTEMS

The singularly perturbed weakly coupled linear dynamical control system has the form consistent with both the singularly perturbed and weakly coupled systems

$$
\begin{bmatrix} \dot{x}_1(t) \\ \dot{x}_2(t) \end{bmatrix} = \begin{bmatrix} A_1 & \varepsilon_1 A_2 \\ \dfrac{\varepsilon_1 A_3}{\varepsilon_2} & \dfrac{A_4}{\varepsilon_2} \end{bmatrix} \begin{bmatrix} x_1(t) \\ x_2(t) \end{bmatrix} + \begin{bmatrix} B_1 & \varepsilon_1 B_2 \\ \dfrac{\varepsilon_1 B_3}{\varepsilon_2} & \dfrac{B_4}{\varepsilon_2} \end{bmatrix} \begin{bmatrix} u_1(t) \\ u_2(t) \end{bmatrix} \tag{4.2}
$$

where $x_i(t) \in \Re^{n_i}$, $i = 1, 2$, are state vectors, $u(t) \in \Re^{m_i}$, $i = 1, 2$, are control inputs and $n = n_1 + n_2$ is the system order. This is a special class of linear dynamical control systems represented, in general, by

$$
\dot{x}(t) = Ax(t) + Bu(t) \tag{4.3}
$$

$$
x(t) = \begin{bmatrix} x_1(t) \\ x_2(t) \end{bmatrix}, \quad u(t) = \begin{bmatrix} u_1(t) \\ u_2(t) \end{bmatrix}
$$

$$
A = \begin{bmatrix} A_1 & \varepsilon_1 A_2 \\ \dfrac{\varepsilon_1 A_3}{\varepsilon_2} & \dfrac{A_4}{\varepsilon_2} \end{bmatrix}, \quad B = \begin{bmatrix} B_1 & \varepsilon_1 B_2 \\ \dfrac{\varepsilon_1 B_3}{\varepsilon_2} & \dfrac{B_4}{\varepsilon_2} \end{bmatrix} \tag{4.4}
$$

Matrices A_i, $i = 1, 2, 3, 4$, and B_i, $i = 1, 2, 3, 4$, are constant and $O(1)$. It is assumed that magnitudes of all the system eigenvalues are $O(1)$, that is, $|\lambda_j| = O(1)$, $j = 1, 2, \ldots, n$, implying that matrices A_1 and A_4 are nonsingular with $\det\{A_1\} = O(1)$ and $\det\{A_4\} = O(1)$. This is the standard assumption for weakly coupled linear systems, which also corresponds to the block diagonal dominance of the system matrix A (Chow and Kokotović 1983). In addition, it is consistent with the standard assumption of singularly perturbed systems that the matrix A_4 is nonsingular (Gajić and Lim 2001). Hence, the main results presented in this chapter are valid under the following assumption.

Assumption 4.1 The system matrices A_i, $i = 1, 2, 3, 4$, and B_i, $i = 1, 2, 3, 4$, are constant and $O(1)$ and magnitudes of all system eigenvalues are $O(1)$, that is, $|\lambda_j| = O(1)$, $j = 1, 2, \ldots, n$, which implies that the matrices A_1, A_4 are nonsingular with $\det\{A_1\} = O(1)$ and $\det\{A_4\} = O(1)$.

A quadratic type functional to be minimized is associated with Equation 4.2 in the form

$$J = \frac{1}{2} \int_0^\infty \left[x^T(t)Qx(t) + u^T(t)Ru(t) \right] dt, \quad Q \geq 0, \quad R > 0 \qquad (4.5)$$

For the purpose of this chapter, we assume that the structures of the matrices Q and R are

$$Q = \begin{bmatrix} Q_1 & \varepsilon_1 Q_2 \\ \varepsilon_1 Q_2^T & Q_3 \end{bmatrix}, \quad R = \begin{bmatrix} R_1 & 0 \\ 0 & R_2 \end{bmatrix} \qquad (4.6)$$

All problem matrices defined in Equations 4.2 through 4.6 are constant and of appropriate dimensions. We will show that in this case one is able to design the optimal controllers by using the reduced-order parallel synchronous algorithms. Even more, the obtained results are applicable under milder conditions than for pure singularly perturbed linear-quadratic control problems. Namely, the stabilizability–detectability conditions are imposed directly on the subsystem matrices A_1, B_1, Q_1 and A_4, B_4, Q_3. For pure singularly perturbed systems the stabilizability–detectability condition is imposed on the slow subsystem matrices, which depend in quite a complicated manner on the original problem matrices. Thus, for pure singularly perturbed systems one is not able to test directly the required stabilizability–detectability conditions.

The optimal problem of minimizing Equation 4.5 along trajectories of Equation 4.2 has the very well-known solution given by

$$u_{opt}[x(t)] = -F_{opt}x(t) = -R^{-1}B^T Px(t) \qquad (4.7)$$

where P is the positive semidefinite stabilizing solution of the algebraic Riccati equation

$$A^T P + PA + Q - PSP = 0, \quad S = BR^{-1}B^T \qquad (4.8)$$

For the development of parallel reduced-order algorithms for solving Equation 4.8, it is very important to discover the proper nature of the solution of Equation 4.8 in terms of small parameters ε_1 and ε_2. By studying partitioned equations (Equation 4.8), it can be shown, under Assumption 4.1 and assuming that weighting matrices Q_j, $j = 1, 2, 3$, R_1, R_2, are also constant and $O(1)$, that the solution of Equation 4.8 is scaled as follows:

$$P = \begin{bmatrix} P_1 & \varepsilon_1 \varepsilon_2 P_2 \\ \varepsilon_1 \varepsilon_2 P_2^T & \sqrt{\varepsilon_1 \varepsilon_2} P_3 \end{bmatrix} \tag{4.9}$$

Partitioning Equation 4.8 compatible to Equations 4.4 through 4.6 and Equation 4.9, we get three matrix algebraic equations

$$0 = P_1 A_1 + A_1^T P_1 + Q_1 - P_1 S_1 P_1 + \varepsilon_1^2 \left(A_3^T P_2^T + P_2 A_3 \right)$$
$$- \varepsilon_1^4 P_2 S_3 P_2^T - \varepsilon_1^2 \left(P_1 S_2 P_1 + P_2 Z^T P_1 + P_1 Z P_2^T + P_2 S_4 P_2^T \right) \tag{4.10}$$

$$P_2 A_4 + Q_2 + P_1 A_2 + \alpha \left(A_3^T P_3 - P_2 S_4 P_3 - P_1 Z P_3 \right) - \varepsilon_1^2 \alpha P_2 S_3 P_3$$
$$+ \varepsilon_2 \left(A_1^T P_2 - P_1 S_1 P_2 \right) - \varepsilon_1^2 \varepsilon_2 (P_1 S_2 P_2 + P_2 Z^T P_2) = 0 \tag{4.11}$$

$$\alpha P_3 A_4 + \alpha A_4^T P_3 + Q_3 - \alpha^2 P_3 S_4 P_3$$
$$+ \varepsilon_1^2 \varepsilon_2 \left(P_2^T A_2 + A_2^T P_2 \right) - \varepsilon_1^2 \varepsilon_2 \alpha \left(P_3 Z^T P_2 + P_2^T Z P_3 \right)$$
$$- \varepsilon_1^2 \varepsilon_2^2 \left(P_2^T S_1 P_2 + \varepsilon_1^2 P_2^T S_2 P_2 \right) - \varepsilon_1^2 \alpha^2 P_3 S_3 P_3 = 0 \tag{4.12}$$

where

$$\begin{aligned} S_1 &= B_1 R_1^{-1} B_1^T \\ S_2 &= B_2 R_2^{-1} B_2^T \\ S_3 &= B_3 R_1^{-1} B_3^T \\ S_4 &= B_4 R_2^{-1} B_4^T \\ Z &= B_1 R_1^{-1} B_3^T + B_2 R_2^{-1} B_4^T \end{aligned} \tag{4.13}$$

with

$$\alpha = \left| \sqrt{\frac{\varepsilon_1}{\varepsilon_2}} \right| \tag{4.14}$$

Since ε_1 and ε_2 are small parameters, we can define $O(\varepsilon)$ approximation of Equations 4.10 through 4.12 as follows:

$$P_1^{(0)} A_1 + A_1^T P_1^{(0)} + Q_1 - P_1^{(0)} S_1 P_1^{(0)} = 0 \tag{4.15}$$

$$P_2^{(0)} \left(A_4 - \alpha S_4 P_3^{(0)} \right) + Q_2 + P_1^{(0)} A_2 + \alpha \left(A_3^T P_3^{(0)} - P_1^{(0)} Z P_3^{(0)} \right) = 0 \tag{4.16}$$

$$\alpha P_3^{(0)} A_4 + \alpha A_4^T P_3^{(0)} + Q_3 - \alpha^2 P_3^{(0)} S_4 P_3^{(0)} = 0 \tag{4.17}$$

Corresponding solution of Equation 4.8 is now given by

$$
P^{(0)} = \begin{bmatrix} P_1^{(0)} & \varepsilon_1\varepsilon_2 P_2^{(0)} \\ \varepsilon_1\varepsilon_2 P_2^{(0)^\mathsf{T}} & \sqrt{\varepsilon_1\varepsilon_2} P_3^{(0)} \end{bmatrix} = P + O(\varepsilon) \tag{4.18}
$$

The unique positive semidefinite stabilizing solution $P^{(0)}$, obtained from Equations 4.15 through 4.17, exists under the assumption that the triples $(A_1, B_1, \mathrm{Chol}(Q_1))$ and $(A_4, \sqrt{\alpha}B_4, \mathrm{Chol}(\frac{1}{\alpha}Q_3))$ are stabilizable–detectable. However, due to the structure of the controllability–observability matrices, we can eliminate the α-dependence so that we need the following assumption.

Assumption 4.2 The triples $(A_1, B_1, \mathrm{Chol}(Q_1))$ and $(A_4, B_4, \mathrm{Chol}(Q_3))$ are stabilizable–detectable.

Defining the approximation errors as

$$
P_j = P_j^{(0)} + \varepsilon E_j, \quad j = 1, 2, 3; \quad \varepsilon = |\sqrt{\varepsilon_1\varepsilon_2}| \tag{4.19}
$$

and using Equation 4.19 in Equations 4.10 through 4.12 and Equations 4.15 through 4.17, we get the following error equations

$$
E_1 D_1 + D_1^\mathsf{T} E_1 = \varepsilon E_1 S_1 E_1 - \varepsilon_1\alpha\left(P_2 A_3 + A_3^\mathsf{T} P_2^\mathsf{T}\right)
$$

$$
+ \varepsilon_1\alpha\left(P_1 S_2 P_1 + P_2 Z^\mathsf{T} P_1 + P_1 Z P_2^\mathsf{T} + P_2 S_4 P_2^\mathsf{T}\right) + \varepsilon_1^3\alpha P_2 S_3 P_2^\mathsf{T} \tag{4.20}
$$

$$
E_2 D_3 + E_1 D_{23} + \alpha D_{21}^\mathsf{T} E_3 = \varepsilon_1 (E_2 S_4 E_3 + E_1 Z E_3)
$$

$$
- \frac{1}{\alpha}(A_1 - S_1 P_1)^\mathsf{T} P_2 + \varepsilon_1\varepsilon(P_1 S_2 P_2 + P_2 Z^\mathsf{T} P_2) + \varepsilon_1\alpha^2 P_2 S_3 P_3 \tag{4.21}
$$

$$
E_3 D_3 + D_3^\mathsf{T} E_3 = \varepsilon_1 E_3 S_4 E_3 + \varepsilon_1\alpha^2 P_3 S_3 P_3
$$

$$
- \varepsilon_1\varepsilon_2\left(A_2^\mathsf{T} P_2 + P_2^\mathsf{T} A_2\right) + \varepsilon_1\varepsilon_2^2\left(P_2^\mathsf{T} S_1 P_2 + \varepsilon_1^2 P_2^\mathsf{T} S_2 P_2\right)
$$

$$
+ \varepsilon_1\varepsilon_2\alpha\left(P_3 Z^\mathsf{T} P_2 + P_2^\mathsf{T} Z P_3\right) \tag{4.22}
$$

where

$$
D_1 = A_1 - S_1 P_1^{(0)}, \quad D_3 = A_4 - \alpha S_4 P_3^{(0)}
$$
$$
D_{23} = A_2 - \alpha Z P_3^{(0)}, \quad D_{21} = A_3 - S_4 P_2^{(0)^\mathsf{T}} - Z^\mathsf{T} P_1^{(0)} \tag{4.23}
$$

Note that all nonlinear terms and all cross-coupling terms in Equations 4.20 through 4.22 are multiplied by small parameters. This fact suggests the following reduced-order parallel synchronous algorithm for solving Equations 4.20 through 4.22 (Gajić and Skatarić 1991).

ALGORITHM 4.1

$$E_1^{(i+1)}D_1 + D_1^T E_1^{(i+1)} = \varepsilon E_1^{(i)} S_1 E_1^{(i)} + \varepsilon_1 \alpha P_2^{(i)} S_4 P_2^{(i)T}$$

$$- \varepsilon_1 \alpha \left(P_2^{(i)} A_3 + A_3^T P_2^{(i)^T} \right) + \varepsilon_1^3 \alpha P_2^{(i)} S_3 P_2^{(i)^T}$$

$$+ \varepsilon_1 \alpha \left(P_1^{(i)} S_2 P_1^{(i)} + P_2^{(i)} Z^T P_1^{(i)} + P_1^{(i)} Z P_2^{(i)^T} \right) \quad (4.24)$$

$$E_2^{(i+1)}D_3 + E_1^{(i+1)}D_{23} + \alpha D_{21}^T E_3^{(i+1)} = \varepsilon_1 \left(E_2^{(i)} S_4 E_3^{(i+1)} + E_1^{(i+1)} Z E_3^{(i+1)} \right)$$

$$+ \varepsilon_1 \alpha^2 P_2^{(i)} S_3 P_3^{(i+1)}$$

$$+ \varepsilon_1 \varepsilon \left(P_1^{(i+1)} S_2 P_2^{(i)} + P_2^{(i)} Z^T P_2^{(i)} \right)$$

$$- \frac{1}{\alpha} \left(A_1 - S_1 P_1^{(i+1)} \right)^T P_2^{(i)} \quad (4.25)$$

$$E_3^{(i+1)}D_3 + D_3^T E_3^{(i+1)} = \varepsilon_1 E_3^{(i)} S_4 E_3^{(i)} + \varepsilon_1 \alpha^2 P_3^{(i)} S_3 P_3^{(i)}$$

$$- \varepsilon_1 \varepsilon_2 \left(A_2^T P_2^{(i)} + P_2^{(i)^T} A_2 \right) + \varepsilon_1 \varepsilon_2 \alpha \left(P_3^{(i)} Z^T P_2^{(i)} + P_2^{(i)^T} Z P_3^{(i)} \right)$$

$$+ \varepsilon_1 \varepsilon_2^2 \left(P_2^{(i)^T} S_1 P_2^{(i)} + \varepsilon_1^2 P_2^{(i)^T} S_2 P_2^{(i)} \right) \quad (4.26)$$

where

$$P_j^{(i)} = P_j^{(0)} + \varepsilon E_j^{(i)}, \quad j = 1, 2, 3, \quad i = 0, 1, 2, 3, \ldots \quad (4.27)$$

with initial conditions

$$E_1^{(0)} = 0, \quad E_2^{(0)} = 0, \quad E_3^{(0)} = 0 \quad (4.28)$$

Note that we have made a parallel algorithm by using the Gauss–Seidel iterations. The similar algorithm could have been derived by using the Jacobi-type iterations, but that algorithm would be slower (Bertsekas and Tsitsiklis 1989, 1991). Since Equations 4.24 through 4.26 are completely decoupled, the solution of Equations 4.20 through 4.22 can be obtained by using three processors working in parallel and exchanging intermediate results after each iteration. The work of these three processors has to be synchronized by a global clock.

The following theorem indicates features of the proposed algorithm Equations 4.24 through 4.28.

THEOREM 4.1

Under conditions stated in Assumptions 4.1 and 4.2, the algorithm Equations 4.24 through 4.28 converges to the exact solution of the error term, and thus to the required solution P, with the rate of convergence of $O(\varepsilon)$, that is

$$\left\| E_j - E_j^{(i+1)} \right\| = O(\varepsilon) \left\| E_j - E_j^{(i)} \right\| \tag{4.29}$$

or equivalently

$$\left\| E_j - E_j^{(i)} \right\| = O\left(\varepsilon^{(i+1)}\right), \quad i = 0, 1, 2, \dots; \quad j = 1, 2, 3 \tag{4.30}$$

Proof The proof of this theorem can be found in Skatarić (1993).

The approximate optimal feedback gain $F^{(i)}$ for the problem under consideration is given by

$$F^{(i)} = -R^{-1} B^{\mathrm{T}} P^{(i)} \tag{4.31}$$

where

$$P^{(i)} = \begin{bmatrix} P_1^{(i)} & \varepsilon_1 \varepsilon_2 P_2^{(i)} \\ \varepsilon_1 \varepsilon_2 P_2^{(i)^{\mathrm{T}}} & \sqrt{\varepsilon_1 \varepsilon_2} P_3^{(i)} \end{bmatrix} \tag{4.32}$$

The approximate criterion is obtained from

$$J_{\mathrm{app}}^{(i)} = \mathrm{tr}\left\{ V^{(i)} \right\} \tag{4.33}$$

where $V^{(i)}$ satisfies

$$\left(A - SP^{(i)} \right)^{\mathrm{T}} V^{(i)} + V^{(i)} \left(A - SP^{(i)} \right) + Q + P^{(i)} SP^{(i)} = 0 \tag{4.34}$$

Many real control systems possess at the same time both singularly perturbed and weakly coupled forms. In this section, we present two of them: a supported beam and a satellite optimal control problem. In Section 4.3, we will study two additional real control systems (turbine governor and fluid catalytic cracker) having the quasi singularly perturbed weakly coupled forms. ∎

4.2.1 CASE STUDY: A SUPPORTED BEAM

The mathematical model of a supported beam in the state space form is given in Hsieh et al. (1989)

TABLE 4.1

Approximate Values for the Criterion

i	J_{app}
0	12.0616
1	a
2	a
3	a
4	11.3799
5	11.3372
6	11.3249
7	11.3248
9	11.3247
10	11.3246
13	11.3245
16	11.3244
20	11.3243
29	$11.3242 = J_{opt}$

[a] The solution of the Riccati equation is indefinite.

$$A = \begin{bmatrix} 0 & 1 & 0 & 0 \\ -1 & -0.01 & 0 & 0 \\ 0 & 0 & 0 & 1 \\ 0 & 0 & -16 & -0.04 \end{bmatrix}, \quad B = \begin{bmatrix} 0 & 0 \\ 0.5878 & -1 \\ 0 & 0 \\ 0.9511 & 2 \end{bmatrix}$$

Weighting matrices Q and R are chosen as identifies. We have solved this problem for $n_1 = 2$ and $\varepsilon_1 = \varepsilon_2 = 0.1$. Simulation results for the optimal and approximate performance criteria are presented in Table 4.1. It is interesting to point out that in this example the proposed algorithm converges rather quickly despite the fact that in iterations 1, 2, and 3 the approximate solution for the algebraic Riccati equation has lost its positive semidefiniteness.

4.2.2 CASE STUDY: A SATELLITE OPTIMAL CONTROL PROBLEM

We demonstrate the result of Theorem 4.1 on a satellite control example (Ackerson and Fu 1970). The system matrices A and B are determined by

$$A = \begin{bmatrix} 0 & 0.667 & 0 & 0 \\ -0.667 & 0 & 0 & 0 \\ 0 & 0 & 0 & 1.53 \\ 0 & 0 & 1.53 & 0 \end{bmatrix}, \quad B = \begin{bmatrix} 0 & 0.2 \\ 1 & 0 \\ 0.4 & 0 \\ 0 & 1 \end{bmatrix}$$

Penalty matrices Q and R are chosen as identities, that is, $Q = I_4$, $R = I_2$. Simulation results for the approximate criterion are presented in Table 4.2, indicating the convergence to the optimal performance criterion with $O(\varepsilon)$. It can be seen that the simulation results are consistent with the statement of Theorem 4.1.

TABLE 4.2

Approximate and Optimal Values
for the Criterion

i	J_{app}
0	13.8580
1	12.0499
2	11.7434
3	11.4806
4	11.4683
5	11.4640
6	11.4557
7	11.4533
8	11.4528
9	11.4529
10	11.4528
11	11.4527
	$11.4527 = J_{opt}$

i	$J_{app} - J_{opt}$
12	5.7729×10^{-6}
16	3.3605×10^{-7}
20	1.0437×10^{-8}
25	1.0225×10^{-10}
30	1.0072×10^{-12}

4.3 QUASI-WEAKLY COUPLED SINGULARLY PERTURBED CONTROL SYSTEMS

The quasi singularly perturbed weakly coupled structures are induced by the system matrix having singularly perturbed weakly coupled form as in Equation 4.4 and by the control input matrix having one of the nonstandard structures, namely

$$\dot{x}(t) = Ax(t) + Bu(t) \qquad (4.35)$$

with

$$(1) \; B = \begin{bmatrix} B_1 \\ 0 \end{bmatrix}$$

$$(2) \; B = \begin{bmatrix} 0 \\ B_2 \end{bmatrix} \qquad (4.36)$$

$$(3) \; B = \begin{bmatrix} B_1 \\ B_2 \end{bmatrix}$$

This implies the existence of only one control agent. All of these three structures appear in the real control systems (see case studies in Section 4.5).

Case 1

It can be shown that the matrix P preserves the structure given by Equation 4.9. With the system matrix A given by Equation 4.4 and the newly defined matrix S as

$$S = \begin{bmatrix} Z_1 & 0 \\ 0 & 0 \end{bmatrix}, \quad Z_1 = \mathbf{B}_1 R^{-1} \mathbf{B}_1^{\mathrm{T}} \tag{4.37}$$

the algebraic Riccati equation (Equation 4.8) is partitioned as

$$P_1 A_1 + A_1^{\mathrm{T}} P_1 + Q_1 - P_1 Z_1 P_1 + \varepsilon_1^2 \left(A_3^{\mathrm{T}} P_2^{\mathrm{T}} + P_2 A_3 \right) = 0 \tag{4.38}$$

$$P_2 A_4 + Q_2 + P_1 A_2 + \alpha A_3^{\mathrm{T}} P_3 + \varepsilon_2 \left(A_1^{\mathrm{T}} P_2 - P_1 Z_1 P_2 \right) = 0 \tag{4.39}$$

$$\alpha P_3 A_4 + \alpha A_4^{\mathrm{T}} P_3 + Q_3 + \varepsilon_1^2 \varepsilon_2 \left(P_2^{\mathrm{T}} A_2 + A_2^{\mathrm{T}} P_2 \right) - \varepsilon^4 P_2^{\mathrm{T}} Z_1 P_2 = 0 \tag{4.40}$$

Following the same arguments as in Section 4.2, the reduced-order solution is obtained as

$$P_1^{(0)} A_1 + A_1^{\mathrm{T}} P_1^{(0)} + Q_1 - P_1^{(0)} Z_1 P_1^{(0)} = 0 \tag{4.41}$$

$$P_2^{(0)} A_4 + Q_2 + P_1^{(0)} A_2 + \alpha A_3^{\mathrm{T}} P_3^{(0)} = 0 \tag{4.42}$$

$$\alpha P_3^{(0)} A_4 + \alpha A_4^{\mathrm{T}} P_3^{(0)} + Q_3 = 0 \tag{4.43}$$

The unique solutions of Equations 4.41 through 4.43 exist under the following assumption.

Assumption 4.3 The triple $(A_1, \mathbf{B}_1, \mathrm{Chol}(Q_1))$ is stabilizable–detectable and the matrix A_4 is stable.

Defining the approximation errors as in Equation 4.19, we get the following expressions for the error equations

$$E_1 \mathbf{D}_1 + \mathbf{D}_1^{\mathrm{T}} E_1 = \varepsilon E_1 Z_1 E_1 - \varepsilon_1 \alpha \left(P_2 A_3 + A_3^{\mathrm{T}} P_2^{\mathrm{T}} \right) \tag{4.44}$$

$$E_2 A_4 + E_1 A_2 + \alpha A_3^{\mathrm{T}} E_3 = -\frac{1}{\alpha} (A_1 - Z_1 P_1)^{\mathrm{T}} P_2 \tag{4.45}$$

$$E_3 A_4 + A_4^{\mathrm{T}} E_3 = -\varepsilon^2 \left(A_2^{\mathrm{T}} P_2 + P_2^{\mathrm{T}} A_2 \right) + \varepsilon_1 \varepsilon_2^2 P_2^{\mathrm{T}} Z_1 P_2 \tag{4.46}$$

where $\mathbf{D}_1 = A_1 - Z_1 P_1^{(0)}$.

The following parallel synchronous algorithm is proposed for solving the error Equations 4.44 through 4.46.

ALGORITHM 4.2

$$E_1^{(i+1)}D_1 + D_1^T E_1^{(i+1)} = \varepsilon_1 E_1^{(i)} Z_1 E_1^{(i)} - \varepsilon_1 \alpha \left(P_2^{(i)} A_3 + A_3^T P_2^{(i)^T} \right) \tag{4.47}$$

$$E_2^{(i+1)}A_4 + E_1^{(i+1)}A_2 + \alpha A_3^T E_3^{(i+1)} = -\frac{1}{\alpha}\left(A_1 - Z_1 P_1^{(i+1)} \right)^T P_2^{(i)} \tag{4.48}$$

$$E_3^{(i+1)}A_4 + A_4^T E_3^{(i+1)} = -\varepsilon^2 \left(A_2^T P_2^{(i)} + P_2^{(i)^T} A_2 \right) + \varepsilon_1 \varepsilon_2^2 P_2^{(i)^T} Z_1 P_2^{(i)} \tag{4.49}$$

with $P_j^{(i)}$ and the initial conditions given in Equations 4.27 and 4.28.

The following theorem summarizes features of the algorithm Equations 4.47 through 4.49.

THEOREM 4.2

Under conditions stated in Assumptions 4.1 and 4.3, the algorithm Equations 4.47 through 4.49 converges to the exact solution of the error term, and thus to the required solution P, with the rate of convergence of $O(\varepsilon)$, that is

$$\left\| E_j - E_j^{(i+1)} \right\| = O(\varepsilon) \left\| E_j - E_j^{(i)} \right\| \tag{4.50}$$

or equivalently

$$\left\| E_j - E_j^{(i)} \right\| = O\left(\varepsilon^{(i+1)}\right), \quad i = 0, 1, 2, \dots; \quad j = 1, 2, 3 \tag{4.51}$$

The proof of Theorem 4.2 follows the ideas of the proof of Theorem 4.1, and is thus omitted.

Case 2

It can be shown that in this case, the matrix P also preserves the structure given by Equation 4.9. With the system matrix A given by Equation 4.4 and the newly defined matrix S given by

$$S = \begin{bmatrix} 0 & 0 \\ 0 & Z_4 \end{bmatrix}, \quad Z_4 = B_2 R^{-1} B_2^T \tag{4.52}$$

The partitioned algebraic Riccati equation (Equation 4.8) becomes

$$P_1 A_1 + A_1^T P_1 + Q_1 - \varepsilon^4 P_2 Z_4 P_2^T + \varepsilon_1^2 \left(A_3^T P_2^T + P_2 A_3 \right) = 0 \tag{4.53}$$

$$P_2 A_4 + Q_2 + P_1 A_2 + \alpha A_3^T P_3 - \varepsilon_2 \varepsilon P_2 Z_4 P_3 + \varepsilon_2 A_1^T P_2 = 0 \tag{4.54}$$

$$\alpha P_3 A_4 + \alpha A_4^T P_3 + Q_3 - \varepsilon^2 P_3 Z_4 P_3 + \varepsilon_1^2 \varepsilon_2 \left(P_2^T A_2 + A_2^T P_2 \right) = 0 \tag{4.55}$$

The reduced-order solution is obtained as

$$P_1^{(0)}A_1 + A_1^T P_1^{(0)} + Q_1 = 0 \tag{4.56}$$

$$P_2^{(0)}A_4 + Q_2 + P_1^{(0)}A_2 + \alpha A_3^T P_3^{(0)} = 0 \tag{4.57}$$

$$\alpha P_3^{(0)}A_4 + \alpha A_4^T P_3^{(0)} + Q_3 = 0 \tag{4.58}$$

The unique solutions of Equations 4.56 through 4.58 exist under the following assumption.

Assumption 4.4 The matrices A_1 and A_4 are stable.

Defining the approximation errors as in Equation 4.19, we get the following expressions for the error equations:

$$E_1 A_1 + A_1^T E_1 = -\varepsilon_1 \alpha \left(P_2 A_3 + A_3^T P_2^T \right) + \varepsilon^3 P_2 Z_4 P_2^T \tag{4.59}$$

$$E_2 A_4 + E_1 A_2 + \alpha A_3^T E_3 = \varepsilon_2 P_2 Z_4 P_3 - \frac{1}{\alpha} A_1^T P_2 \tag{4.60}$$

$$E_3 A_4 + A_4^T E_3 = -\varepsilon^2 \left(A_2^T P_2 + P_2^T A_2 \right) + \varepsilon_2 E_3 Z_4 E_3 \tag{4.61}$$

The following parallel synchronous algorithm is proposed for solving the error equations.

ALGORITHM 4.3

$$E_1^{(i+1)} A_1 + A_1^T E_1^{(i+1)} = \varepsilon^3 P_2^{(i)} Z_4 P_2^{(i)^T} - \varepsilon_1 \alpha \left(P_2^{(i)} A_3 + A_3^T P_2^{(i)^T} \right) \tag{4.62}$$

$$E_2^{(i+1)} A_4 + E_1^{(i+1)} A_2 + \alpha A_3 E_3^{(i+1)} = \varepsilon_2 P_2^{(i)} Z_4 P_3^{(i+1)} - \frac{1}{\alpha} A_1^T P_2^{(i)} \tag{4.63}$$

$$E_3^{(i+1)} A_4 + A_4^T E_3^{(i+1)} = -\varepsilon^2 \left(A_2^T P_2^{(i)} + P_2^{(i)^T} A_2 \right) + \varepsilon_2 P_3^{(i)} Z_4 P_3^{(i)} \tag{4.64}$$

with $P_j^{(i)}$ and the initial conditions are defined in Equations 4.27 and 4.28.

The following theorem summarizes the features of algorithm Equations 4.62 through 4.64.

THEOREM 4.3

Under conditions stated in Assumptions 4.1 and 4.4, the algorithm Equations 4.62 through 4.64 converges to the exact solution of the error term, and thus to the required solution P, with the rate of convergence of $O(\varepsilon)$, that is

$$\left\| E_j - E_j^{(i+1)} \right\| = O(\varepsilon) \left\| E_j - E_j^{(i)} \right\| \tag{4.65}$$

or equivalently

$$\left\| E_j - E_j^{(i)} \right\| = O(\varepsilon^{(i+1)}), \quad i = 0, 1, 2, \ldots; \quad j = 1, 2, 3 \tag{4.66}$$

The proof of this theorem is similar to the proof of Theorem 4.1, and it can be found in Skatarić (1993).

Case 3

It can be shown that the matrix P preserves the structure given by Equation 4.9. With the system matrix A given by Equation 4.4 and the newly defined matrix S as

$$S = \begin{bmatrix} Z_1 & Z_2 \\ Z_2^T & Z_4 \end{bmatrix}, \quad Z_2 = \mathbf{B_1} R^{-1} \mathbf{B_2^T} \tag{4.67}$$

the algebraic Riccati equation (Equation 4.8) is partitioned as

$$\begin{aligned} & P_1 A_1 + A_1^T P_1 + Q_1 - P_1 Z_1 P_1 + \varepsilon_1^2 \left(A_3^T P_2^T + P_2 A_3 \right) \\ & - \varepsilon^2 \left(P_2 Z_2^T P_1 + P_1 Z_2 P_2^T \right) - \varepsilon^4 P_2 Z_4 P_2^T = 0 \end{aligned} \tag{4.68}$$

$$\begin{aligned} & P_2 A_4 + Q_2 + P_1 A_2 + \alpha \left(A_3^T P_3 - \varepsilon_2^2 P_2 Z_4 P_3 \right) - \frac{1}{\alpha} P_1 Z_2 P_3 \\ & + \varepsilon_2 \left(A_1^T P_2 - P_1 Z_1 P_2 \right) - \varepsilon_1 \varepsilon_2^2 P_2 Z_2^T P_2 = 0 \end{aligned} \tag{4.69}$$

$$\begin{aligned} & \alpha P_3 A_4 + \alpha A_4^T P_3 + Q_3 - \varepsilon^2 P_3 Z_4 P_3 + \varepsilon_1^2 \varepsilon_2 \left(P_2^T A_2 + A_2^T P_2 \right) \\ & - \varepsilon^3 \left(P_3 Z_2^T P_2 + P_2^T Z_2 P_3 \right) - \varepsilon^4 P_2^T Z_1 P_2 = 0 \end{aligned} \tag{4.70}$$

Since ε_1 and ε_2 are small parameters, we can define $O(\varepsilon)$ approximation of Equations 4.68 through 4.70 as follows:

$$P_1^{(0)} A_1 + A_1^T P_1^{(0)} + Q_1 - P_1^{(0)} Z_1 P_1^{(0)} = 0 \tag{4.71}$$

$$P_2^{(0)} A_4 + Q_2 + P_1^{(0)} A_2 + \alpha A_3^T P_3^{(0)} - \frac{1}{\alpha} P_1^{(0)} Z_2 P_3^{(0)} = 0 \tag{4.72}$$

$$\alpha P_3^{(0)} A_4 + \alpha A_4^T P_3^{(0)} + Q_3 = 0 \tag{4.73}$$

The unique positive semidefinite stabilizing solution $P^{(0)}$, obtained from Equations 4.71 through 4.73, exists under Assumption 4.3.

Defining the approximation errors as before, and using the same logic, we get the following error equations:

$$\begin{aligned} E_1 \mathbf{D_1} + \mathbf{D_1^T} E_1 = {} & \varepsilon E_1 Z_1 E_1 - \varepsilon_1 \alpha \left(P_2 A_3 + A_3^T P_2^T \right) \\ & + \varepsilon \left(P_2 Z_2^T P_1 + P_1 Z_2 P_2^T \right) + \varepsilon^3 P_2 Z_4 P_2^T \end{aligned} \tag{4.74}$$

$$E_2 A_4 + E_1 A_2 + \alpha A_3^T E_3 = \varepsilon_2 E_1 Z_2 E_3 - \frac{1}{\alpha} A_1^T P_2 + \varepsilon_2 P_2 Z_4 P_3$$

$$+ \varepsilon_2^2 \alpha P_2 Z_2^T P_2 + \frac{1}{\alpha} P_1 Z_1 P_2 + \frac{1}{\alpha} \left(E_1 Z_2 P_3^{(0)} + P_1^{(0)} Z_2 E_3 \right)$$

$$(4.75)$$

$$E_3 A_4 + A_4^T E_3 = - \varepsilon^2 \left(A_2^T P_2 + P_2^T A_2 \right) + \varepsilon_2 P_3 Z_4 P_3$$

$$+ \varepsilon_2^2 \alpha \left(P_3 Z_2^T P_2 + P_2^T Z_2 P_3 \right) + \varepsilon_1 \varepsilon_2^2 P_2^T Z_1 P_2 \qquad (4.76)$$

Let us propose the following reduced-order parallel synchronous algorithm for solving Equations 4.74 through 4.76.

ALGORITHM 4.4

$$E_1^{(i+1)} \mathbf{D_1} + \mathbf{D_1^T} E_1^{(i+1)} = \varepsilon E_1^{(i)} Z_1 E_1^{(i)} - \varepsilon_1 \alpha \left(P_2^{(i)} A_3 + A_3^T P_2^{(i)T} \right) + \varepsilon^3 P_2^{(i)} Z_4 P_2^{(i)T}$$

$$+ \varepsilon \left(P_2^{(i)} Z_2^T P_1^{(i)} + P_1^{(i)} Z_2 P_2^{(i)T} \right) \qquad (4.77)$$

$$E_2^{(i+1)} A_4 + E_1^{(i+1)} A_2 + \alpha A_3^T E_3^{(i+1)} = \varepsilon_2 E_1^{(i+1)} Z_2 E_3^{(i+1)} - \frac{1}{\alpha} A_1^T P_2^{(i)}$$

$$+ \varepsilon_2 P_2^{(i)} Z_4 P_3^{(i+1)} + \varepsilon_2^2 \alpha P_2^{(i)} Z_2^T P_2^{(i)}$$

$$+ \frac{1}{\alpha} P_1^{(i+1)} Z_1 P_2^{(i)} + \frac{1}{\alpha} \left(E_1^{(i+1)} Z_2 P_3^{(0)} + P_1^{(0)} Z_2 E_3^{(i+1)} \right)$$

$$(4.78)$$

$$E_3^{(i+1)} A_4 + A_4^T E_3^{(i+1)} = - \varepsilon^2 \left(A_2^T P_2^{(i)} + P_2^{(i)T} A_2 \right) + \varepsilon_2 P_3^{(i)} Z_4 P_3^{(i)}$$

$$+ \varepsilon_2^2 \alpha \left(P_3^{(i)} Z_2^T P_2^{(i)} + P_2^{(i)T} Z_2 P_3^{(i)} \right) + \varepsilon_1 \varepsilon_2^2 P_2^{(i)T} Z_1 P_2^{(i)} \qquad (4.79)$$

with $P_j^{(i)}$ and the initial conditions given in Equations 4.27 and 4.28.

The following theorem indicates the features of the proposed algorithm Equations 4.77 through 4.79.

THEOREM 4.4

Under conditions stated in Assumptions 4.1 and 4.3, the algorithm Equations 4.77 through 4.79 converges to the exact solution of the error term, and thus to the required solution P, with the rate of convergence of $O(\varepsilon)$, that is

$$\left\| E_j - E_j^{(i+1)} \right\| = O(\varepsilon) \left\| E_j - E_j^{(i)} \right\| \qquad (4.80)$$

or equivalently

$$\left|E_j - E_j^{(i)}\right| = O\left(\varepsilon^{(i+1)}\right), \quad i = 0, 1, 2, \ldots; \quad j = 1, 2, 3 \tag{4.81}$$

Proof The proof of this theorem is omitted. It can be found in Skatarić (1993, 2005). In the next section, we present several case studies of quasi singularly perturbed weakly coupled systems. ∎

4.3.1 CASE STUDIES

Case Study 1
The design of turbine governors of the power system considered in Arnautovic and Skatarić (1991) is represented by the state space model of the form

$$A = \begin{bmatrix} -0.71 & 0 & 0 & 0 & 0 \\ 0 & -2 & 0 & 0 & 0 \\ 0.61 & 1.28 & -1.46 & 0.566 & 0 \\ -0.18 & -0.37 & 0.56 & -0.594 & -0.23 \\ 0 & 0 & 0 & 314.16 & 0 \end{bmatrix}, \quad B = \begin{bmatrix} 0.71 \\ 2 \\ 0 \\ 0 \\ 0 \end{bmatrix}$$

This system is partitioned with $n_1 = 3$. The penalty matrices are chosen as $Q = 0.1 \times I_5$, $R = 1$. Obtained results are presented in Table 4.3.

TABLE 4.3

Approximate and Optimal Values for the Criterion

i	J_{app}
0	a
1	107.4983
2	83.9686
3	79.5014
4	79.4011
5	77.6671
6	77.0594
7	77.0955
8	76.9988
9	76.9211
10	76.9153
11	76.9157
12	76.9090
13	76.9059
14	76.9061
15	76.9059

(continued)

TABLE 4.3 (continued)
Approximate and Optimal Values
for the Criterion

i	J_{app}
16	76.9054
17	76.9053
18	76.9053
19	76.9052
20	$76.9052 = J_{opt}$

[a] $P^{(0)}$ is indefinite.

Case Study 2

In order to demonstrate the efficiency of the proposed algorithm, we have run a fifth-order real world example, an industrially important reactor considered in Arkun and Ramakrishnan (1983). Matrices A, B, C, Q, and R are given by

$$A = \begin{bmatrix} -16.11 & -0.39 & 27.2 & 0 & 0 \\ 0.01 & -16.99 & 0 & 0 & 12.47 \\ 15.11 & 0 & -53.6 & -16.57 & 71.78 \\ -53.36 & 0 & 0 & -107.2 & 232.11 \\ 2.27 & 69.1 & 0 & 0 & -102.99 \end{bmatrix}$$

$$B^T = \begin{bmatrix} 11.12 & -3.61 & -21.91 & -53.6 & 69.1 \\ -12.6 & 3.36 & 0 & 0 & 0 \end{bmatrix}$$

$$C = \begin{bmatrix} 0 & 0 & 0 & 0 & 1 \\ 0 & 1 & 0 & 0 & 0 \end{bmatrix}, \quad Q = I_5, \quad R = I_2$$

The eigenvalues of the matrix A are -2.8, -7.7, -74, -82, and -129. The small parameter is chosen as $\varepsilon = 0.1$, which is roughly the ratio of 7.7 and 74. The matrices A and B are partitioned with $n_1 = 3$. The penalty matrices are chosen as $Q = 10 \times I_5$, $R = I_2$. Obtained results are presented in Table 4.4.

In both case studies we have seen very good convergence properties of the proposed parallel reduced-order algorithms.

Exercise 4.1: Derive the reduced-order parallel algorithm for the optimal control for the standard Draper/RPL satellite considered in Keel and Bhattacharyya (1990).

TABLE 4.4

Approximate and Optimal Values for the Criterion

i	J_{app}
0	2.1738
1	1.7840
2	1.7747
3	1.7463
4	1.7462
5	1.7422
6	1.7419
7	1.7415
8	1.7414
9	1.7413
10	1.7413
11	1.7412
	$1.7412 = J_{opt}$

i	$J_{app} - J_{opt}$
12	2.5376×10^{-5}
15	6.0143×10^{-6}
20	6.2229×10^{-7}
25	6.2168×10^{-8}

4.4   CONCLUSION

In this chapter, we have shown that the large-scale systems containing small parameters in the sense of singular perturbations and weak coupling are inherently parallel in nature, and thus, very well suited for parallel and distributed computation. Corresponding parallel synchronous algorithms are developed. The extension of these results to the asynchronous parallel algorithms is possible. It is well known that the asynchronous algorithms generate the required solution faster than the synchronous ones, but they have problems with convergence.

This chapter has presented optimal control theory results for a class of linear dynamic systems that are simultaneously weakly coupled and singularly perturbed. Such systems, under different assumptions, have been studied also within the concept of multimodeling (Khalil and Kokotović 1978) and using the slow coherency method (Chow 1982). The assumptions imposed here allow us solve this class of singularly perturbed weakly coupled systems using the standard techniques of the recursive approach to weakly coupled optimal control systems and facilitate reduced-order computations. The assumptions used in the multimodeling and slow coherency techniques make them more challenging for investigation.

REFERENCES

Ackerson, G. and K. Fu, On the state estimation in switching environments, *IEEE Transactions on Automatic Control*, AC-15, 10–17, 1970.

Arkun, Y. and S. Ramakrishnan, Bounds of the optimum quadratic cost of structure constrained regulators, *IEEE Transactions Automatic Control*, AC-28, 924–927, 1983.

Arnautovic, D. and D. Skatarić, Suboptimal design of hydroturbine governors, *IEEE Transactions on Energy Conversion*, 6, 438–444, 1991.

Bertsekas, D. and J. Tsitsiklis, *Parallel and Distributed Computation: Numerical Methods*, Prentice-Hall, Englewood Cliffs, NJ, 1989.

Bertsekas, D. and J. Tsitsiklis, Some aspects of parallel and distributed iterative algorithms—A survey, *Automatica*, 27, 3–21, 1991.

Chow, J., *Time-Scale Modeling of Dynamic Networks with Applications to Power Systems*, Springer Verlag, New York, 1982.

Chow, J. and P. Kokotović, Sparsity and time scales, *Proceedings of the American Control Conference*, San Francisco, pp. 656–661, 1983.

Coumarbatch, C. and Z. Gajić, Exact decomposition of the algebraic Riccati equation of deterministic multimodeling optimal control problems, *IEEE Transactions on Automatic Control*, 44, 790–794, 2000a.

Coumarbatch, C. and Z. Gajić, Parallel optimal Kalman filtering for stochastic systems in multimodeling form, *Transactions of ASME Journal Dynamic Systems, Measurement and Control*, 122, 542–550, 2000b.

Gajić, Z., The existence of a unique and bounded solution of the algebraic Riccati equation of the multimodel estimation and control problems, *Systems and Control Letters*, 10, 185–190, 1988.

Gajić, Z. and H. Khalil, Multimodel strategies under random disturbances and imperfect partial observations, *Automatica*, 22, 121–125, 1986.

Gajić, Z. and M. Lim, *Optimal Control of Singularly Perturbed Linear Systems and Applications: High Accuracy Techniques*, Marcel Dekker, New York, 2001.

Gajić, Z. and X. Shen, *Parallel Algorithms for Optimal Control of Large Scale Linear Systems*, Springer Verlag, London, 1993.

Gajić, Z. and D. Skatarić, Singularly perturbed weakly coupled linear control systems, *Proceedings European Control Conference*, Grenoble, France, pp. 1607–1612, 1991.

Hsieh, C., R. Skelton, and F. Dampa, Minimum energy controllers with inequality constraints on output variances, *Optimal Control Applications and Methods*, 10, 347–366, 1989.

Keel, L. and S. Bhattacharyya, A matrix equation approach to the design of low-order regulators, *SIAM Journal on Matrix Analysis and Applications*, 11, 180–199, 1990.

Khalil, H., Multi-model design of a Nash strategy, *Journal of Optimization Theory and Applications*, 31, 553–564, 1980.

Khalil, H. and P. Kokotović, Control strategies for decision makers using different models of the same system, *IEEE Transactions on Automatic Control*, AC-23, 289–298, 1978.

Khalil, H. and P. Kokotović, Control of linear systems with multiparameter singular perturbations, *Automatica*, 15, 197–207, 1979a.

Khalil, H. and P. Kokotović, D-stability and multiparameter singular perturbations, *SIAM Journal on Control and Optimization*, 17, 56–65, 1979b.

Kokotoric, P., H. Khalil, and J. O'Reilly, *Singular Perturbation Method in Control: Analysis and Design*, Academic Press, Orlando, 1986.

Mukaidani, H. Near-optimal control for multimodeling systems, *Transactions of the Society of Instrument and Control Engineers*, 37, 960–969, 2001.

Mukaidani, H., Near-optimal Kalman filters for multiparameter singularly perturbed linear systems, *IEEE Transactions on Circuits and Systems—I: Fundamental Theory and Applications*, 50, 717–721, 2003.

Mukaidani, H., A new approach to robust guaranteed cost controller for uncertain multi-modeling systems, *Automatica*, 41, 1055–1062, 2005.

Mukaidani, H., T. Shimomura, and K. Mizukami, Algebraic expansions and a new numerical algorithm of the algebraic Riccati equation for multiparameter singularly perturbed systems, *Journal of Mathematical Analysis and Its Applications*, 267, 209–234, 2002a.

Mukaidani, H., T. Shimomura, and H. Xu, Near-optimal control of linear multiparameter singularly perturbed systems, *IEEE Transactions on Automatic Control*, 47, 2051–2057, 2002b.

Mukaidani, H., H. Xua, and K. Mizukami, New results for near-optimal control of linear multiparameter singularly perturbed systems, *Automatica*, 39, 2151–2167, 2003.

Saksena, V. and J. Cruz, A multimodel approach to stochastic Nash games, *Automatica*, 17, 295–305, 1981a.

Saksena, V. and J. Cruz, Nash strategies in decentralized control of multiparameter singularly perturbed large scale systems, *Large Scale Systems*, 2, 219–234, 1981b.

Saksena, V. and T. Basar, A multimodel approach to stochastic team problems, *Automatica*, 18, 713–720, 1982.

Saksena, V., J. Cruz, W. Perkins, and T. Basar, Information induced multimodel solution in multiple decision make problems, *IEEE Transactions on Automatic Control*, AC-28, 716–728, 1983.

Skatarić, D., *Parallel Algorithms for Reduced-Order Optimal Control of Quasi Singularly Perturbed and Weakly Coupled Systems,* Doctoral Dissertation, University of Novi Sad, Novi Sad, 1993.

Skatarić, D., *Optimal Control of Quasi Singularly Perturbed and Weakly Coupled Systems,* Planeta Print, Belgrade, 2005.

Zhuang, J. and Z. Gajić, Stochastic multimodel strategy with perfect measurements, *Control—Theory and Advanced Technology*, 7, 173–182, 1991.

5 Decoupling Transformation, Lyapunov Equation, and Boundary Value Problem

Decoupling transformations play very important roles in control systems containing small parameters. Under certain, usually very mild conditions, these transformations allow the linear system decomposition into independent reduced-order subsystems. The decoupling transformation for weakly coupled linear systems was introduced in Gajić and Shen (1989). In Qureshi (1992), another version of the transformation obtained by Gajić and Shen (1989) was derived. These transformations were obtained for weakly coupled linear systems composed of two subsystems. In Gajić and Borno (1995), a general transformation that decouples a weakly coupled system composed of N subsystems into N independent subsystems was presented.

In this chapter, we first present the main results of Gajić and Shen (1989) for continuous-time invariant linear weakly coupled linear systems (see Section 5.1). Numerical techniques for solving algebraic equations comprising the transformation for weakly coupled systems are also discussed. The decoupling transformation for weakly coupled linear systems obtained in Qureshi (1992) is also presented in this section. In Section 5.2, we consider the transformation for decoupling N weakly coupled linear subsystems by following the results of Gajić and Borno (1995). The transformation for decoupling N weakly coupled linear systems is also considered in Prljaca and Gajić (2007).

The transformation developed for decoupling two linear weakly coupled subsystems, applied to the Lyapunov differential equation of weakly coupled linear systems, produces the complete decompositions of this equation into three Lyapunov equations of the reduced order, which was demonstrated in Section 5.3.

In Section 5.4, we have solved the general boundary value problem of continuous-time varying weakly coupled linear systems producing its decomposition into two-reduced-order initial value problems by following the results of Qureshi and Gajić (1991). This section demonstrates how the transformation developed in Section 5.1 can be used to solve the basic mathematical problem of linear-quadratic optimal control theory. The discrete-time version of the results obtained in Section 5.4 are presented in Section 5.5.

5.1 DECOUPLING TRANSFORMATION OF GAJIĆ AND SHEN

The linear weakly coupled system is represented by (Kokotović et al., 1969)

$$\dot{x}(t) = A_1 x(t) + \varepsilon A_2 z(t) + B_1 u_1(t) + \varepsilon B_2 u_2(t) \tag{5.1}$$

$$\dot{z}(t) = \varepsilon A_3 x(t) + A_4 z(t) + \varepsilon B_3 u_1(t) + B_4 u_2(t) \tag{5.2}$$

where

  $x(t) \in \Re^{n_1}$, $z(t)$ $\Re^{n_2}$, $n_1 + n_2 = n$ are subsystem states
  $u_i(t) \in \Re^{m_i}$ are subsystem controls $i = 1, 2$
  ε is a small coupling parameter

Matrices A_i, $i = 1, 2, 3, 4$, and B_i, $i = 1, 2, 3, 4$, are constant and $O(1)$. It is also assumed that magnitudes of all the system eigenvalues are $O(1)$, that is $|\lambda_j| = O(1)$, $j = 1, 2, \ldots, n$, implying that matrices A_1 and A_4 are nonsingular with $\det\{A_1\} = O(1)$ and $\det\{A_4\} = O(1)$, which is the standard assumption for weakly coupled linear control systems and also corresponds to block diagonal dominance of the system matrix (Chow and Kokotović 1983). Hence, the main results presented in this section are valid under the following assumption.

Assumption 5.1   Matrices, A_i, $i = 1, 2, 3, 4$, are constant and $O(1)$. In addition, magnitudes of all system eigenvalues are $O(1)$, that is, $|\lambda_j| = O(1)$, $j = 1, 2, \ldots, n$, which implies that the matrices A_1, A_4 are nonsingular with $\det\{A_1\} = O(1)$ and $\det\{A_4\} = O(1)$.

  Note that when this assumption is not satisfied, the system defined in Equations 5.1 and 5.2 in addition of weak coupling also displays multiple timescale phenomena (singular perturbations) (Phillips and Kokotović 1981; Kokotović and Khalil 1986).

  In this section, we derive the nonsingular transformation that completely decouples linear weakly coupled systems (filters or estimators first of all) by following the results of Gajić and Shen (1989).

  Introducing the change of variables

$$x(t) = \eta(t) + \varepsilon L_1 z(t) \tag{5.3}$$

the original system (Equation 5.1) is transformed into

$$\dot{\eta}(t) = A_{10}\eta(t) + \varepsilon \Phi_1(L_1)z(t) + B_{10}u_1(t) + \varepsilon B_{20}u_2(t) \tag{5.4}$$

where

$$\begin{aligned}
A_{10} &= A_1 - \varepsilon^2 L_1 B_3 \\
B_{10} &= B_1 - \varepsilon^2 L_1 B_3 \\
B_{20} &= B_2 - L_1 B_4
\end{aligned} \tag{5.5}$$

and

$$\Phi_1(L_1) = A_1 L_1 - L_1 A_4 + A_2 - \varepsilon^2 L_1 A_3 L_1 \tag{5.6}$$

Assuming that matrix L_1 can be chosen such that $\Phi_1(L_1) = 0$, Equation 5.4 represents a completely independent (decoupled) subsystem

$$\dot{n}(t) = A_{10}\eta(t) + B_{10}u_1(t) + \varepsilon B_{20}u_2(t) \tag{5.7}$$

As a matter of fact, Equations 5.2 and 5.7 form a triangular system (after elimination of $x(t)$ from Equation 5.2 using Equation 5.3).

Introducing the second change of variables as

$$\zeta(t) = z(t) + \varepsilon H_1 \eta(t) \tag{5.8}$$

Equation 5.2 becomes

$$\dot{\zeta}(t) = A_{40}\zeta(t) + \varepsilon \Phi_2(H_1)\eta(t) + \varepsilon B_{30}u_1(t) + B_{40}u_2(t) \tag{5.9}$$

with

$$\begin{aligned} A_{40} &= A_4 + \varepsilon^2 A_3 L_1 \\ B_{30} &= B_3 + H_1 B_{10} \\ B_{40} &= B_4 + \varepsilon^2 H_1 B_{20} \end{aligned} \tag{5.10}$$

and

$$\Phi_2(H_1) = H_1 A_{10} - A_{40} H_1 + A_3 \tag{5.11}$$

In addition, if matrix H_1 can be chosen such that $\Phi_2(H_1) = 0$, we have

$$\dot{\zeta}(t) = A_{40}\zeta(t) + \varepsilon B_{30}u_1(t) + B_{40}u_2(t) \tag{5.12}$$

so that Equations 5.7 and 5.12 represent two completely decoupled linear subsystems. Notice that the weakly coupled structure of the control inputs in Equations 5.1 and 5.2 is preserved in the new coordinates, that is, in Equations 5.7 and 5.12. This means that the proposed transformation is applicable to the feedback structure of Equations 5.1 and 5.2 also. Thus, applying the nonsingular transformation

$$\begin{bmatrix} \eta(t) \\ \zeta(t) \end{bmatrix} = \begin{bmatrix} I_{n_1} & -\varepsilon L_1 \\ \varepsilon H_1 & I_{n_2} - \varepsilon^2 H_1 L_1 \end{bmatrix} \begin{bmatrix} x(t) \\ x(t) \end{bmatrix} = \mathbf{T_1} \begin{bmatrix} x(t) \\ z(t) \end{bmatrix} \tag{5.13}$$

with

$$\mathbf{T_1^{-1}} = \begin{bmatrix} I_{n_1} - \varepsilon^2 L_1 H_1 & \varepsilon L_1 \\ -\varepsilon H_1 & I_{n_2} \end{bmatrix} \tag{5.14}$$

the linear weakly coupled system (Equations 5.1 and 5.2) is completely decoupled and uniquely determined by its subsystems (Equations 5.7 and 5.12).

Obviously, the transformation $\mathbf{T_1}$ is uniquely defined if unique solutions of the following two algebraic equations exist:

$$A_1 L_1 - L_1 A_4 + A_2 - \varepsilon^2 L_1 A_3 L_1 = 0 \tag{5.15}$$

$$H_1(A_1 - \varepsilon^2 L_1 A_3) - (A_4 + \varepsilon^2 A_3 L_1)H_1 + A_3 = 0 \tag{5.16}$$

It is important to notice that at $\varepsilon = 0$ we have

$$A_1 L_1^{(0)} - L_1^{(0)} A_4 + A_2 = 0 \tag{5.17}$$

$$H_1^{(0)} A_1 - A_4 H_1^{(0)} + A_3 = 0 \tag{5.18}$$

so that

$$\begin{aligned} L_1 &= L_1^{(0)} + O(\varepsilon^2) \\ H_1 &= H_1^{(0)} + O(\varepsilon^2) \end{aligned} \tag{5.19}$$

Equations 5.17 and 5.18 are Sylvester equations and their unique solutions exist if matrices A_1 and A_4 have no eigenvalues in common (Lancaster and Tismenetsky 1985). Thus, the presented results will be valid under the following assumption.

Assumption 5.2   Matrices A_1 and A_4 have no common eigenvalues.

By the implicit function theorem (Ortega and Rheinboldt 1970), for a sufficiently small $\varepsilon \in (0, \varepsilon_1]$ there exists a unique solution of weakly nonlinear algebraic equation (Equation 5.15). Under the assumption that A_1 and A_4 have no eigenvalues in common and by the fact that the eigenvalues are continuous functions of matrix elements (Kato 1980), there exists ε_2 small enough such that for any $\varepsilon \in (0, \varepsilon_2]$ matrices A_{10} and A_{40} will not have eigenvalues in common and thus, the unique solution of Equation 5.16 will exist.

In summary, we have established the following theorem.

THEOREM 5.1

Under Assumptions 5.1 and 5.2, there exists a small parameter $\varepsilon \in (0, \min(\varepsilon_1, \varepsilon_2)]$ such that the unique solutions of Equations 5.15 and 5.16 exist.

Trajectories of the transformed (decoupled) system are $O(\varepsilon)$ close to trajectories of the original system. If the coupling parameter ε is extremely small, or if in the design procedure the accuracy of $O(\varepsilon)$ is sufficient, there is no need for the decomposition. However, if $O(\varepsilon)$ is not very small, or if the high accuracy is required, then one needs methods that will produce any desired accuracy, that is, the accuracy of $O(\varepsilon^k)$, $k = 2, 3, 4, \ldots$. Thus, the method proposed in this section is very useful for the intermediate values of ε and for systems with high accuracy requirements. In addition, the importance of the proposed transformation is in the design of linear filters and observers—dynamical systems built by the designer. Apparently, it is much easier and less expensive to build two dynamical systems of order n_1 and n_2, than one dynamical system of order $n_1 + n_2$.

Note that transformations (Equations 5.3 and 5.8) can be used independently to put the system in either the lower or upper triangular form. For some applications, this might be sufficient.

Numerical solutions for L_1 and H_1 can be obtained by using the fixed point type recursive algorithms similar to those developed by Petrović and Gajić (1988) and Harkara et al. (1989). In the case of Equations 5.15 and 5.16 the corresponding algorithm is given by

$$A_1 L_1^{(i+1)} - L_1^{(i+1)} A_4 + A_2 - \varepsilon^2 L_1^{(i)} A_3 L_1^{(i)} = 0 \tag{5.20}$$

with $i = 0, 1, 2, \ldots, N - 1$, and $L_1^{(0)}$ obtained from Equation 3.26

$$H_1^{(N)} A_{10}^{(N)} - A_{40}^{(N)} H_1^{(N)} + A_3 = 0 \tag{5.21}$$

where

$$A_{10}^{(N)} = A_1 - \varepsilon^2 L_1^{(N)} A_3, \quad A_{40}^{(N)} = A_4 + \varepsilon^2 A_3 L_1^{(N)} \tag{5.22}$$

Using the results of the references given above, it can be shown that

$$L_1 = L_1^{(N)} + O(\varepsilon^{2N}) \tag{5.23}$$

and

$$H_1 = H_1^{(N)} + O(\varepsilon^{2N}) \tag{5.24}$$

Hence, the algorithm (Equation 5.20) converges with the rate of convergence of $O(\varepsilon^2)$.

Example 5.1

In order to demonstrate the efficiency of the proposed algorithm (Equation 5.20), we have run a sixth-order example. Matrices A_i, $i = 1, 2, 3, 4$, are chosen randomly (standard deviation equals 1 and the mean value equals 0 for A_1, A_2, and A_3; standard deviation equals 2 and the mean value equals 0 for A_4).

$$A_1 = \begin{bmatrix} -1.720 & -0.999 & -0.592 \\ -1.434 & 0.799 & 0.856 \\ -0.729 & 0.105 & 0.867 \end{bmatrix}, \quad A_2 = \begin{bmatrix} -1.614 & -1.429 & 0.516 \\ 0.225 & 1.928 & 0.310 \\ -0.332 & 0.067 & 0.329 \end{bmatrix}$$

$$A_3 = \begin{bmatrix} -1.398 & 1.039 & 0.557 \\ 1.298 & 1.349 & -0.891 \\ -0.472 & -0.610 & -0.873 \end{bmatrix}, \quad A_4 = \begin{bmatrix} -2.956 & 1.219 & 2.269 \\ -0.038 & -2.240 & 2.296 \\ -0.873 & -2.020 & 2.344 \end{bmatrix}$$

TABLE 5.1

Number of Iterations for the Fixed Point Method

ε	Number of Required Iterations Such That $\|L_1 - L_1^{(i)}\|_\infty < 10^{-10}$
0.8	*
0.7	28
0.6	17
0.5	12
0.3	9
0.1	5
0.05	3
0.01	2

The simulation results for different values of the coupling parameter ε are given in Table 5.1.

Table 5.1 strongly support the necessity for the existence of the recursive scheme for solving Equation 5.15, since unless ε is very small, the zeroth and first-order approximations are far from the optimal solution.

In Table 5.2, we show the propagations of the error per iteration when $\varepsilon = 0.1$. We notice that the rate of convergence of the proposed algorithm (Equation 5.20) is $O(\varepsilon^2) = O(10^{-2})$.

The algorithm (Equation 5.20) is based on the fixed point iterations, and it will converge as long as the small parameter ε is small enough so that the radius of convergence is $p(\varepsilon) < 1$ at each iteration.

An alternative way of solving Equation 5.15 is by using the Newton method where solution of Equation 5.17 plays the role of the initial condition. The Newton method for the similar type of algebraic equations has been presented by Grodt and Gajić (1988). The Newton algorithm for Equation 5.15 can be constructed by setting

TABLE 5.2

Error Propagation for the Fixed Point Method

i $\varepsilon = 0.1$	$\|L_1 - L_1^{(i)}\|_\infty$
0	4.129×10^{-2}
1	7.4645×10^{-4}
2	1.6401×10^{-5}
3	2.1149×10^{-7}
4	2.0989×10^{-10}

$L_1^{(i+1)} = L_1^{(i)} + \Delta L_1^{(i)}$ and neglecting $O((\Delta L_1)^2)$ terms. This will produce a Sylvester-type equation of the form

$$D_1^{(i)} L_1^{(i+1)} + L_1^{(i+1)} D_2^{(i)} = Q^{(i)}, \quad i = 0, 1, 2, \dots \tag{5.25}$$

where

$$\begin{aligned}
D_1^{(i)} &= A_1 - \varepsilon^2 L_1^{(i)} A_3 \\
D_2^{(i)} &= -\left(A_4 + \varepsilon^2 A_3 L_1^{(i)}\right) \\
Q^{(i)} &= -\left(A_2 + \varepsilon^2 L_1^{(i)} A_3 L_1^{(i)}\right)
\end{aligned} \tag{5.26}$$

with the initial condition $L_1^{(0)}$ obtained from Equation 5.17.

Example 5.2

The Newton method is demonstrated by solving the same example. For the different values of ε the results are presented in Table 5.3.

It can be seen, that for this particular example, the Newton method converges much faster than the fixed point iteration algorithm. It is a well-known fact that the Newton method converges quadratically in the neighborhood of the sought solution and that its main problem lies in the choice of the initial guess. For algebraic equation (Equation 5.15) the initial guess is easily obtained with the accuracy of $O(\varepsilon^2)$, and the Newton method, if it converges, will produce a sequence $O(\varepsilon^4)$, $O(\varepsilon^8)$, $O(\varepsilon^{16})$, close to the exact solution. However, in some cases the Newton method does not converge at all (bad initial guess) and one needs to have some other efficient techniques available. The fixed point method presented earlier in this section is one of them, since its rate of convergence of $O(\varepsilon^2)$ is remarkable. Another method for solving Equation 5.15 is the eigenvector method of Kecman (2006). This method is presented in Chapter 10.

TABLE 5.3

Number of Iterations for the Newton Method

ε	Number of Required Iterations Such That $\|L_1 - L_1^{(i)}\|_\infty < 10^{-10}$
0.8	5
0.7	5
0.6	4
0.5	4
0.3	3
0.1	2
0.05	2
0.01	1

The importance of the decoupling transformation introduced for weakly coupled linear systems is in the decomposition of the linear Kalman filters and observers. Namely, they are dynamical systems built by the control engineers. It is much easier and cheaper to build two filters of order n_1 and n_2, than one filter of order $n_1 + n_2$. The reduced-order filters are much faster. Due to parallelism, the online computations are considerably reduced at every time instant.

In the case of time varying linear weakly coupled systems represented by

$$\dot{x}_1(t) = A_1(t)x_1(t) + \varepsilon A_2(t)x_2(t) \tag{5.27}$$

$$\dot{x}_2(t) = \varepsilon A_3(t)x_1(t) + A_4(t)x_2(t) \tag{5.28}$$

where matrices A_1, A_2, A_3, and A_4 are bounded functions of time t with all matrix elements bounded by $O(1)$, we can use the same idea and obtain the decoupling transformation as

$$\mathbf{T}_2(t) = \begin{bmatrix} I_{n_1} & -\varepsilon L_2(t) \\ \varepsilon H_2(t) & I_{n_2} - \varepsilon^2 H_2(t)L_2(t) \end{bmatrix} \tag{5.29}$$

where $L_2(t)$ and $H_2(t)$ can be obtained from the following two coupled differential equations

$$\dot{L}_2(t) = A_1(t)L_2(t) - L_2(t)A_4(t) + A_2(t) - \varepsilon^2 L_2(t)A_3(t)L_2(t) \tag{5.30}$$

$$\dot{H}_2(t) = H_2(t)\big(A_1(t) - \varepsilon^2 L_2(t)A_3(t)\big) \\ - \big(A_4(t) + \varepsilon^2 A_3(t)L_2(t)\big)H_2(t) + A_3(t) \tag{5.31}$$

This difficulty in solving Equations 5.30 and 5.31 is in the fact that Equation 5.31 can be solved only after the results of Equation 5.30 are available. Therefore, computation must be done sequentially. Furthermore, two different algorithms are needed: one for Equation 5.30 and the other for Equation 5.31. In this section, this difficulty is overcome by introducing another transformation that decouples the original systems as well as the transformation equations (Qureshi 1992). This will enable us to compute $L_2(t)$ and $H_2(t)$ in parallel, and by using only one algorithm. The proposed transformation is extremely efficient, from the numerical point of view, in the case of time varying systems since corresponding differential equations are completely decoupled.

5.1.1 DECOUPLING TRANSFORMATION OF QURESHI

Introducing the change of variables like in Gajić and Shen (1989)

$$\eta(t) = x_1(t) - \varepsilon L_3(t)x_2(t) \tag{5.32}$$

and differentiating both sides, we obtain

$$\dot{\eta}(t) = \dot{x}_1(t) - \varepsilon \dot{L}_3(t)x_2(t) - \varepsilon L_3(t)\dot{x}_2(t) \tag{5.33}$$

Substituting for $\dot{x}_1(t)$ and $\dot{x}_2(t)$ and simplifying, we get

$$\dot{\eta}(t) = A_{10}(t)\eta(t) - \varepsilon\Phi_3\big(L_3(t), \dot{L}_3(t)\big)x_2(t) \tag{5.34}$$

where

$$A_{10}(t) = A_1(t) - \varepsilon^2 L_3(t)A_3(t) \tag{5.35}$$

and

$$\Phi_3\big(L_3(t), \dot{L}_3(t)\big) = \dot{L}_3(t) - A_1(t)L_3(t) + L_3(t)A_4(t)$$
$$- A_2(t) + \varepsilon^2 L_3(t)A_3(t)L_3(t) \tag{5.36}$$

Assuming that matrix $L_3(t)$ can be chosen so that $\Phi_3(L_3(t), \dot{L}_3(t)) = 0$ (Equation 5.34) will represent a completely independent (decoupled) system.

Introducing the second change of variables as

$$\zeta(t) = -\varepsilon H_3(t)x_1(t) + x_2(t) \tag{5.37}$$

and following similar calculations, we get

$$\dot{\zeta}(t) = A_{40}(t)\zeta(t) - \varepsilon\Phi_4\big(H_3(t), \dot{H}_3(t)\big)x_1(t) \tag{5.38}$$

where

$$A_{40}(t) = A_4(t) - \varepsilon^2 H_3(t)A_2(t) \tag{5.39}$$

and

$$\Phi_4\big(H_3(t), \dot{H}_3(t)\big) = \dot{H}_3(t) - A_4(t)H_3(t) + H_3(t)A_1(t)$$
$$- A_3(t) + \varepsilon^2 H_3(t)A_2(t)H_3(t) \tag{5.40}$$

Assuming that matrix $H_3(t)$ can be chosen such that the following equation is satisfied for every t, $\Phi_4(H_3(t), \dot{H}_3(t)) = 0$, Equations 5.34 and 5.38 will represent a completely decoupled linear system. Note that the initial conditions for differential Equations 5.36 and 5.40 are arbitrary. Thus, applying the transformation

$$\begin{bmatrix} \eta(t) \\ \zeta(t) \end{bmatrix} = \begin{bmatrix} I_{n_1} & -\varepsilon L_3(t) \\ -\varepsilon H_3(t) & I_{n_2} \end{bmatrix} \begin{bmatrix} x_1(t) \\ x_2(t) \end{bmatrix} = T_3(t) \begin{bmatrix} x_1(t) \\ x_2(t) \end{bmatrix} \tag{5.41}$$

where

$$T_3^{-1}(t) = \begin{bmatrix} I_{n_1} + \varepsilon^2 L_3(t)M(t)H_3(t) & \varepsilon L_3(t)M(t) \\ \varepsilon M(t)H_3(t) & M(t) \end{bmatrix} \tag{5.42}$$

with $M(t) = (I_{n_1} - \varepsilon^2 H_3(t)L_3(t))^{-1}$ the linear weakly coupled system is completely decoupled and uniquely determined by its subsystems (Equations 5.34 and 5.38). The nonsingularity of the transformation $\mathbf{T}_3(t)$ can be noticed by the fact that the off-diagonal elements are of the order of ε, while the blocks on the main diagonal are identity matrices. Therefore, for a sufficiently small ε, $\mathbf{T}_3(t)$ is strictly diagonally dominant and hence nonsingular.

Note that exactly the same transformation can be applied to the continuous-time invariant systems. In that case differential Equations 5.36 and 5.40 reduce to the algebraic ones, so that the matrix $\mathbf{T}_3$ is constant.

We can apply transformation Equation 5.41 to the discrete-time invariant weakly coupled linear system defined by

$$\begin{aligned} x_1(n+1) &= A_{11}x_1(n) + \varepsilon A_{12}x_2(n) \\ x_2(n+1) &= \varepsilon A_{21}x_1(n) + A_{22}x_2(n) \end{aligned} \tag{5.43}$$

Following similar calculations, the transformed block diagonal system is obtained as follows

$$\begin{aligned} \eta(n+1) &= [A_{11} - \varepsilon^2 L_3 A_{21}]\eta(n) \\ \zeta(n+1) &= [A_{22} - \varepsilon^2 H_3 A_{12}]\zeta(n) \end{aligned} \tag{5.44}$$

where L_3 and H_3 satisfy the following decoupled algebraic equations:

$$A_{11}L_3 - L_3 A_{22} - \varepsilon^2 L_3 A_{21} L_3 + A_{12} = 0 \tag{5.45}$$

$$A_{22}H_3 - H_3 A_{11} - \varepsilon^2 H_3 A_{12} H_3 + A_{21} = 0 \tag{5.46}$$

Note that unique solutions of Equations 5.45 and 5.46 exist for sufficiently small values of ε under the assumption that matrices A_{11} and A_{22} have no eigenvalues in common (Assumption 5.1).

Exercise 5.1: Derive Equations 5.44 through 5.46. Then show that a similar transformation is applicable to the decomposition of the discrete-time varying linear control system represented by

$$\begin{aligned} x_1(n+1) &= A_{11}(n)x_1(n) + \varepsilon A_{12}(n)x_2(n) \\ &\quad + B_{11}(n)u_1(n) + \varepsilon B_{12}(n)u_2(n) \\ x_2(n+1) &= \varepsilon A_{12}(n)x_1(n) + A_{22}(n)x_2(n) \\ &\quad + \varepsilon B_{21}(n)u_1(n) + B_{22}(n)u_2(n) \end{aligned}$$

Find expressions for the system matrices in the new coordinates and the difference equations whose solutions comprise the required transformation.

The advantage of this transformation over the one of Gajić and Shen (1989) is that transformation equations for L_3 and H_3, namely Equations 5.45 and 5.46, respectively, have exactly the same form, and they are independent of each other. Therefore, we can use the same algorithm to solve both L_3 and H_3 equations. Moreover, due to the fact that they are independent from each other, the computations can be done in parallel. However, a price is paid for this convenience when we go back to the original variables $x_1(n)$ and $x_2(n)$. This step requires computation of the matrix inverse, that is finding $(I_{n_1} - \varepsilon^2 L_3 H_3)^{-1}$.

5.2   DECOUPLING TRANSFORMATION FOR N WEAKLY COUPLED SUBSYSTEMS

In this section, a transformation is introduced for exact decomposition (block-diagonalization) of linear weakly coupled systems composed of N-subsystems. This transformation can also be used for block diagonalization of block-diagonally dominant matrices, and under certain assumptions it can be applied for block diagonalization of nearly completely decomposable Markov chains (Phillips and Kokotović 1981). A 12th-order real-world power system example is included at the end of this section to demonstrate the efficiency of the proposed method.

A decoupling transformation that exactly decomposes weakly coupled linear systems composed of two subsystems into independent subsystems was presented in Section 5.1. In this section, we extend the results of Gajić and Shen (1989) and Qureshi (1992) to the general case of weakly coupled linear systems composed of N subsystems, and establish conditions under which such a transformation is feasible. The estimate of the rate of convergence of the corresponding algorithm used for decomposition of N weakly coupled linear subsystems is given and compared to the case of two weakly coupled linear subsystems.

Consider a continuous-time systems consisting of n states clustered into N groups of strongly interacting states. Weak interactions among different groups are expressed in terms of a small perturbation parameter ε. The dynamics of such systems are represented by the differential equation

$$\frac{dx(t)}{dt} = Ax(t) \tag{5.47}$$

where $x(t)$ is n-dimensional state vector partitioned consistently with N subsystems as $x(t) = \left[x_1^T(t) \; x_2^T(t) \; \cdots \; x_N^T(t)\right]^T$, $\dim\{x_i(t)\} = n_i$. The constant matrix A is partitioned as

$$A = \begin{bmatrix} A_{11} & \varepsilon A_{12} & \cdots & \varepsilon A_{1N} \\ \varepsilon A_{21} & A_{22} & \cdots & \cdots \\ \cdots & \cdots & \cdots & \cdots \\ \varepsilon A_{N1} & \varepsilon A_{N2} & \cdots & A_{NN} \end{bmatrix} \tag{5.48}$$

where ε is a small parameter. Each block, A_{ii}, is of dimensions, $n_i \times n_i$, hence $\sum_{i=1}^{N} n_i = n$. All submatrices A_{ij} in matrix A are assumed to be constant and $O(1)$.

In addition, it is assumed that magnitudes of all system eigenvalues are $O(1)$, that is $|\lambda_j| = O(1)$, $j = 1, 2, \ldots, n$. This implies that the matrices A_{ii} are nonsingular with $\det\{A_{ii}\} = O(1)$, $i = 1, 2, \ldots, N$, which is the standard assumption for weakly coupled linear control systems (Chow and Kokotović 1983), and also corresponds to block diagonal dominance of the system matrix A. Thus, the main results presented in this section are valid under the following assumption.

Assumption 5.3   All matrices A_{ij} are constant and $O(1)$, and magnitudes of all system eigenvalues are $O(1)$, that is, $|\lambda_j| = O(1)$, $j = 1, 2, \ldots, n$, which implies that the matrices A_{ii}, $i = 1, 2, \ldots, N$ are nonsingular with $\det\{A_{ii}\} = O(1)$.

As indicated in Section 5.1, when this assumption is not satisfied, the system Equations 5.47 and 5.48, in addition to weak coupling also displays multiple timescale phenomena.

In the cases when a linear weakly coupled system is not in its explicit form defined by Equations 5.47 and 5.48, one can use the methodology of Sezer and Siljak (1986, 1991) in order to achieve the desired weakly coupled structure.

Our goal is to find a transformation that makes the matrix A block diagonal. Consider the following change of the state variables, which represents a generalization of the transformation derived in Qureshi (1992) for two subsystems

$$\eta_i(t) = x_i(t) + \varepsilon \sum_{j=1, j \neq i}^{N} L_{ij} x_j(t), \quad i = 1, \ldots, N \tag{5.49}$$

This leads to

$$\dot{\eta}_i(t) = \dot{x}_i(t) + \varepsilon \sum_{j=1, j \neq i}^{N} L_{ij} \dot{x}_j(t), \quad i = 1, 2, \ldots, N \tag{5.50}$$

Eliminating $\dot{x}_i(t)$, $i = 1, 2, \ldots, N$, from Equation 5.50 using Equations 5.47 and 5.48, that is

$$\dot{x}_i(t) = A_{ii} x_i(t) + \varepsilon \sum_{j=1, j \neq i}^{N} A_{ij} x_j(t), \quad i = 1, 2, \ldots, N \tag{5.51}$$

and using Equation 5.49, we get

$$\dot{\eta}_i(t) = \left(A_{ii} + \varepsilon^2 \sum_{j=1, j \neq i}^{N} L_{ij} A_{ji} \right) \eta_i(t)$$

$$+ \varepsilon \sum_{j=1, j \neq i}^{N} \Phi_{ij}(L_{ij}, \varepsilon) x_j(t), \quad i = 1, 2, \ldots, N \tag{5.52}$$

where

$$\Phi_{ij}(L_{ij}, \varepsilon) = L_{ij}A_{jj} - A_{ii}L_{ij} + A_{ij} + \varepsilon \left(\sum_{k=1, k \neq i,j}^{N} L_{ik}A_{kj} \right)$$

$$- \varepsilon^2 \left(\sum_{k=1, k \neq i}^{N} L_{ik}A_{ki} \right) L_{ij} \tag{5.53}$$

$$i, j = 1, 2, \ldots, N, \quad i \neq j$$

In order to achieve complete decoupling the matrices, L_{ij} must be chosen such that

$$\Phi_{ij}(L_{ij}, \varepsilon) = 0, \quad \forall i, j = 1, 2, \ldots, N \tag{5.54}$$

Assuming that Equation 5.54 is satisfied, we get in the new coordinates a set of completely decomposed N subsystems, that is

$$\dot{\eta}_i(t) = \Omega_i \eta_i(t), \quad i = 1, 2, \ldots, N \tag{5.55}$$

with

$$\Omega_i = A_{ii} + \varepsilon^2 \sum_{j=1, j \neq i}^{N} L_{ij}A_{ji}, \quad j = 1, 2, \ldots, N \tag{5.56}$$

Let $\eta = \left[\eta_1^T \eta_2^T, \ldots, \eta_N^T \right]^T$, then

$$\dot{\eta}(t) = \Omega \eta(t) \tag{5.57}$$

where $\Omega = \mathrm{diag}\{\Omega_1, \Omega_2, \ldots, \Omega_N\}$.

The transformation matrix, that relates the original weakly coupled linear system and the set of completely decoupled subsystems in the new coordinates is given by

$$\eta(t) = \Gamma x(t) \tag{5.58}$$

where

$$\Gamma(\varepsilon) = \begin{bmatrix} I & \varepsilon L_{12} & \cdots & \varepsilon L_{1N} \\ \varepsilon L_{21} & I & \cdots & \varepsilon L_{2N} \\ \vdots & \cdots & \ddots & \vdots \\ \varepsilon L_{N_1} & \cdots & \varepsilon L_{N(N-1)} & I \end{bmatrix} = 1 + \varepsilon \Psi \tag{5.59}$$

with the obvious definition of Ψ.

Note that $\Gamma(\varepsilon)$ is invertible for sufficiently small values of ε. This transformation offers the advantage that it exactly decomposes a high-order linear system into N completely decoupled, reduced-order subsystems that can be solved independently. The state of the system in the original coordinates can be determined by the inverse transformation as

$$x(t) = \Gamma^{-1}(\varepsilon)\eta(t) \tag{5.60}$$

The main problem that we are faced with is the solution of the system of algebraic equation (Equation 5.54). This system has the form

$$L_{ij}A_{jj} - A_{ii}L_{ij} + A_{ij} + \varepsilon\left(\sum_{k=1,\,k\neq i,\,j}^{N} L_{ik}A_{kj}\right)$$

$$- \varepsilon^2\left(\sum_{k=1,\,k\neq i}^{N} L_{ik}A_{ki}\right)L_{ij} = 0 \tag{5.61}$$

$$i, j = 1, 2, \ldots, N, \quad i \neq j$$

It represents a system of nonlinear algebraic equations. However, the nonlinear (quadratic) terms are nicely multiplied by the squares of the small perturbation parameter ε. Solving the system of algebraic equations (Equation 5.61) will be the focus of the next section.

In the following, we present three iterative algorithms for computing matrices L_{ij} by performing iterations on a set of linear algebraic equations.

ALGORITHM 5.1　Fixed-Point Algorithm

The first algorithm that can be used to efficiently solve the set of nonlinear algebraic equation (Equation 5.61) is based on fixed point iterations. The algorithms is given in two steps.

Step 1: Set $\varepsilon = 0$ in Equation 5.61, and solve the $O(\varepsilon)$ perturbed set of completely decoupled reduced-order algebraic Sylvester equations

$$L_{ij}^{(0)}A_{jj} - A_{ii}L_{ij}^{(0)} + A_{ij} = 0, \quad i, j = 1, 2, \ldots, N, \quad i \neq j \tag{5.62}$$

Equations 5.62 have unique solutions under the assumption that matrices A_{jj} and A_{ii} have no eigenvalues in common (Gajić and Qureshi 1995), thus we have to impose the following assumption.

Assumption 5.4   The matrices A_{jj} and A_{ii} have no eigenvalues in common for every $i, j, i \neq j$.

This step produces on $O(\varepsilon)$ approximation for the desired solution, that is, $\left\| L_{ij} \right\| = \left\| L_{ij}^{(0)} \right\| + O(\varepsilon)$. Note that under Assumptions 5.3 and 5.4 we have $\left\| L_{ij}^{(i)} \right\| = O(1)$ and $\left\| L_{ij} \right\| = O(1)$.

Step 2: In order to improve the required solution accuracy up to any arbitrary order, we propose the following fixed point iterations scheme with $L_{ij}^{(0)}$, obtained in Step 1, playing the role of the initial conditions

$$L_{ij}^{(m+1)} A_{jj} - A_{ii} L_{ij}^{(m+1)} + A_{ij} + \varepsilon \left(\sum_{k=1, k \neq i,j}^{N} L_{ik}^{(m)} A_{kj} \right)$$

$$- \varepsilon^2 \left(\sum_{k=1, k \neq i}^{N} L_{ik}^{(m)} A_{ki} \right) L_{ij}^{(m)} = 0 \qquad (5.63)$$

$$i, j = 1, 2, \ldots, N, \quad i \neq j; \quad m = 0, 1, 2, \ldots$$

Algorithm 5.1 has the advantage that it operates on the *linear decoupled Sylvester's equations to solve the set of nonlinear coupled algebraic equations* (*Equation 5.61*).

The convergence proof of the fixed point algorithm (Equation 5.63) can be obtained under Assumptions 5.3 and 5.4 by generalizing the corresponding proofs of Gajić and Shen (1989) and Qureshi (1992) to N subsystems. Note that under Assumptions 5.3 and 5.4, the system of nonlinear algebraic equations (Equation 5.61) has unique solutions for sufficiently small values of ε since the corresponding Jacobian is nonsingular at $\varepsilon = 0$. This also implies that $\left\| L_{ij}^{(0)} \right\| = O(1)$. By generalizing the results of Gajić and Shen (1989), it can be established that the rate of convergence of the algorithm (Equation 5.63) is $O(\varepsilon)$, hence $\left\| L_{ij} - L_{ij}^{(m)} \right\| = O(\varepsilon^m)$, where m is the number of iterations. It is interesting to point out that in the case of algorithms considered in Gajić and Shen (1989), that is, for $N = 2$ the convergence rate of the corresponding algorithms is much faster, that is, it is equal to $O(\varepsilon^{2m})$.

ALGORITHM 5.2   Newton Algorithm

Since the Equations 5.62 produce initial guesses that are only $O(\varepsilon)$ apart from the exact solutions, it seems that the Newton method is an excellent candidate for solving nonlinear algebraic equations (Equation 5.61). In the following, the Newton algorithm for solving the set of nonlinear coupled algebraic equation (Equations 5.61) is derived. The Newton method is known for its quadratic rate of convergence, hence this algorithm will converge to the solutions of the algebraic equation (Equations 5.61) faster than the fixed point Algorithm 5.1, which has linear convergence. The Newton algorithm can be basically derived by replacing L_{ij} with $L_{ij}^{(m+1)}$, $\forall i, j$, substituting $L_{ij}^{(m+1)} = L_{ij}^{(m)} + \Delta_{ij}$, $\forall i, j$, $i \neq j$ into quadratic terms and neglecting quadratic terms with respect to Δ_{ij}. This yields the following algorithm:

$$L_{ij}^{(m+1)} A_{ii} - \left\{ A_{ii} + \varepsilon^2 \left(\sum_{k=1, k\neq i}^{N} L_{ik}^{(m)} A_{ki} \right) \right\} L_{ij}^{(m+1)}$$

$$+ \varepsilon \left(\sum_{k=1, k\neq i,j}^{N} L_{ik}^{(m+1)} A_{kj} \right) - \varepsilon^2 \left(\sum_{k=1, k\neq i}^{N} L_{ik}^{(m+1)} A_{ki} \right) L_{ij}^{(m)}$$

$$+ A_{ij} + \varepsilon^2 \left(\sum_{k=1, k\neq i}^{N} L_{ik}^{(m)} A_{ki} \right) L_{ij}^{(m)} = 0 \qquad (5.64)$$

$$i, j = 1, 2, \ldots, N, \quad i \neq j; \quad m = 0, 1, 2, \ldots$$

ALGORITHM 5.3   Hybrid Newton/Fixed Point Iterations Algorithm

It can be seen that the Newton method leads to a set of linear equations coupled by the terms $O(\varepsilon)$ and $O(\varepsilon^2)$. These equations can be solved in terms of the decoupled linear equations by using the fixed point iterations as in Algorithm 5.1. Since $\left\| L_{ij}^{(m+1)} - L_{ij}^{(m)} \right\| = O(\varepsilon)$ we can replace $L_{ij}^{(m+1)}$ in the third and the fourth terms of Equation 5.64 by $L_{ij}^{(m)}$ without affecting the corresponding fixed point type algorithm which now has the form

$$L_{ij}^{(m+1)} A_{jj} - \left\{ A_{ii} + \varepsilon^2 \left(\sum_{k=1, k\neq i}^{N} L_{ik}^{(m)} A_{ki} \right) \right\} L_{ij}^{(m+1)} + A_{ij}$$

$$+ \varepsilon \left(\sum_{k=1, k\neq i,j}^{N} L_{ik}^{(m)} A_{kj} \right) = 0 \qquad (5.65)$$

$$i, j = 1, 2, \ldots, N, \quad i \neq j, \quad m = 0, 1, 2, \ldots$$

This algorithm can be called the hybrid Newton-fixed-point iterations algorithm.

    The Newton method requires an initial guess that has to be quite close to the exact solution, otherwise the Newton method does not converge. In such cases when the initial guess is not good (small parameter ε is not very small), one has to use the fixed point iterations presented in Algorithm 5.1.

    Note that the Sylvester equation (Equation 5.62) has unique solutions if the square matrices A_{ii} and A_{jj}, $i, j = 1, 2, \ldots, N$, $i \neq j$, have one or more common eigenvalues. However, any solution of Equation 5.62 and subsequently any solution of Equation 5.63 will produce the desired transformation. Namely, the fact that there are several solutions of Equation 5.61 implies that there are several transformations having the form of Equation 5.59 that block diagonalize the considered weakly coupled system composed of N subsystems. This is particularly important for nearly decomposable Markov chains for which Assumptions 5.3 and 5.4 are not satisfied. For block iterative methods for Markov chains the reader is referred to the excellent book by Stewart (1994); here we just give the known result for solvability of Equations 5.62 needed when Assumption 5.4 is not satisfied. Solvability condition of the Sylvester equations is given by the following lemma (Flanders and Wimmer 1997).

LEMMA 5.1

Equation 5.62 has a solution if and only if matrices

$$\begin{bmatrix} A_{jj} & 0 \\ 0 & A_{ii} \end{bmatrix} \quad \text{and} \quad \begin{bmatrix} A_{jj} & A_{ji} \\ 0 & A_{ii} \end{bmatrix} \tag{5.66}$$

are similar.

Example 5.3: Power System

In order to demonstrate the efficiency of the presented method we have run a twelfth-order power system example composed of three machines playing the role of three subsystems. Each machine is modeled as a fourth-order subsystem consisting of a third-order synchronous machine and a first-order exciter regulator system (Delacour et al. 1978). The system matrix is defined by

$$A = \begin{bmatrix} A_{11} & A_{12} & A_{13} \\ A_{21} & A_{22} & A_{23} \\ A_{31} & A_{32} & A_{33} \end{bmatrix}$$

Matrices A_{ij}, $i, j = 1, 2, 3$, are given by

$$A_{11} = \begin{bmatrix} 0 & 1 & -0.266 & -0.009 \\ -2.75 & -2.78 & -1.36 & -0.037 \\ 0 & 0 & 0 & 1 \\ -4.95 & 0 & -55.5 & -0.039 \end{bmatrix}$$

$$A_{12} = \begin{bmatrix} 0.024 & 0 & -0.087 & 0.002 \\ -0.158 & 0 & 1.11 & -0.011 \\ 0 & 0 & 0 & 0 \\ 0.222 & 0 & 8.17 & 0.004 \end{bmatrix}$$

$$A_{13} = \begin{bmatrix} 0.072 & 0 & -0.25 & 0.003 \\ -0.46 & 0 & 2.8 & -0.02 \\ 0 & 0 & 0 & 0 \\ 0.924 & 0 & 17.5 & 0.02 \end{bmatrix}$$

$$A_{21} = \begin{bmatrix} 0.021 & 0 & 0.121 & 0.003 \\ -1.1 & 0 & -1.62 & -0.015 \\ 0 & 0 & 0 & 0 \\ -2.43 & 0 & 1.37 & -0.034 \end{bmatrix}$$

$$A_{22} = \begin{bmatrix} -0.21 & 1 & -1.6 & -0.005 \\ -1.9 & -1.8 & 9.3 & -0.12 \\ 0 & 0 & 0 & 1 \\ -3.1 & 0 & -56 & 0.032 \end{bmatrix}$$

$$A_{23} = \begin{bmatrix} 0.06 & 0 & 0.46 & 0.002 \\ -1 & 0 & 1.49 & -0.04 \\ 0 & 0 & 0 & 0 \\ 0.12 & 0 & 29.8 & -0.028 \end{bmatrix}$$

$$A_{31} = \begin{bmatrix} -0.002 & 0 & 0.083 & 0 \\ -6.78 & 0 & -10.1 & 0.09 \\ 0 & 0 & 0 & 0 \\ -1.24 & 0 & 0.498 & -0.017 \end{bmatrix}$$

$$A_{32} = \begin{bmatrix} 0.011 & 0 & 0.22 & 0 \\ -2.1 & 0 & 1.7 & -0.123 \\ 0 & 0 & 0 & 0 \\ -0.07 & 0 & 6.37 & -0.011 \end{bmatrix}$$

$$A_{33} = \begin{bmatrix} -0.197 & 1 & -1.2 & -0.003 \\ -54.4 & -20 & 70.1 & -2.37 \\ 0 & 0 & 0 & 1 \\ -3.4 & 0 & -21 & -0.017 \end{bmatrix}$$

The control input matrix B is

$$B^T = \begin{bmatrix} 0 & 36.1 & 0 & 0 & 0 & 0 & 0 & 0 & 0 & 0 & 0 & 0 \\ 0 & 0 & 0 & 0 & 0 & 78.9 & 0 & 0 & 0 & 0 & 0 & 0 \\ 0 & 0 & 0 & 0 & 0 & 0 & 0 & 0 & 0 & 1000 & 0 & 0 \end{bmatrix}$$

The eigenvalues and determinants of the diagonal blocks satisfy both Assumptions 5.3 and 5.4 since

$$\det\{A_{11}\} = 142.2$$
$$\lambda(A_{11}) = \{-0.0362 \pm j7.4534, \ -1.3733 \pm j0.8211\}$$
$$\det\{A_{22}\} = 147.5$$
$$\lambda(A_{22}) = \{-0.0241 \pm j7.44461, \ -0.9649 \pm j1.3148\};$$
$$\det\{A_{33}\} = 1381.9$$
$$\lambda(A_{33}) = \{-16.6128, \ -3.9337, \ -0.1660 \pm j4.5956\}$$

The new decoupled system matrix is obtained as

$$\Omega = \begin{bmatrix} \Omega_1 & 0 & 0 \\ 0 & \Omega_2 & 0 \\ 0 & 0 & \Omega_3 \end{bmatrix}$$

with

$$\Omega_1 = \begin{bmatrix} 0.0000 & 1 & -0.2660 & -0.0090 \\ -2.7498 & -2.78 & -1.3601 & -0.0370 \\ -0.0011 & 0 & 0.0006 & 1.0000 \\ -4.9447 & 0 & -55.5026 & -0.0389 \end{bmatrix}$$

$$\Omega_2 = \begin{bmatrix} -0.21 & 1 & -1.5998 & -0.005 \\ -1.90 & -1.8 & 9.2999 & -0.120 \\ 0.00 & 0 & -0.0007 & 1.000 \\ -3.10 & 0 & -55.9992 & 0.032 \end{bmatrix}$$

$$\Omega_3 = \begin{bmatrix} -0.197 & 1 & -1.2001 & -0.003 \\ -54.400 & -20 & 70.1000 & -2.370 \\ 0.000 & 0 & 0.0000 & 1.000 \\ -3.400 & 0 & -20.9994 & -0.017 \end{bmatrix}$$

All elements in matrix Ω denoted by "0" are zeros with the accuracy of at least 10^{-14}. Using Algorithm 5.1 with $\varepsilon = 0.01$, we have obtained the results presented in Table 5.4.

The accuracy of the solutions obtained is measured by using the MATLAB function norm in the following sense error$(m) = \max_{i,j} \left\| \Phi_{ij}(L_{ij}^{(m)}, \varepsilon) \right\|$. Note that this error estimate is a conservative measure so that the results presented in the table are

---

TABLE 5.4

Error Propagation per Iteration

Iteration (m)	Error (m)
0	6.2458×10^0
1	2.5816×10^{-1}
2	7.0832×10^{-3}
3	6.7502×10^{-4}
4	2.6511×10^{-5}
5	7.7792×10^{-7}
6	6.4017×10^{-8}
7	2.1459×10^{-9}
8	1.4050×10^{-10}
9	5.2846×10^{-12}
10	2.5311×10^{-13}

slightly worse than predicted by the rate of convergence of the presented algorithm. In addition, matrices A_{11} and A_{22} have a pair of complex conjugate eigenvalues close to each other causing numerical ill-conditioning. In achieving higher order of accuracy, we have experienced some problems with MATLAB® function 1yap. However, by using MATLAB function 1yap2 those problems have been eliminated.

The transformation introduced can be used for decomposition of weakly coupled linear dynamic controllers, observers, and Kalman filters, that is, for block-diagonal control and filtering of weakly coupled linear deterministic and stochastic systems. It can be also used to simplify computations of large systems of linear and nonlinear algebraic equations displaying block diagonal dominance. The transformation is very useful for parallel processing of information and computations on parallel computers. In addition, this matrix block diagonalization transformation might simplify many problems of linear algebra such as the problem of finding matrix eigenvalues.

Another transformation that can be used for decoupling N weakly coupled subsystems has been recently considered in Prljaca and Gajić (2007). In fact that paper recommends that a weakly coupled system composed of N subsystems be decoupled using a composition of transformations presented in Section 5.1 for decoupling a weakly coupled system composed of two subsystems. This gives some flexibility in choosing two subsystems to be decoupled and facilitating the condition (given in Assumption 5.4) under which such a decomposition is feasible.

5.3 DECOMPOSITIONS OF THE DIFFERENTIAL LYAPUNOV EQUATION

In the following, we first show that the introduced decoupling transformation for weakly coupled linear systems also completely decouples the Lyapunov matrix differential equation corresponding to weakly coupled systems.

Consider the Lyapunov matrix differential equation of weakly coupled systems

$$\dot{P}(t) = A^{\mathrm{T}} P(t) + P(t)A + Q, \quad Q = Q^{\mathrm{T}}, \quad P(t_0) = P_0 \tag{5.67}$$

where the given matrices A and Q are partitioned as

$$A = \begin{bmatrix} A_1 & \varepsilon A_2 \\ \varepsilon A_3 & A_4 \end{bmatrix}, \quad Q = \begin{bmatrix} Q_1 & \varepsilon Q_2 \\ \varepsilon Q_2^{\mathrm{T}} & Q_3 \end{bmatrix} \tag{5.68}$$

Due to assumed structure for A and Q, the matrix P is properly scaled as Kokotović et al. (1969)

$$P(t) = \begin{bmatrix} P_1(t) & \varepsilon P_2(t) \\ \varepsilon P_2^{\mathrm{T}}(t) & P_3(t) \end{bmatrix} \tag{5.69}$$

Multiplying Equation 5.67 from the left by $\mathbf{T}_1^{-\mathrm{T}}$ and from the right-hand side by $\mathbf{T}_1^{-1}$ we get

$$\mathbf{T}_1^{-\mathrm{T}} \dot{P}(t) \mathbf{T}_1^{-1} = \mathbf{T}_1^{-\mathrm{T}} A^{\mathrm{T}} P(t) \mathbf{T}_1^{-1} + \mathbf{T}_1^{-\mathrm{T}} P(t) A \mathbf{T}_1^{-1} + \mathbf{T}_1^{-\mathrm{T}} Q \mathbf{T}_1^{-1} \tag{5.70}$$

which can be written as

$$\dot{K}(t) = a^T K(t) + K(t)a + q, \quad K(t_0) = K_0 \tag{5.71}$$

where

$$a = \mathbf{T_1} A \mathbf{T_1}^{-1} = \begin{bmatrix} A_{10} & 0 \\ 0 & A_{40} \end{bmatrix}$$

$$q = \mathbf{T_1}^{-T} Q \mathbf{T_1} = \begin{bmatrix} q_1 & \varepsilon q_2 \\ \varepsilon q_2^T & q_3 \end{bmatrix} \tag{5.72}$$

$$K(t) = \mathbf{T_1}^{-T} P(t) \mathbf{T_1}^{-1} = \begin{bmatrix} K_1(t) & \varepsilon K_2(t) \\ \varepsilon K_2^T(t) & K_3(t) \end{bmatrix}, \quad K(t_0) = \mathbf{T_1}^{-T} P_0 \mathbf{T_1}^{-1}$$

Partitioning Equation 5.71 we can note a completely decoupled form among elements of $K(t)$, that is

$$\dot{K}_1(t) = K_1(t) A_{10} + A_{10}^T K_1(t) + q_1 \tag{5.73}$$

$$\dot{K}_2(t) = K_2(t) A_{40} + A_{10}^T K_2(t) + q_2 \tag{5.74}$$

$$\dot{K}_3(t) = K_3(t) A_{40} + A_{40}^T K_3(t) + q_3 \tag{5.75}$$

Having obtained $K_i(t)$'s from Equations 5.73 through 5.75, we can get the solution of the original Lyapunov differential equation as

$$P(t) = \mathbf{T_1}^T K(t) \mathbf{T_1} \tag{5.76}$$

which follows from Equation 5.72.

It is left to the reader to find the corresponding decomposition of the difference Lyapunov equation of weakly coupled systems, which has been formulated as an exercise.

Exercise 5.2: Following the methodology presented in Section 5.3 find the decomposition for the weakly coupled matrix difference Lyapunov equation.

5.4   BOUNDARY VALUE PROBLEM OF LINEAR CONTINUOUS SYSTEMS

In this section, we study the general boundary value problem of linear time varying weakly coupled systems. We show how the decoupling transformation developed in Section 5.1 can be used to solve this fundamental problem of linear-quadratic optimal control theory. The existence of the solution in terms of the reduced-order completely decoupled dynamical systems is established.

Consider the general boundary value problem for the linear weakly coupled system

$$\dot{x}(t) = A_1(t)x(t) + \varepsilon A_2(t)z(t) \tag{5.77}$$

$$\dot{z}(t) = \varepsilon A_3(t)x(t) + A_4(t)z(t) \tag{5.78}$$

with boundary conditions

$$M(\varepsilon)\begin{bmatrix} x(0,\ \varepsilon) \\ z(0,\ \varepsilon) \end{bmatrix} + N(\varepsilon)\begin{bmatrix} x(1,\ \varepsilon) \\ z(1,\ \varepsilon) \end{bmatrix} = \begin{bmatrix} c_1(\varepsilon) \\ c_2(\varepsilon) \end{bmatrix} \tag{5.79}$$

on the interval $0 \le t \le 1$, where $x(t) \in \mathfrak{R}^{n_1}$, $z(t) \in \mathfrak{R}^{n_2}$, $c_1 \in \mathfrak{R}^{n_1}$, $c_2 \in \mathfrak{R}^{n_2}$, and $A_1(t)$, $A_2(t)$, $A_3(t)$, $A_4(t)$, $M(\varepsilon)$, $N(\varepsilon)$ are matrices of appropriate dimensions. A small parameter ε couples the states $x(t)$ and $z(t)$. Note that for weakly coupled systems the following standard assumption is imposed.

Assumption 5.5

$$\det\{A_1(t)\} = O(1) \quad \text{and} \quad \det\{A_4(t)\} = O(1), \quad \forall t \in [0,\ 1]$$

In addition, the following assumption is made in this section.

Assumption 5.6   The matrices $A_1(t)$, $A_2(t)$, $A_3(t)$, and $A_4(t)$ are continuous function for $0 \le t \le 1$, and $M(\varepsilon) = M(0) + O(\varepsilon)$, $N(\varepsilon) = N(0) + O(\varepsilon)$, $c_i(\varepsilon) = c_i(0) + O(\varepsilon)$, $i = 1, 2$.
   In the following, we will use the transformation derived in Section 5.1 to block diagonalize Equations 5.77 and 5.78. The obtained results simplify the analytical and computational treatment of Equations 5.77 and 5.79 in terms of the reduced-order completely decoupled dynamical systems.
   It is shown in Section 5.1.1 that the following transformation

$$\begin{bmatrix} n \\ \zeta \end{bmatrix} = \begin{bmatrix} I_{n_1} & -\varepsilon L_3(t) \\ -\varepsilon H_3(t) & I_{n_2} \end{bmatrix}\begin{bmatrix} x_1 \\ x_2 \end{bmatrix} = \mathbf{T}_3(t,\ \varepsilon)\begin{bmatrix} x_1 \\ x_2 \end{bmatrix} \tag{5.80}$$

produces the block diagonal form of Equations 5.77 and 5.78, that is,

$$\dot{\eta}(t) = A_{10}(t)\eta(t) \tag{5.81}$$

$$\dot{\zeta}(t) = A_{40}(t)\zeta(t) \tag{5.82}$$

where

$$A_{10}(t) = A_1(t) - \varepsilon^2 L_3(t)A_3(t) \tag{5.83}$$

$$A_{40}(t) = A_4(t) - \varepsilon^2 H_3(t)A_2(t) \tag{5.84}$$

The transformation matrices $L_3(t)$ and $H_3(t)$ satisfy the following Riccati-type differential equations:

$$\dot{L}_3(t) = A_1(t)L_3(t) - L_3(t)A_4(t) + A_2(t) - \varepsilon^2 L_3(t)A_3(t)L_3(t) \tag{5.85}$$

$$\dot{H}_3(t) = A_4(t)H_3(t) - H_3(t)A_1(t) + A_3(t) - \varepsilon^2 H_3(t)A_2(t)H_3(t) \tag{5.86}$$

Note that for sufficiently small ε, the matrix $\mathbf{T}_3(t, \varepsilon)$ is diagonally dominant, and hence, nonsingular. In addition, no initial or terminal conditions are imposed on Equations 5.85 and 5.86.

Since the differential Riccati equation, in general, has a finite escape time (Sasagawa, 1982), we have to establish the existence of bounded solutions for Equations 5.85 and 5.86. The only assumption we impose is that $A_1(t)$, $A_2(t)$, $A_3(t)$, and $A_4(t)$ are continuous function for $0 \le t \le 1$. Due to the same structure of Equations 5.85 and 5.86, it suffices to show that either one of them possesses a bounded solution. The following lemma is proved for the existence of such a solution.

LEMMA 5.2

There exists $\varepsilon_0 > 0$ such that Equations 5.85 and 5.86 have solutions $L_3(t)$ and $H_3(t)$, respectively, which are uniformly bounded for $0 \le t \le 1$ in the region $0 < \varepsilon \le \varepsilon_0$.

Proof Let $X(t)$ be the fundamental matrix of $\dot{x}(t) = A_1(t)x(t)$, and $Z(t)$ be the fundamental matrix of $\dot{z}(t) = A_4(t)z(t)$. Since $A_1(t)$ and $A_4(t)$ are continuous on [0, 1], and therefore bounded on [0, 1], there exist $\sigma_1 > 0$ and $\sigma_2 > 0$ such that

$$\|A_1(t)\| < \sigma_1 = O(1)$$
$$\|A_4(t)\| < \sigma_2 = O(1)$$

which implies (Chang 1972)

$$\|X(t)X^{-1}(s)\| \le K_1 \exp(\sigma_1|t - s|)0 \le t, s \le 1$$
$$\|Z(t)Z^{-1}(s)\| \le K_2 \exp(\sigma_2|t - s|)0 \le t, s \le 1 \tag{5.87}$$

where K_1 and K_2 are positive constants.

It can be verified by differentiation that

$$L_3(t) = \int_0^t X(t)X^{-1}(s)\left[A_2(s) - \varepsilon^2 L_3(s)A_3(s)L_3(s)\right]Z(s)Z^{-1}(t)ds \tag{5.88}$$

is a solution of Equation 5.85. We will show that this is a bounded solution, that is, $\|L_3(t)\| \leq \rho$, for some ρ. From Equation 5.88, we have

$$\|L_3(t)\| \leq \int_0^t K_1 K_2 e^{\sigma_1|t-s|} e^{\sigma_2|t-s|} ds\left(\|A_2(t)\| + \varepsilon^2 \rho^2 \|A_3(t)\|\right)$$

Since $s \leq t$, then $|t-s| = t-s$, so that we obtain

$$\|L_3(t)\| \leq \frac{K_1 K_2}{\sigma_1 + \sigma_2}\left[e^{(\sigma_1+\sigma_2)t} - 1\right]\left(\|A_2(t)\| + \varepsilon^2 \rho^2 \|A_3(t)\|\right)$$

This inequality is valid for $t = 1$, giving

$$\|L_3(t)\| \leq \alpha\left(\|A_2(t)\| + \varepsilon^2 \rho^2 \|A_3(t)\|\right)$$

where

$$\alpha = \frac{K_1 K_2}{\sigma_1 + \sigma_2}\left[e^{(\sigma_1+\sigma_2)t} - 1\right]$$

Now we have to find ρ and ε_0 such that $\|L_3(t)\| \leq \rho$. Pick

$$\rho = 2\alpha\|A_2(t)\|$$

so that

$$\|L_3(t)\| \leq \frac{\rho}{2} + \alpha\varepsilon^2 \rho^2 \|A_3(t)\|$$

Pick ε_0 such that

$$\alpha\varepsilon^2 \rho^2 \|A_3(t)\| \leq \frac{\rho}{2}$$

or

$$\varepsilon_0^2 \leq \frac{1}{4\alpha^2 \|A_2(t)\| \|A_3(t)\|}$$

Thus

$$\|L_3(t)\| \leq \frac{\rho}{2} + \frac{\rho}{2} \leq \rho$$

Therefore, $L_3(t)$ has a bounded solution, that is, $\|L_3(t)\| \leq \rho$. Since $[0, 1]$ is a compact interval, the solution is uniformly bounded. On the same lines, it can be proved that $H_3(t)$ also has a uniformly bounded solution.   ∎

Consequently, the change of variables (Equation 5.80) transforms the system Equations 5.77 and 5.78 to Equations 5.81 and 5.82. Applying Equation 5.80 to the boundary conditions of Equation 5.79, we obtain

$$\hat{M}(\varepsilon) \begin{bmatrix} \eta(0, \varepsilon) \\ \zeta(0, \varepsilon) \end{bmatrix} + \hat{N}(\varepsilon) \begin{bmatrix} \eta(1, \varepsilon) \\ \zeta(1, \varepsilon) \end{bmatrix} = \begin{bmatrix} c_1(\varepsilon) \\ c_2(\varepsilon) \end{bmatrix} \tag{5.89}$$

where $\hat{M}(\varepsilon) = M(\varepsilon)\mathbf{T}_3^{-1}(0, \varepsilon)$, $\hat{N}(\varepsilon) = N(\varepsilon)\mathbf{T}_3^{-1}(1, \varepsilon)$ and $\mathbf{T}_3^{-1}(t, \varepsilon)$ is given by

$$\mathbf{T}_3^{-1}(t, \varepsilon) = \begin{bmatrix} I_{n_1} + \varepsilon^2 L_3(t)M(t)H_3(t) & \varepsilon L_3(t)M(t) \\ \varepsilon M(t)H(t) & M(t) \end{bmatrix}$$

where $M(t) = (I_{n_2} - \varepsilon^2 H(t)L(t))^{-1}$. Therefore, the transformed boundary value problem is described by Equations 5.81, 5.82, and 5.89.

Let $\xi(t, \varepsilon)$ be the transition matrix of Equation 5.81, and $\chi(t, \varepsilon)$ be the transition matrix of Equation 5.82; then we have the following solution for Equations 5.81 and 5.82

$$\begin{bmatrix} \eta(t, \varepsilon) \\ \zeta(t, \varepsilon) \end{bmatrix} = \begin{bmatrix} \xi(t, \varepsilon) & 0 \\ 0 & \chi(t, \varepsilon) \end{bmatrix} \begin{bmatrix} \alpha_1(\varepsilon) \\ \alpha_2(\varepsilon) \end{bmatrix} \tag{5.90}$$

where $\alpha_1(\varepsilon)$ and $\alpha_2(\varepsilon)$ are arbitrary constants. It remains to choose $\alpha_1(\varepsilon)$ and $\alpha_2(\varepsilon)$ to satisfy the boundary condition Equation 5.89. Substituting Equation 5.90 into Equation 5.89 yields

$$\Delta(\varepsilon_1) \begin{bmatrix} \alpha_1(\varepsilon) \\ \alpha_2(\varepsilon) \end{bmatrix} = \begin{bmatrix} c_1(\varepsilon) \\ c_2(\varepsilon) \end{bmatrix}$$

where

$$\Delta(\varepsilon) = \hat{M}(\varepsilon) + \hat{N}(\varepsilon) \begin{bmatrix} \xi(1, \varepsilon) & 0 \\ 0 & \chi(1, \varepsilon) \end{bmatrix} \tag{5.91}$$

If $\Delta^{-1}(\varepsilon)$ exist, then

$$\begin{bmatrix} a_1(\varepsilon) \\ a_2(\varepsilon) \end{bmatrix} = \Delta^{-1}(\varepsilon) \begin{bmatrix} c_1(\varepsilon) \\ c_2(\varepsilon) \end{bmatrix}$$

Thus Equation 5.90 represents a solution to the boundary value problem.

Note that as $\varepsilon \to 0$, $\mathbf{T}_3^{-1}(t, \varepsilon) \to I$, $\lim_{\varepsilon \to 0} \hat{M}(\varepsilon) = M(0)$ and $\lim_{\varepsilon \to 0} \hat{N}(\varepsilon) = N(0)$

$$\lim_{e \to 0} \Delta(\varepsilon) = M(0) + N(0) \begin{bmatrix} \xi(1, 0) & 0 \\ 0 & \chi(1, 0) \end{bmatrix}$$

where $\xi(t, 0) = \lim_{e \to 0} \xi(t, \varepsilon)$ and $\chi(t, 0) = \lim_{e \to 0} \chi(t, \varepsilon)$. Hence for sufficiently small ε, the inverse $\Delta^{-1}(\varepsilon)$ exists under the following assumption.

Assumption 5.7   The matrix $\Delta(0) = \lim_{\varepsilon \to 0} \Delta(\varepsilon)$ in nonsingular.

Note that Assumption 5.7 is always satisfied for the linear filtering and control problem, Qureshi (1992).

Consequently, the following theorem summarizes the results.

THEOREM 5.2

Let Assumptions 5.6 and 5.7 hold, then for sufficiently small ε the original weakly coupled boundary value problems 5.77 through 5.79 has the solution given by

$$\begin{bmatrix} x(t, \varepsilon) \\ z(t, \varepsilon) \end{bmatrix} = \mathbf{T}_3^{-1}(t, \varepsilon) \begin{bmatrix} \xi(t, \varepsilon) & 0 \\ 0 & \chi(t, \varepsilon) \end{bmatrix} \Delta^{-1}(\varepsilon) \begin{bmatrix} c_1(\varepsilon) \\ c_2(\varepsilon) \end{bmatrix}$$

for $0 \le t \le 1$.

5.5   BOUNDARY VALUE PROBLEM OF LINEAR DISCRETE SYSTEMS

In the procedure of solving the general *continuous-time* boundary value problem of linear weakly coupled systems the transition matrices of two time-varying subsystems are required. Since, in general, these matrices cannot be found analytically, it would be interesting, from the practical point of view, to develop the corresponding results for the weakly coupled time-varying discrete boundary value problem, where the corresponding *discrete-time* transition matrices can be found easily.

Consider the general boundary value problem of the discrete-time linear weakly coupled system

$$x(k + 1) = A_1(k)x(k) + \varepsilon A_2(k)z(k) \tag{5.92}$$

$$z(k + 1) = \varepsilon A_3(k)x(k) + A_4(k)z(k) \tag{5.93}$$

with boundary conditions

$$M(\varepsilon) \begin{bmatrix} x(0, \varepsilon) \\ z(0, \varepsilon) \end{bmatrix} + N(\varepsilon) \begin{bmatrix} x(n, \varepsilon) \\ z(n, \varepsilon) \end{bmatrix} = \begin{bmatrix} c_1(\varepsilon) \\ c_2(\varepsilon) \end{bmatrix} \tag{5.94}$$

on the interval $0 \le k \le n$, with basic dimensions $x(k) \in \Re^{n_1}$, $z(k) \in \Re^{n_2}$, $c_1 \in \Re^{n_1}$, $c_2 \in \Re^{n_2}$, and $A_1(k)$, $A_2(k)$, $A_3(k)$, $A_4(k)$, $M(\varepsilon)$, $N(\varepsilon)$ being of compatible dimensions. A small parameter ε couples the states $x(k)$ and $z(k)$. It is assumed that $M(\varepsilon) = M(0) + O(\varepsilon)$, $N(\varepsilon) = N(0) + O(\varepsilon)$ $c_i(\varepsilon) = c_i(0) + O(\varepsilon)$, $i = 1, 2$. Note that for weakly coupled systems

$$\det\{A_1(k)\} = O(1) \text{ and } \det\{A_4(k)\} = O(1) \text{ for all } 0 \le k \le n \text{ (see Assumption 5.1).}$$

The corresponding boundary value problem for continuous-time weakly coupled systems is studied in the previous section. In this section, we will use the same transformation to block diagonalize Equations 5.92 and 5.93, which simplifies the analytical and computational treatment of Equations 5.92 through 5.94 in terms of the reduced-order completely decoupled dynamical systems.

The following transformation

$$\begin{bmatrix} \eta(k) \\ \zeta(k) \end{bmatrix} = \begin{bmatrix} I & -\varepsilon L(k) \\ -\varepsilon H(k) & I \end{bmatrix} \begin{bmatrix} x(k) \\ z(k) \end{bmatrix}$$
$$= \mathbf{T}_3(k, \varepsilon) \begin{bmatrix} x(k) \\ z(k) \end{bmatrix} \tag{5.95}$$

produces a block diagonal form of Equations 5.92 and 5.93, that is

$$\eta(k + 1) = A_{10}(k)\eta(k) \tag{5.96}$$

$$\zeta(k + 1) = A_{40}(k)\zeta(k) \tag{5.97}$$

where

$$A_{10}(k) = A_1(k) - \varepsilon^2 L(k)A_3(k) \tag{5.98}$$

$$A_{40}(k) = A_4(k) - \varepsilon^2 H(k)A_2(k) \tag{5.99}$$

The transformation matrices $L(k)$ and $H(k)$ satisfy the following difference equations:

$$A_1(k)L(k) - L(k+1)A_4(k) + A_2(k)$$
$$- \varepsilon^2 L(k+1)A_3(k)L(k) = 0 \tag{5.100}$$

$$A_4(k)H(k) - H(k+1)A_1(k) + A_3(k)$$
$$- \varepsilon^2 H(k+1)A_2(k)H(k) = 0 \tag{5.101}$$

Note that for sufficiently small ε, the matrix $\mathbf{T}_3(k, \varepsilon)$ in Equation 5.95 is diagonally dominant, and hence nonsingular. In addition, no initial or terminal conditions are imposed on Equations 5.100 and 5.101.

In order to have a bounded solution for the block diagonalized system, Equations 5.100 and 5.101, $\|A_{10}(k)\|$ and $\|A_{40}(k)\|$ must be bounded. First of all, we make the following assumption on the matrices $A_1(k)$, $A_2(k)$, $A_3(k)$, and $A_4(k)$.

Assumption 5.8   The matrices $A_i(k)$, $i = 1, 2, 3, 4$, are bounded, that is, $\|A_i(k)\| \leq K_i$, $i = 1, 2, 3, 4$, where K_i are bounded constants. Furthermore, $A_1(k)$ and $A_4(k)$ are invertible.

Under Assumption 5.8, it follows that Equations 5.100 and 5.101 have bounded solutions, and hence, the systems 5.96 and 5.97 have bounded solutions. Thus, we have a dual lemma to Lemma 5.2 stated as following.

LEMMA 5.3

Under Assumption 5.8, there exist $\varepsilon_0 > 0$ such that Equations 5.100 and 5.101 have solutions $L(k)$ and $H(k)$, respectively, which are bounded for $0 \leq k \leq n$ in the region $0 < \varepsilon \leq \varepsilon_0$.

Consequently, the change of variable (Equation 5.95) transforms the system Equations 5.92 and 5.93 to Equations 5.96 and 5.97. Applying Equation 5.95 to the boundary conditions (Equation 5.94), we obtain

$$\hat{M}(\varepsilon)\begin{bmatrix} \eta(0, \varepsilon) \\ \zeta(0, \varepsilon) \end{bmatrix} + \hat{N}(\varepsilon)\begin{bmatrix} \eta(n, \varepsilon) \\ \zeta(n, \varepsilon) \end{bmatrix} = \begin{bmatrix} c_1(\varepsilon) \\ c_2(\varepsilon) \end{bmatrix} \tag{5.102}$$

where $\hat{M}(\varepsilon) = M(\varepsilon)\mathbf{T}_3^{-1}(0, \varepsilon)$, $\hat{N}(\varepsilon) = N(\varepsilon)\mathbf{T}_3^{-1}(n, \varepsilon)$ with $\mathbf{T}_3(k, \varepsilon)$ defined in Equation 5.95. Therefore the transformed boundary value problem is described by Equations 5.96, 5.97, and 5.102.

Let $\mathcal{E}(k, \varepsilon)$ be the transition matrix of Equation 5.96, and $\mathcal{Z}(k, \varepsilon)$ be the transition matrix of Equation 5.97, then we have the following solutions for Equations 5.96 and 5.97

$$\begin{bmatrix} \eta(k, \varepsilon) \\ \zeta(k, \varepsilon) \end{bmatrix} = \begin{bmatrix} \mathcal{E}(k, \varepsilon) & 0 \\ 0 & \mathcal{Z}(k, \varepsilon) \end{bmatrix}\begin{bmatrix} \alpha_1(\varepsilon) \\ \alpha_2(\varepsilon) \end{bmatrix} \tag{5.103}$$

where $\alpha_1(\varepsilon)$ and $\alpha_2(\varepsilon)$ are arbitrary constants. It remains to choose $\alpha_1(\varepsilon)$ and $\alpha_2(\varepsilon)$ to satisfy the boundary conditions (Equation 5.102). Substituting Equation 5.103 into Equation 5.102 yields

$$\Delta(\varepsilon)\begin{bmatrix} \alpha_1(\varepsilon) \\ \alpha_2(\varepsilon) \end{bmatrix} = \begin{bmatrix} c_1(\varepsilon) \\ c_2(\varepsilon) \end{bmatrix}$$

where

$$\Delta(\varepsilon) = \hat{M}(\varepsilon) + \hat{N}(\varepsilon)\begin{bmatrix} \mathcal{E}(n, \varepsilon) & 0 \\ 0 & \mathcal{Z}(k, \varepsilon) \end{bmatrix} \tag{5.104}$$

If $\Delta^{-1}(\varepsilon)$ exists, then

$$\begin{bmatrix} \alpha_1(\varepsilon) \\ \alpha_2(\varepsilon) \end{bmatrix} = \Delta^{-1}(\varepsilon)\begin{bmatrix} c_1(\varepsilon) \\ c_2(\varepsilon) \end{bmatrix}$$

and thus Equation 5.103 represents a solution of the considered boundary value problem.

It is easy to see that as $\varepsilon \to 0$, $\mathbf{T}_3^{-1}(k, \varepsilon) \to I$, $\lim_{\varepsilon \to 0} \hat{M}(\varepsilon) = M(0)$ and $\lim_{\varepsilon \to 0} \hat{N}(\varepsilon) = N(0)$. Therefore

$$\lim_{\varepsilon \to 0} \Delta(\varepsilon) = M(0) + N(0) \begin{bmatrix} \mathcal{E}(n, 0) & 0 \\ 0 & \mathcal{Z}(n, 0) \end{bmatrix} \tag{5.105}$$

where $\mathcal{E}(k, 0) = \lim_{\varepsilon \to 0} \mathcal{E}(k, \varepsilon)$ and $\mathcal{Z}(k, 0) = \lim_{\varepsilon \to 0} \mathcal{Z}(k, \varepsilon)$. Hence for sufficiently small ε, the unique solutions for $\alpha_1(\varepsilon)$ and $\alpha_2(\varepsilon)$ exist under the following assumption.

Assumption 5.9   The matrix (Equation 5.104) is nonsingular.

Note that as opposed to the continuous-time systems, the transition matrices and hence, $\Delta(\varepsilon)$ for discrete systems can be found analytically. This gives us the explicit solution for the boundary value problem of discrete weakly coupled linear systems.

Consequently, the following theorem summarizes the results.

THEOREM 5.3

Let Assumption 5.9 hold, then under conditions stated in Lemma 5.3 for sufficiently small ε, the original boundary value problems 5.92 and 5.93 has the solution given by

$$\begin{bmatrix} x(k, \varepsilon) \\ z(k, \varepsilon) \end{bmatrix} = \mathbf{T}_3^{-1}(k, \varepsilon) \begin{bmatrix} \mathcal{E}(k, \varepsilon) & 0 \\ 0 & \mathcal{Z}(k, \varepsilon) \end{bmatrix} \begin{bmatrix} \alpha_1(\varepsilon) \\ \alpha_2(\varepsilon) \end{bmatrix} \tag{5.106}$$

for $0 \le k \le n$.

REFERENCES

Chang, K., Singular perturbations of a general boundary value problem, *SIAM Journal of Mathematical Analysis*, 3, 520–526, 1972.

Chow, J. and P. Kokotović, Sparsity and time scales, *Proceedings of the American Control Conference*, San Francisco, CA, pp. 656–661, 1983.

Delacour, J.D., M. Darwish, and J. Fantin, Control strategies for large-scale power systems, International Journal of Control, 27, 753–767, 1978.

Flanders, H. and H. Wimmer, On the matrix equations $AX - XB = C$ and $AX - YB = C$, *SIAM Journal of Applied Mathematics*, 32, 707–710, 1977.

Gajić, Z. and I. Borno, Lyapunov iterations for optimal control of jump linear systems at steady state, *IEEE Transactions on Automatic Control*, 40, 1971–1975, 1995.

Gajić, Z. and M. Qureshi, *Lyapunov Matrix Equation in Systems Stability and Control*, Academic Press, San Diego, CA, 1995.

Gajić, Z. and X. Shen, Decoupling transformation for weakly coupled linear systems, *International Journal of Control*, 50, 1517–1523, 1989.

Grodt, T. and Z. Gajić, The recursive reduced order numerical solution of the singularly perturbed matrix differential Riccati equation, *IEEE Transactions on Automatic Control*, AC-33, 751–754, 1988.

Harkara, N., D. Petkovski, and Z. Gajić, The recursive algorithm for the optimal static output feedback control problem of linear weakly coupled systems, *International Journal of Control*, 50, 1–11, 1989.

Kato, T., *Perturbation Theory of Linear Operators*, Springer-Verlag, New York, 1980.

Kecman, V., Eigenvector approach for reduced-order optimal control problems of weakly coupled systems, *Dynamics of Continuous Discrete and Impulsive Systems*, 13, 569–588, 2006.

Kokotović, P. and H. Khalil, *Singular Perturbations in Systems and Control*, IEEE Press, New York, 1986.

Kokotović, P., W. Perkins, J. Cruz, and G. D'Ans, ε—coupling approach for near-optimum design of large scale linear systems, *Proceeding of IEE, Part D.*, 116, 889–892, 1969.

Lancaster, P. and M. Tismenetsky, *The Theory of Matrices*, Academic Press, Orlando, FL, 1985.

Ortega, J. and W. Rheinboldt, *Iterative Solution of Nonlinear Equations in Several Variables*, Academic Press, New York, 1970.

Petrović, B. and Z. Gajić, The recursive solution of linear quadratic Nash games for weakly interconnected systems, *Journal of Optimization Theory and Applications*, 56, 463–477, 1988.

Phillips, R. and P. Kokotović, A singular perturbation approach to modeling and control of Markov chains, *IEEE Transactions on Automatic Control*, AC-26, 1087–1094, 1981.

Prljaca, N. and Z. Gajić, A transformation for block diagonalization of weakly coupled systems composed of N subsystems, *WSEAS Transactions on Systems*, 6, 848–851, 2007.

Qureshi, M., Parallel algorithms for discrete singularly perturbed and weakly coupled filtering and control problems, PhD dissertation, Rutgers University, 1992.

Qureshi, M. and Z. Gajić, Boundary value problem of linear weakly coupled systems, *Proceedings of the Allerton Conference on Communication, Control and Computing*, Urbana, IL, 455–462, 1991.

Sasagawa, T., On the finite escape phenomena for matrix Riccati equation, *IEEE Transactions on Automatic Control*, AC-27, 977–979, 1982.

Sezer, M. and D. Siljak, Nested epsilon decomposition and clustering of complex systems, *Automatica*, 22, 59–702, 1986.

Sezer, M. and D. Siljak, Nested epsilon decomposition of linear systems: Weakly coupled and overlapping blocks, *SIAM Journal of Matrix Analysis and Applications*, 12, 521–533, 1991.

Stewart, W., *Introduction to Numerical Solution of Markov Chains*, Princeton University Press, Princeton, NJ, 1994.

6 Stochastic Linear Weakly Coupled Systems

In this chapter, we present first approaches for decomposition and approximation of linear-quadratic Gaussian (LQG) estimation and control problems for weakly coupled continuous-time linear systems. The global Kalman filter of linear weakly coupled continuous systems is decomposed into separate reduced-order local filters via the use of the decoupling transformation introduced in Chapter 5. A near-optimal control law is derived by approximating the coefficients of the optimal control law. The order of approximation of the optimal performance is $O(\varepsilon^k)$, where k is the order of approximation of the coefficients. The electrical power system example demonstrates the failures of the $O(\varepsilon^2)$ and $O(\varepsilon^4)$ theories and the necessity for the existence of the $O(\varepsilon^k)$ theory (Shen and Gajić 1990a). The proposed method produces the reduction in both off-line and online computational requirements and converges under mild assumptions.

The results obtained for continuous systems are extended to the discrete-time domain in Section 6.2. To that end, the near-optimum steady-state regulator is derived for the discrete stochastic weakly coupled system and applied to a fifth-order distillation column (Shen and Gajić 1990c). The proposed methods allow parallel processing of information and reduce considerably the size of required off-line and online computations, since they introduce full parallelism in the design procedures.

The static output feedback of discrete quasi-weakly coupled stochastic systems is studied in Section 6.3. A parallel reduced-order algorithm is developed which is very efficient, since the algorithm decomposes a high-order system into a low-order system. The low-order system is represented by six Lyapunov equations that may be solved in parallel to reduce computational time. The required solution is obtained up to an arbitrary order of accuracy, $O(\varepsilon^{2k})$, where ε is a weak coupling parameter and k represents the number of iterations. The efficiency of the proposed method is demonstrated on two real aircraft that possess the quasi-weakly coupled structure under prescribed degree of stability assumption (Hogan and Gajić 1994).

In Section 6.4, we consider parallel algorithms for optimal control of weakly coupled jump parameter linear stochastic systems. These algorithms yield arbitrary orders of accuracy. The algorithms operate on reduced-order algebraic equations, hence, they are numerically very efficient. An example is included to demonstrate the procedure. The results of this section are based on the work of Borno and Gajić (1995).

6.1 CONTINUOUS WEAKLY COUPLED STOCHASTIC LINEAR CONTROL SYSTEMS

The LQG control problem of the weakly coupled continuous-time systems is studied in this section by using the results reported in Shen and Gajić (1990a). Although the duality of the regulator and filter Riccati equations can be used, together with the results presented in Chapter 2, to obtain corresponding approximations to the regulator and filter gains, such approximations will not be sufficient, because they only reduce off-line computations and do not help online computations of implementing the Kalman filter that will be of the same order as the overall weakly coupled system. The weakly coupled structure of the global Kalman filter is exploited in this section such that it may be replaced by two lower order local filters. This has been achieved via the use of a decoupling transformation introduced in Gajić and Shen (1989).

In this section, we present the approach to the decomposition and approximation of the LQG control problem of weakly coupled systems by treating the decomposition and approximation tasks separately. The decoupling transformation of Gajić and Shen (1989) is used for the exact block diagonalization of the global Kalman filter. The approximate feedback control law is then obtained by approximating the coefficients of the optimal local filters with the accuracy of $O(\varepsilon^N)$. The resulting feedback control law is shown to be a near-optimal solution of the LQG by studying the corresponding closed-loop system as a system driven by white noise. It is shown that the order of approximation of the optimal performance is $O(\varepsilon^N)$, and the order of approximation of the optimal system trajectories is $O(\varepsilon^{2N})$. All required coefficients of the desired accuracy are easily obtained by using the recursive fixed point type numerical techniques developed in Chapter 2. Given numerical algorithms converge to the required coefficients with the rate of convergence of $O(\varepsilon^2)$. In addition, only low-order subsystems are involved in algebraic computations and no analyticity requirements are imposed on the system coefficients—which is the standard assumption in the power-series expansion method. As a consequence of these properties, under very mild conditions (coefficients are bounded functions of a small coupling parameter over $\varepsilon \in [0, \varepsilon_1]$), in addition to the standard stabilizability–detectability subsystem assumptions, we have achieved the reduction in both off-line and online computational requirements.

This section is organized as follows. At first, we study the approximation of weakly coupled systems driven by white noise. It is shown that an Nth-order approximation in which the system coefficients are $O(\varepsilon^N)$ close to the exact ones is a valid approximation in the sense that the differences between the exact and approximate solutions are $O(\varepsilon^{2N})$. Then, we use these results in the study of the LQG problem. A decoupling nonsingular transformation is used to transform the Kalman filter in new coordinates in which local filters are completely decoupled. An Nth-order approximate feedback control law is defined by approximating coefficients by $O(\varepsilon^N)$. A study of the corresponding closed-loop system shows that the absolute increase in the performance criterion over its optimal value is $O(\varepsilon^N)$.

Consider the linear time-invariant weakly coupled system driven by white noise

$$\begin{bmatrix} \dot{x}_1(t) \\ \dot{x}_2(t) \end{bmatrix} = \begin{bmatrix} A_{11}(\varepsilon) & \varepsilon A_{12}(\varepsilon) \\ \varepsilon A_{21}(\varepsilon) & A_{22}(\varepsilon) \end{bmatrix} \begin{bmatrix} x_1(t) \\ x_2(t) \end{bmatrix} + \begin{bmatrix} G_{11}(\varepsilon) & \varepsilon G_{12}(\varepsilon) \\ \varepsilon G_{21}(\varepsilon) & G_{22}(\varepsilon) \end{bmatrix} \begin{bmatrix} w_1(t) \\ w_2(t) \end{bmatrix} \quad (6.1)$$

where

$$x_i(t) \in \Re^{n_i}$$
$$w_i(t) \in \Re^{r_i}, \quad i = 1, 2, \ldots$$

ε is a small parameter

The system matrices are bounded functions of ε (Petrović and Gajić 1988; Harkara et al. 1989; Gajić et al. 1990) of appropriate dimensions. The inputs $w_i(t)$ are zero mean, stationary, Gaussian uncorrelated white noise processes with intensities $W_i > 0$, $i = 1, 2$. It is well known that the variance of the linear systems driven by white noise is given by the Lyapunov equation (Gajić and Qureshi 1995). In order to assure the existence of its solution we assume that $A_{ii}(\varepsilon)$, $i = 1, 2$, are stable matrices. Hence, in this section we will need the following assumption.

Assumption 6.1 $A_{ii}(\varepsilon)$, $i = 1, 2$, are stable matrices.

In addition, the standard weak coupling assumption used throughout this book to define linear weakly coupled systems and eliminate slow and fast phenomena, that is, to provide pure weak coupling is applicable in this chapter also. It is given as follows.

Assumption 6.2a   Magnitudes of all system eigenvalues are $O(1)$, that is, $|\lambda_j| = O(1)$, $j = 1, 2, \ldots, n$, $n = n_1 + n_2$, which implies that the matrices A_{11}, A_{22} are nonsingular with $\det\{A_{11}\} = O(1)$ and $\det\{A_{22}\} = O(1)$.

The purpose of this section is to study approximations of $x_i(t)$, $i = 1, 2$, when ε is small. We are interested in approximations $x_i^N(t)$ which are defined by the following equations:

$$\begin{bmatrix} \dot{x}_1^N(t) \\ \dot{x}_2^N(t) \end{bmatrix} = \begin{bmatrix} A_{11}^N(\varepsilon) & \varepsilon A_{12}^N(\varepsilon) \\ \varepsilon A_{21}^N(\varepsilon) & A_{22}^N(\varepsilon) \end{bmatrix} \begin{bmatrix} x_1^N(t) \\ x_2^N(t) \end{bmatrix} + \begin{bmatrix} G_{11}^N(\varepsilon) & \varepsilon G_{12}^N(\varepsilon) \\ \varepsilon G_{21}^N(\varepsilon) & G_{22}^N(\varepsilon) \end{bmatrix} \begin{bmatrix} w_1(t) \\ w_2(t) \end{bmatrix} \quad (6.2)$$

where

$$A_{ij}(\varepsilon) - A_{ij}^N(\varepsilon) = O(\varepsilon^N)$$
$$G_{ij}(\varepsilon) - G_{ij}^N(\varepsilon) = O(\varepsilon^N), \quad i, j = 1, 2 \quad (6.3)$$

The quantities of interest are the variances of the errors

$$e_i(t) = x_i(t) - x_i^N(t), \quad i = 1, 2 \quad (6.4)$$

at steady state. We study the impact of the steady-state errors on a quadratic form given by

$$
\sigma = \mathrm{tr}\left\{ \begin{bmatrix} H^{\mathrm{T}}(\varepsilon)H(\varepsilon) & H^{\mathrm{T}}(\varepsilon)J(\varepsilon) \\ J^{\mathrm{T}}(\varepsilon)H(\varepsilon) & J^{\mathrm{T}}(\varepsilon)J(\varepsilon) \end{bmatrix} \times E\begin{bmatrix} x_1(t)x_1^{\mathrm{T}}(t) & x_1(t)x_2^{\mathrm{T}}(t) \\ x_2(t)x_1^{\mathrm{T}}(t) & x_2(t)x_2^{\mathrm{T}}(t) \end{bmatrix} \right\} \tag{6.5}
$$

where $H(\varepsilon)$ and $J(\varepsilon)$ are also bounded functions of ε. Such a quadratic form appears in the steady-state LQG control problem. We examine the approximation of σ by σ^N defined by

$$
\sigma^N = \mathrm{tr}\left\{ \begin{bmatrix} H^{N^{\mathrm{T}}}(\varepsilon)H^N(\varepsilon) & H^{N^{\mathrm{T}}}(\varepsilon)J^N(\varepsilon) \\ J^{N^{\mathrm{T}}}(\varepsilon)H^N(\varepsilon) & J^{N^{\mathrm{T}}}(\varepsilon)J^N(\varepsilon) \end{bmatrix} \times E\begin{bmatrix} x_1^N(t)x_1^{N^{\mathrm{T}}}(t) & x_1^N(t)x_2^{N^{\mathrm{T}}}(t) \\ x_2^N(t)x_1^{N^{\mathrm{T}}}(t) & x_2^N(t)x_2^{N^{\mathrm{T}}}(t) \end{bmatrix} \right\} \tag{6.6}
$$

where

$$
H^N(\varepsilon) - H(\varepsilon) = O(\varepsilon^N), \quad J^N(\varepsilon) - J(\varepsilon) = O(\varepsilon^N) \tag{6.7}
$$

For the purpose of estimating the order of quantities involved in derivations, we need the following assumption about the order of the quantities involved in calculations.

Assumption 6.2b   Matrices A_{ij}, A_{ij}^N, G_{ij}, G_{ij}^N, $i, j = 1, 2$, and H, J, H^N, J^N are $O(1)$.
   In the following we will suppress the ε-dependence of the problem matrices in order to simplify notation.
   The main results of this section are given in the following two theorems.

THEOREM 6.1

Under Assumptions 6.1 and 6.2 imposed on A_{ii}, $i = 1, 2$; the approximation errors at steady state satisfy

$$
\mathrm{Var}\{e_i(t)\} = \mathrm{Var}\{x_i(t) - x_i^N(t)\} = O(\varepsilon^{2N}), \quad i = 1, 2
$$
$$
\mathrm{Cov}\{e_1(t), e_2(t)\} = O(\varepsilon^{2N}) \tag{6.8}
$$

THEOREM 6.2

Under Assumptions 6.1 and 6.2, the quadratic forms (Equations 6.5 and 6.6) at steady state satisfy

$$
\Delta\sigma = \sigma - \sigma^N = O(\varepsilon^N) \tag{6.9}
$$

Proof The proof of these two theorems can be obtained by studying the following augmented system driven by white noise:

$$
\begin{bmatrix} \dot{x}_1(t) \\ \dot{e}_1(t) \\ \dot{x}_2(t) \\ \dot{e}_2(t) \end{bmatrix} = \begin{bmatrix} A_{11} & 0 & \varepsilon A_{11} & 0 \\ O(\varepsilon^N) & A_{11}^N & O(\varepsilon^{N+1}) & \varepsilon A_{12}^N \\ \varepsilon A_{21} & 0 & A_{22} & 0 \\ O(\varepsilon^{N+1}) & \varepsilon A_{21}^N & O(\varepsilon^N) & A_{22}^N \end{bmatrix} \begin{bmatrix} x_1(t) \\ e_1(t) \\ x_2(t) \\ e_2(t) \end{bmatrix} + \begin{bmatrix} G_{11} & \varepsilon G_{12} \\ O(\varepsilon^N) & O(\varepsilon^{N+1}) \\ \varepsilon G_{21} & G_{22} \\ O(\varepsilon^{N+1}) & O(\varepsilon^N) \end{bmatrix} \begin{bmatrix} w_1(t) \\ w_2(t) \end{bmatrix}
$$

$$(6.10)$$

For shorthand notation, Equation 6.10 is written as

$$\dot{z}(t) = Az(t) + \Gamma w(t) \tag{6.11}$$

with obvious definitions of $z(t)$, $\omega(t)$, Λ, and Γ. The variance of $z(t)$ at steady state is given by the algebraic Lyapunov equation (Gajić and Qureshi 1995)

$$0 = \Lambda Q + Q \Lambda^{\mathrm{T}} + \Gamma W \Gamma^{\mathrm{T}} \tag{6.12}$$

where

$$W = \begin{bmatrix} W_1 & 0 \\ 0 & W_2 \end{bmatrix}$$

The variance of $z(t)$ is partitioned as

$$Q = \begin{bmatrix} Q_{11} & Q_{12} & \varepsilon Q_{13} & \varepsilon Q_{14} \\ Q_{12}^{\mathrm{T}} & Q_{22} & \varepsilon Q_{23} & \varepsilon Q_{24} \\ \varepsilon Q_{13}^{\mathrm{T}} & \varepsilon Q_{23}^{\mathrm{T}} & Q_{33} & Q_{34} \\ \varepsilon Q_{14}^{\mathrm{T}} & \varepsilon Q_{24}^{\mathrm{T}} & Q_{34}^{\mathrm{T}} & Q_{44} \end{bmatrix}$$

Studying the partitioned form of Equation 6.12 will produce (after lengthy calculations)

$$Q_{ij} = O(1), \quad ij = 11, 13, 33 \tag{6.13a}$$

$$Q_{ij} = O(\varepsilon^{2N}), \quad ij = 22, 24, 44 \tag{6.13b}$$

$$Q_{ij} = O(\varepsilon^N), \quad ij = 12, 14, 23, 34 \tag{6.13c}$$

which proves Theorem 6.1.

Quadratic forms defined in Equations 6.5 and 6.6 can be now expressed in terms of the elements of the matrix Q as

$$\sigma = \mathrm{tr}\{H^{\mathrm{T}} H Q_{11} + J^{\mathrm{T}} J Q_{33} + 2\varepsilon J^{\mathrm{T}} H Q_{13}\} \tag{6.14}$$

and

$$\sigma^N = \text{tr}\left\{2\varepsilon J^{N^T} H^N(Q_{13} - Q_{23} - Q_{14} + Q_{24})\right\}$$
$$+ \text{tr}\left\{H^{N^T} H^N(Q_{11} - 2Q_{12} + Q_{22}) + J^{N^T} J^N(Q_{33} - 2Q_{34} + Q_{44})\right\} \quad (6.15)$$

From Equations 6.14 and 6.15 and estimates for Q_{ij}, $i, j = 1, 2, 3, 4$, one has

$$\Delta\sigma = \sigma - \sigma^N$$
$$= O(\varepsilon^N)\text{tr}\left\{(Q_{11} + Q_{23}) + (H^{N^T} H^N + J^{N^T} J^N)\right\} + O(\varepsilon^{N+1}) \quad (6.16)$$

Since Q_{11}, Q_{33} and H^N, J^N are $O(1)$ quantities, one can conclude that $\Delta\sigma = O(\varepsilon^N)$, which completes the proof of Theorem 6.2. ∎

At this point, we can introduce the LQG control problem of weakly coupled systems and study its approximation and decomposition by utilizing results from Theorems 6.1 and 6.2.

Consider the weakly coupled linear system

$$\begin{bmatrix} \dot{x}_1(t) \\ \dot{x}_2(t) \end{bmatrix} = \begin{bmatrix} A_{11}(\varepsilon) & \varepsilon A_{12}(\varepsilon) \\ \varepsilon A_{21}(\varepsilon) & A_{22}(\varepsilon) \end{bmatrix} \begin{bmatrix} x_1(t) \\ x_2(t) \end{bmatrix}$$
$$+ \begin{bmatrix} B_{11}(\varepsilon) & \varepsilon B_{12}(\varepsilon) \\ \varepsilon B_{21}(\varepsilon) & B_{22}(\varepsilon) \end{bmatrix} \begin{bmatrix} u_1(t) \\ u_2(t) \end{bmatrix} + \begin{bmatrix} G_{11}(\varepsilon) & \varepsilon G_{12}(\varepsilon) \\ \varepsilon G_{21}(\varepsilon) & G_{22}(\varepsilon) \end{bmatrix} \begin{bmatrix} w_1(t) \\ w_2(t) \end{bmatrix}$$
$$(6.17)$$

$$\begin{bmatrix} y_1(t) \\ y_2(t) \end{bmatrix} = \begin{bmatrix} C_{11}(\varepsilon) & e C_{12}(\varepsilon) \\ \varepsilon C_{21}(\varepsilon) & C_{22}(\varepsilon) \end{bmatrix} \begin{bmatrix} x_1(t) \\ x_2(t) \end{bmatrix} + \begin{bmatrix} v_1(t) \\ v_2(t) \end{bmatrix} \quad (6.18)$$

where $x_i(t) \in \mathcal{R}^{n_i}$, $u_i(t) \in \mathcal{R}^{m_i}$, $y_i(t) \in \mathcal{R}^{r_i}$, $i = 1, 2$ are state, control, and measurement vectors, respectively, and $w_i(t) \in \mathcal{R}^{s_i}$, $v_i(t) \in \mathcal{R}^{r_i}$, $i = 1, 2$, are independent zero-mean stationary white Gaussian noise processes with intensities W_i, V_i, $i = 1, 2$. The degree of interaction between subsystems is measured by a small parameter ε. With Equations 6.17 and 6.18, consider the performance criterion

$$J = \text{tr}\left\{D^T D E \begin{bmatrix} x_1(t)x_1^T(t) & x_1(t)x_2^T(t) \\ x_2(t)x_1^T(t) & x_2(t)x_2^T(t) \end{bmatrix}\right\}$$
$$+ \text{tr}\left\{R E \begin{bmatrix} u_1(t)u_1^T(t) & u_1(t)u_2^T(t) \\ u_2(t)u_1^T(t) & u_2(t)u_2^T(t) \end{bmatrix}\right\} \quad (6.19)$$

with positive definite R. Matrices D^TD and R have the weakly coupled structure, that is, they are given by

$$D^TD = \begin{bmatrix} D_1^T D_1 & \varepsilon D_1^T D_2 \\ \varepsilon D_2^T D_1 & D_2^T D_2 \end{bmatrix} \quad R = \begin{bmatrix} R_1 & 0 \\ 0 & R_2 \end{bmatrix}$$

where

$$R_i \in \Re^{m_i \times m_i}$$
$$D_i^T D_i \in \Re^{n_i \times n_i}, \; i = 1, 2$$

The optimal control law has the very well-known form (Kwakernaak and Sivan 1972)

$$\begin{bmatrix} u_1(t) \\ u_2(t) \end{bmatrix} = - \begin{bmatrix} F_{11} & \varepsilon F_{12} \\ \varepsilon F_{21} & F_{22} \end{bmatrix} \begin{bmatrix} \hat{x}_1(t) \\ \hat{x}_2(t) \end{bmatrix} \tag{6.20}$$

$$\begin{bmatrix} \dot{\hat{x}}_1(t) \\ \dot{\hat{x}}_2(t) \end{bmatrix} = \begin{bmatrix} A_{11} & \varepsilon A_{12} \\ \varepsilon A_{21} & A_{22} \end{bmatrix} \begin{bmatrix} \hat{x}_1(t) \\ \hat{x}_2(t) \end{bmatrix} + \begin{bmatrix} B_{11} & \varepsilon B_{12} \\ \varepsilon B_{21} & B_{22} \end{bmatrix} \begin{bmatrix} u_1(t) \\ u_2(t) \end{bmatrix}$$
$$\times \begin{bmatrix} K_{11} & \varepsilon K_{12} \\ \varepsilon K_{21} & K_{22} \end{bmatrix} \begin{bmatrix} y_1(t) - C_{11}\hat{x}_1(t) - \varepsilon C_{12}\hat{x}_2(t) \\ y_2(t) - \varepsilon C_{21}\hat{x}_1 - C_{22}\hat{x}_2(t) \end{bmatrix} \tag{6.21}$$

Introducing the notation

$$A = \begin{bmatrix} A_{11} & \varepsilon A_{12} \\ \varepsilon A_{21} & A_{22} \end{bmatrix}, \quad B = \begin{bmatrix} B_{11} & \varepsilon B_{12} \\ \varepsilon B_{21} & B_{22} \end{bmatrix}, \quad G = \begin{bmatrix} G_{11} & \varepsilon G_{12} \\ \varepsilon G_{21} & G_{22} \end{bmatrix}$$

$$C = \begin{bmatrix} C_{11} & \varepsilon C_{12} \\ \varepsilon C_{21} & C_{22} \end{bmatrix}, \quad F = \begin{bmatrix} F_{11} & \varepsilon F_{12} \\ \varepsilon F_{21} & F_{22} \end{bmatrix}, \quad K = \begin{bmatrix} K_{11} & \varepsilon K_{12} \\ \varepsilon K_{21} & K_{22} \end{bmatrix}$$

$$W = \begin{bmatrix} W_1 & 0 \\ 0 & W_2 \end{bmatrix}, \quad V = \begin{bmatrix} V_1 & 0 \\ 0 & V_2 \end{bmatrix}$$

the regulator and filter gains are obtained from

$$F = R^{-1}B^T P, \quad K = QC^T V^{-1} \tag{6.22}$$

where P and Q are positive semidefinite stabilizing solutions of the algebraic Riccati equations

$$A^T P + PA - PS_R P + D^T D = 0, \quad S_R = BR^{-1}B^T \tag{6.23}$$

$$AQ + QA^T - QS_F Q + GWG^T = 0, \quad S_F = C^T V^{-1} C \tag{6.24}$$

To be able to give an estimate of the order of the quantities to appear in the derivations that follow, we need to impose an assumption on the order of the quantities that define the optimization problem (Equations 6.17 through 6.22).

Assumption 6.3   Matrices A_{ij}, B_{ij}, C_{ij}, $i, j = 1, 2$, and matrices D_l, R_l, W_l, V_l, $l = 1, 2$, are $O(1)$.

Due to the weakly coupled structure of all coefficients in Equations 6.23 and 6.24, and based on Assumption 6.3, the solutions of algebraic Riccati equations have the forms

$$P = \begin{bmatrix} P_1 & \varepsilon P_2 \\ \varepsilon P_2^T & P_3 \end{bmatrix}, \quad Q = \begin{bmatrix} Q_1 & \varepsilon Q_2 \\ \varepsilon Q_2^T & Q_3 \end{bmatrix} \qquad (6.25)$$

Solutions of Equations 6.23 and 6.24 can be found in terms of the reduced-order problems by imposing standard stabilizability–detectability assumptions on subsystems. The efficient fixed point algorithms for solving Equations 6.23 and 6.24 are obtained in Section 2.2. The algorithms for solving regulator and filter algebraic Riccati equations of weakly coupled systems are convergent under the following assumptions.

Assumption 6.4   Triples (A_{ii}, B_{ii}, D_{ii}), $i = 1, 2$, are stabilizable and detectable.

Assumption 6.5   Triples (A_{ii}, C_{ii}, G_{ii}), $i = 1, 2$, are stabilizable and detectable.

Getting approximate solutions for P and Q in terms of the reduced-order problems will produce savings in off-line computations. However, in the case of stochastic systems, where an additional dynamic system—filter—has to be built, one is particularly interested in the reduction of online computations. We will achieve this by using the decoupling transformation presented in Section 5.1.

The Kalman filter (Equation 6.21) is viewed as a system driven by the innovation process. However, one might study the filter form when it is driven by both measurements and controls. The filter form under consideration is obtained from Equation 6.21 as

$$\begin{bmatrix} \dot{\hat{x}}_1(t) \\ \dot{\hat{x}}_2(t) \end{bmatrix} = \begin{bmatrix} K_{11} & \varepsilon K_{12} \\ \varepsilon K_{21} & K_{22} \end{bmatrix} \begin{bmatrix} v_1(t) \\ v_2(t) \end{bmatrix}$$
$$+ \begin{bmatrix} (A_{11} - B_{11}F_{11} - \varepsilon^2 B_{12}F_{21}) & \varepsilon(A_{12} - B_{11}F_{12} - B_{12}F_{22}) \\ \varepsilon(A_{21} - B_{21}F_{11} - B_{22}F_{21}) & (A_{22} - B_{22}F_{22} - \varepsilon^2 B_{21}F_{12}) \end{bmatrix} \begin{bmatrix} \hat{x}_1(t) \\ \hat{x}_2(t) \end{bmatrix}$$
$$\qquad (6.26)$$

with innovation processes

$$v_1(t) = y_1(t) - C_{11}\hat{x}_1(t) - \varepsilon C_{12}\hat{x}_2(t)$$
$$v_2(t) = y_2(t) - \varepsilon C_{21}\hat{x}_1(t) - C_{22}\hat{x}_2(t) \qquad (6.27)$$

The nonsingular state transformation from Section 5.1 will block diagonalize Equation 6.26 under the condition that matrices $(A_{11} - B_{11}F_{11} - \varepsilon^2 B_{12}F_{12})$ and $(A_{22} - B_{22}F_{22} - \varepsilon^2 B_{21}F_{12})$ have no eigenvalues in common. This transformation is given by

$$\begin{bmatrix} \hat{\eta}_1(t) \\ \hat{\eta}_2(t) \end{bmatrix} = \begin{bmatrix} I_{n_1} - \varepsilon^2 LH & -\varepsilon L \\ \varepsilon H & I_{n_2} \end{bmatrix} \begin{bmatrix} \hat{x}_1(t) \\ \hat{x}_2(t) \end{bmatrix} = T_1^{-1} \begin{bmatrix} \hat{x}_1(t) \\ \hat{x}_2(t) \end{bmatrix} \qquad (6.28)$$

with

$$T_1 = \begin{bmatrix} I_{n_1} & \varepsilon L \\ -\varepsilon H & I_{n_2} - \varepsilon^2 HL \end{bmatrix} \tag{6.29}$$

where matrices L and H satisfy equations given in Section 5.1. The optimal feedback control expressed in the new coordinates has the form

$$u_1(t) = -f_{11}\hat{\eta}_1(t) - \varepsilon f_{12}\hat{\eta}_2(t) \tag{6.30a}$$

$$u_2(t) = -\varepsilon f_{21}\hat{\eta}_1(t) - f_{22}\hat{\eta}_2(t) \tag{6.30b}$$

with

$$\dot{\hat{\eta}}_1(t) = \alpha_1\hat{\eta}_1(t) + \beta_{11}v_1(t) + \varepsilon\beta_{12}v_2(t) \tag{6.31a}$$

$$\dot{\hat{\eta}}_2(t) = \alpha_2\hat{\eta}_2(t) + \varepsilon\beta_{21}v_1(t) + \beta_{22}v_2(t) \tag{6.31b}$$

where

$$\begin{aligned}
f_{11} &= F_{11} - \varepsilon^2 F_{12}H, \quad F_{12} = F_{12} + (F_{11} - \varepsilon^2 F_{12}H)L \\
f_{21} &= F_{21} - F_{22}H, \quad f_{22} = F_{22} + \varepsilon^2(F_{21} - F_{22}H)L \\
\alpha_1 &= \alpha_{11} - \varepsilon^2 a_{12}H, \quad \alpha_2 = a_{22} + \varepsilon^2 H a_{12} \\
\beta_{11} &= K_{11} - \varepsilon^2(LH + LK_{21}), \quad \beta_{12} = K_{12} - LK_{22} - \varepsilon^2 LHK_{12} \\
\beta_{21} &= HK_{11} + K_{21}, \quad \beta_{22} = K_{22} + \varepsilon^2 HK_{12}
\end{aligned} \tag{6.32}$$

and

$$\begin{aligned}
a_{11} &= A_{11} - B_{11}F_{11} - \varepsilon^2 B_{12}F_{21}, \quad a_{12} = A_{12} - B_{11}F_{12} - B_{12}F_{22} \\
a_{21} &= A_{21} - B_{21}F_{11} - B_{22}F_{21}, \quad a_{22} = A_{22} - B_{22}F_{22} - \varepsilon^2 B_{21}F_{12}
\end{aligned}$$

The innovation processes $v_1(t)$ and $v_2(t)$ are now given by

$$v_1(t) = y_1(t) - d_{11}\hat{\eta}_1(t) - \varepsilon d_{12}\hat{\eta}_2(t) \tag{6.33a}$$

$$v_2(t) = y_2(t) - \varepsilon d_{21}\hat{\eta}_1(t) - d_{22}\hat{\eta}_2(t) \tag{6.33b}$$

where

$$\begin{aligned}
d_{11} &= C_{11} - \varepsilon^2 C_{12}H, \quad d_{12} = C_{11}L + C_{12} - \varepsilon^2 C_{12}HL \\
d_{21} &= C_{21} - C_{22}H, \quad d_{22} = C_{22} + \varepsilon^2(C_{21} - C_{22}H)L
\end{aligned}$$

Approximate control laws are defined by perturbing coefficients F_{ij}, K_{ij}, $i, j = 1, 2$; L and H by $O(\varepsilon^k)$, $k = 1, 2, \ldots$, in other words by using kth approximations for these coefficients, where k stands for the required order of accuracy, that is,

$$u_1^{(k)}(t) = -f_{11}^{(k)} \hat{\eta}_1^{(k)}(t) - \varepsilon f_{12}^{(k)} \hat{\eta}_2^{(k)}(t) \tag{6.34a}$$

$$u_2^{(k)}(t) = -\varepsilon f_{21}^{(k)} \hat{\eta}_1^{(k)}(t) - f_{22}^{(k)} \hat{\eta}_2^{(k)}(t) \tag{6.34b}$$

with

$$\dot{\hat{\eta}}_1^{(k)}(t) = \alpha_1^{(k)} \hat{\eta}_1^{(k)}(t) + \beta_{11}^{(k)} v_1^{(k)}(t) + \varepsilon \beta_{12}^{(k)} v_2^{(k)}(t) \tag{6.35a}$$

$$\dot{\hat{\eta}}_2^{(k)}(t) = \alpha_2^{(k)} \hat{\eta}_2^{(k)}(t) + \varepsilon \beta_{21}^{(k)} v_1^{(k)}(t) + \beta_{22}^{(k)} v_2^{(k)}(t) \tag{6.35b}$$

where

$$v_1^{(k)}(t) = y_1(t) - d_{11}^{(k)} \hat{\eta}_1^{(k)}(t) - \varepsilon d_{12}^{(k)} \hat{\eta}_2^{(k)}(t) \tag{6.36a}$$

$$v_2^{(k)}(t) = y_2(t) - \varepsilon d_{21}^{(k)} \hat{\eta}_1^{(k)}(t) - d_{22}^{(k)} \hat{\eta}_2^{(k)}(t) \tag{6.36b}$$

and

$$f_{ij}^{(k)} = f_{ij} + O(\varepsilon^k), \quad d_{ij}^{(k)} = d_{ij} + O(\varepsilon^k)$$
$$\alpha_{ij}^{(k)} = \alpha_{ij} + O(\varepsilon^k), \quad \beta_{ij}^{(k)} = \beta_{ij} + O(\varepsilon^k) \quad i, j = 1, 2 \tag{6.37}$$

The near-optimality of the proposed control law (Equation 6.34) is established in the following theorem.

THEOREM 6.3

Let $x_1(t)$ and $x_2(t)$ be the optimal trajectories and J be the optimal value of the performance criterion. Let $x_1^{(k)}(t)$, $x_2^{(k)}(t)$, and $J^{(k)}$ be the corresponding quantities under the approximate control law $u^{(k)}(t)$, then under Assumptions 6.3 through 6.5, we have

$$J - J^{(k)} = O(\varepsilon^k) \tag{6.38}$$

$$\text{Var}\{(x_i(t) - x_i^{(k)}(t))\} = O(\varepsilon^{2k}), \quad k = 0, 1, 2, \ldots \tag{6.39}$$

Proof The results of Theorems 6.1 and 6.2 are employed by studying a system of equations driven by white noise. For the truly optimal control consider the equations

$$\begin{bmatrix} \dot{\eta}_1(t) \\ \dot{e}_1(t) \\ \dot{\eta}_2(t) \\ \dot{e}_2(t) \end{bmatrix} = \begin{bmatrix} A_{11} & \varepsilon A_{12} \\ \varepsilon A_{21} & A_{22} \end{bmatrix} \begin{bmatrix} \hat{\eta}_1(t) \\ e_1(t) \\ \hat{\eta}_2(t) \\ e_2(t) \end{bmatrix} + \begin{bmatrix} \Theta_{11} & \varepsilon \Theta_{12} \\ \varepsilon \Theta_{21} & \Theta_{22} \end{bmatrix} \begin{bmatrix} v_1(t) \\ w_1(t) \\ v_2(t) \\ w_2(t) \end{bmatrix} \tag{6.40}$$

where $e_i(t) = \eta_i(t) - \hat{\eta}_i(t)$, $i = 1, 2$, are estimation errors. The corresponding equation for the approximate control is

$$
\begin{bmatrix} \dot{\hat{\eta}}_1^N(t) \\ \dot{e}_1^N(t) \\ \dot{\hat{\eta}}_2^N(t) \\ \dot{e}_2^N(t) \end{bmatrix} = \begin{bmatrix} A_{11}^N & \varepsilon A_{12}^N \\ \varepsilon A_{21}^N & A_{22}^N \end{bmatrix} \begin{bmatrix} \dot{\hat{\eta}}_1^N(t) \\ \dot{e}_1^N(t) \\ \dot{\hat{\eta}}_2^N(t) \\ \dot{e}_2^N(t) \end{bmatrix} + \begin{bmatrix} \Theta_{11}^N & \varepsilon \Theta_{12}^N \\ \varepsilon \Theta_{21}^N & \Theta_{22}^N \end{bmatrix} \begin{bmatrix} v_1(t) \\ w_1(t) \\ v_2(t) \\ w_2(t) \end{bmatrix} \tag{6.41}
$$

where $e_i^N(t) = \eta_i(t) - \hat{\eta}_i^N(t)$ are corresponding estimation errors. The matrices A_{ij}, Θ_{ij}, and A_{ij}^N, Θ_{ij}^N in Equations 6.40 and 6.41 are obtained in an obvious way. It can be verified that

$$
A_{ij} - A_{ij}^N = O(\varepsilon^N), \quad \Theta_{ij} - \Theta_{ij}^N = O(\varepsilon^N), \quad i, j = 1, 2
$$

and $A_{ii}(0)$, $i = 1, 2$, are given by

$$
A_{ii}(0) = \begin{bmatrix} A_{ii} - B_{ii} F_{ii} & K_{ii} C_{ii} \\ 0 & A_{ii} - K_{ii} C_{ii} \end{bmatrix} \tag{6.42}
$$

which by stabilizability–detectability assumptions imposed on the triples (A_{ii}, B_{ii}, D_i) and (A_{ii}, C_{ii}, G_{ii}), $i = 1, 2$, guarantees the stability of matrices $A_{ii}(0)$. The results of Theorems 6.1 and 6.2 can now be directly used to establish Equations 6.38 and 6.39.

Results obtained in this section are along the lines of those obtained in Kokotović and Cruz (1969). It is shown in Kokotović and Cruz (1969) that an $O(\varepsilon^N)$ approximation of coefficients for a deterministic linear-quadratic regulator implies the $O(\varepsilon^{2N})$ approximation of the corresponding performance criterion. In this section, we show that for the weakly coupled linear-quadratic *stochastic* control an $O(\varepsilon^N)$ approximation of coefficients implies the absolute error of the performance criterion of *only* $O(\varepsilon^N)$—Theorem 6.3.   ■

6.1.1 CASE STUDY: ELECTRIC POWER SYSTEM

In order to demonstrate the numerical behavior of the near-optimum design of weakly coupled LQG regulator, we present results for an LQG controller of a power system composed of two interconnected areas (Geromel and Pres 1985). The system model is given by

$$
A = \begin{bmatrix}
0 & 0.55 & 0.0 & 0.0 & 0.0 & -0.55 & 0.0 & 0.0 & 0.0 \\
0 & 0.0 & 1.0 & 0.0 & 0.0 & 0.0 & 0.0 & 0.0 & 0.0 \\
0 & -3.3 & -0.05 & 6.0 & 0.0 & 3.3 & 0.0 & 0.0 & 0.0 \\
0 & 0.0 & 0.0 & -3.3 & 3.3 & 0.0 & 0.0 & 0.0 & 0.0 \\
0 & 0.0 & -5.2 & 0.0 & -13 & 0.0 & 0.0 & 0.0 & 0.0 \\
0 & 0.0 & 0.0 & 0.0 & 0.0 & 0.0 & 1.0 & 0.0 & 0.0 \\
0 & 3.3 & 0.0 & 0.0 & 0.0 & -3.3 & -0.05 & 6.0 & 0.0 \\
0 & 0.0 & 0.0 & 0.0 & 0.0 & 0.0 & 0.0 & -3.3 & 3.3 \\
0 & 0.0 & 0.0 & 0.0 & 0.0 & 0.0 & -5.2 & 0.0 & -13
\end{bmatrix}
$$

$$B = \begin{bmatrix} 0 & 0 & 0 & 0 & 13 & 0 & 0 & 0 & 0 \\ 0 & 0 & 0 & 0 & 0 & 0 & 0 & 0 & 13 \end{bmatrix}$$

$$C = \begin{bmatrix} 1 & 0.43 & 0 & 0 & 0 & 0.0 & 0 & 0 & 0 \\ 0 & 0.0 & 0 & 1 & 0 & 0.0 & 0 & 0 & 0 \\ -1 & 0.0 & 0 & 0 & 0 & 0.43 & 0 & 0 & 0 \\ 0 & 0.0 & 0 & 0 & 0 & 0.0 & 0 & 1 & 0 \end{bmatrix}$$

$$D^T D = \begin{bmatrix} 1.0 & 0.0 & 0.0 & 0.0 & 0.0 & 0.0 & 0.0 & 0.0 & 0.0 \\ 0.0 & 1.3 & 0.0 & 0.0 & 0.0 & -0.3 & 0.0 & 0.0 & 0.0 \\ 0.0 & 0.0 & 1.0 & 0.0 & 0.0 & 0.0 & 0.0 & 0.0 & 0.0 \\ 0.0 & 0.0 & 0.0 & 0.0 & 0.0 & 0.0 & 0.0 & 0.0 & 0.0 \\ 0.0 & 0.0 & 0.0 & 0.0 & 0.0 & 0.0 & 0.0 & 0.0 & 0.0 \\ 0.0 & -0.3 & 0.0 & 0.0 & 0.0 & 1.3 & 0.0 & 0.0 & 0.0 \\ 0.0 & 0.0 & 0.0 & 0.0 & 0.0 & 0.0 & 1.0 & 0.0 & 0.0 \\ 0.0 & 0.0 & 0.0 & 0.0 & 0.0 & 0.0 & 0.0 & 0.0 & 0.0 \\ 0.0 & 0.0 & 0.0 & 0.0 & 0.0 & 0.0 & 0.0 & 0.0 & 0.0 \end{bmatrix}$$

$$R = \begin{bmatrix} 1.0 & 0.0 \\ 0.0 & 1.0 \end{bmatrix}$$

It is assumed that $G = B$, and that the noise intensity matrices are given by

$$W_1 = 0.1, \quad W_2 = 0.1, \quad V_1 = I_2, \quad V_2 = I_2$$

We can note relatively big elements in the cross-coupling matrices A_{12}, A_{21}, and C_{21}. The small parameter ε is built in the problem. The value for ε should be estimated from the problem's strongest coupled matrix, in this case matrix C. It seems from our experience that the formula

$$\varepsilon = \frac{\max(\|C_{12}\|, \|C_{21}\|)}{\max(\|C_{11}\|, \|C_{22}\|)} = \frac{1}{1.43} = 0.699 \tag{6.43}$$

produces quite good estimate for ε, where $\| \ \|$ is any suitable norm. In this example we have used the infinity norm.

It is important to notice that there is no known method in the literature which produces an upper bound for the small parameter ε. This is true for the entire theory of small parameters (weak coupling and singular perturbations). It happens that in this particular example, despite the relatively large value for the small parameter ε, the proposed method converges, since the radius of convergence of all algorithms used is less than one at each iteration. The simulation results are presented in the following table.

The small parameter ε is relatively big in this example, that is, $\varepsilon = 0.7$. Since $O(0.7^{26}) \approx 10^{-4}$, it will require 24 terms in order to get the accuracy of 10^{-4} if the

TABLE 6.1

Approximate Values for Criterion

k	$J^{(k)}$	$J^{(k)} - J$	$(0.7)^k$
2	∞	∞	*
4	∞	∞	*
6	5.9415	0.9645	0.11765
10	5.1111	0.1341	0.02825
18	4.9788	0.0018	0.00163
26	4.9770	$<10^{-4}$	9.4×10^{-5}
Optimal	4.9770	*	*

power-series expansion method is used—which is not feasible. On the other hand, the fixed pointed method scheme used in this section will demand 12 iterations (rate of convergence is $O(\varepsilon^2)$) of the presented algorithms—which can easily be achieved. Moreover, the $O(\varepsilon^2)$ and $O(\varepsilon^4)$ approximate filters in this problem do not stabilize the plant-filter augmented system, and the approximate filter has to be found with an accuracy of at least $O(\varepsilon^6)$.

Table 6.1 verifies the result of Theorem 6.3, namely $J - J^{(k)} = O(\varepsilon^k)$, and supports the formula (Equation 6.43) for the estimate of the weak coupling parameter.

6.2   DISCRETE WEAKLY COUPLED STOCHASTIC LINEAR CONTROL SYSTEMS

In this section, we study the LQG control problem of weakly coupled discrete-time systems. The partitioned form of the main equation of the optimal linear control theory, the Riccati equation, has a very complicated form in the discrete-time domain. In Chapter 2, this problem is overcome by using a bilinear transformation which is applicable under quite mild assumptions, so that the reduced-order solution of the discrete algebraic Riccati equation of weakly coupled systems can be obtained up to any order of accuracy by using known reduced-order results for the corresponding continuous-time algebraic Riccati equation.

Although the duality of the filter and regulator Riccati equations can be used together with results reported in Shen and Gajić (1990b) to obtain corresponding approximations to the filter and regulator gains, such approximations will not be sufficient because they only reduce the off-line computations of implementing the Kalman filter that will be of the same order as the overall weakly coupled system. The weakly coupled structure of the global Kalman filter is exploited in this section such that it may be replaced by two lower order local filters. This has been achieved via the use of a decoupling transformation introduced in Gajić and Shen (1989).

The decoupling transformation of Gajić and Shen (1989) is used for the exact block diagonalization of the global Kalman filter. The approximate feedback control law is then obtained by approximating the coefficients of the optimal local filters and regulators with the accuracy of $O(\varepsilon^N)$. The resulting feedback control law is shown to be a near-optimal solution of the LQG by studying the corresponding closed-loop system as a system driven by white noise. It is shown that the order of approximation of the optimal performance is $O(\varepsilon^N)$, and the order of approximation of the optimal system trajectories is $O(\varepsilon^{2N})$. All required coefficients of desired accuracy are easily obtained by using the recursive reduced-order fixed point type numerical techniques developed in Chapter 2. The obtained numerical algorithms converge to the required optimal coefficients with the rate of convergence of $O(\varepsilon^2)$. In addition, only low-order subsystems are involved in the algebraic computations and no analyticity requirements are imposed on the system coefficients—which is the standard assumption in the power-series expansion method. As a consequence of these properties, under very mild conditions (coefficients are bounded functions of a small coupling parameter), in addition to the standard stabilizability–detectability subsystem assumptions, we have achieved the reduction in both off-line and online computational requirements.

The results presented in this section are mostly based on the results of Shen (1990) and Shen and Gajić (1990b,c).

Consider the linear discrete weakly coupled stochastic system

$$
\begin{aligned}
x_1(n+1) &= A_{11}x_1(n) + \varepsilon A_{12}x_2(n) + B_{11}u_1(n) + \varepsilon B_{12}u_2(n) \\
&\quad + G_{11}w_1(n) + \varepsilon G_{12}w_2(n) \\
x_2(n+1) &= \varepsilon A_{21}x_1(n) + A_{22}x_2(n) + \varepsilon B_{21}u_1(n) + B_{22}u_2(n) \\
&\quad + \varepsilon G_{21}w_1(n) + G_{22}w_2(n) \\
y_1(n) &= C_{11}x_1(n) + \varepsilon C_{12}x_2(n) + v_1(n) \\
y_2(n) &= \varepsilon C_{21}x_1(n) + C_{22}x_2(n) + v_2(n)
\end{aligned}
\tag{6.44}
$$

with the performance criterion

$$
J = \frac{1}{2}E\left\{\sum_{n=0}^{\infty}\left[z^T(n)z(n) + u_1^T(n)R_1u_1(n) + u_2^T(n)R_2u_2(n)\right]\right\}
\tag{6.45}
$$

where
$x_i(n) \in \mathfrak{R}^{n_i}$, $i = 1, 2$, comprise state vectors
$u_i(n) \in \mathfrak{R}^{m_i}$, $i = 1, 2$, are control inputs
$y_i(n) \in \mathfrak{R}^{l_i}$, $i = 1, 2$, are observed outputs
$w_i(n) \in \mathfrak{R}^{r_i}$ and $v_i(n) \in \mathfrak{R}^{l_i}$ are independent zero-mean stationary Gaussian mutually uncorrelated white noise processes with intensities $W_i > 0$ and $V_i > 0$, respectively

$z_i(n) \in \mathfrak{R}^{s_i}$, $i = 1, 2$, are the controlled outputs given by

$$
\begin{aligned}
z_1(n) &= D_{11}x_1(n) + \varepsilon D_{12}x_2(n) \\
z_2(n) &= \varepsilon D_{21}x_1(n) + D_{22}x_2(n)
\end{aligned}
\tag{6.46}
$$

All matrices are bounded functions of a small coupling parameter ε and have appropriate dimensions. It is assumed that R_i, $i = 1, 2$, are positive definite matrices. In addition, the standard weak coupling Assumption 6.2a is assumed to be satisfied.

The optimal control law in the discrete-time domain is given by (Kwakernaak and Sivan 1972)

$$u(n) = -F\hat{x}(n) \tag{6.47}$$

with

$$\hat{x}(n + 1) = A\hat{x}(n) + Bu(n) + K[y(n) - C\hat{x}(n)] \tag{6.48}$$

where

$$A = \begin{bmatrix} A_{11} & \varepsilon A_{12} \\ \varepsilon A_{21} & A_{22} \end{bmatrix}, \quad B = \begin{bmatrix} B_{11} & \varepsilon B_{12} \\ \varepsilon B_{21} & B_{22} \end{bmatrix}, \quad C = \begin{bmatrix} C_{11} & \varepsilon C_{12} \\ \varepsilon C_{21} & C_{22} \end{bmatrix}$$

$$K = \begin{bmatrix} K_{11} & \varepsilon K_{12} \\ \varepsilon K_{21} & K_{22} \end{bmatrix}, \quad F = \begin{bmatrix} F_{11} & \varepsilon F_{12} \\ \varepsilon F_{21} & F_{22} \end{bmatrix} \tag{6.49}$$

The regulator gain F and filter gain K are obtained from

$$F = (R + B^\mathsf{T} PB)^{-1} B^\mathsf{T} PA \tag{6.50}$$

$$K = AQC^\mathsf{T} (V + CQC^\mathsf{T})^{-1} \tag{6.51}$$

where P and Q are positive semidefinite stabilizing solutions of the discrete-time algebraic regulator and filter Riccati equations, respectively, given by

$$P = D^\mathsf{T} D + A^\mathsf{T} PA - A^\mathsf{T} PB(R + B^\mathsf{T} PB)^{-1} B^\mathsf{T} PA \tag{6.52}$$

$$Q = AQA^\mathsf{T} - AQC^\mathsf{T} (V + CQC^\mathsf{T})^{-1} CQA^\mathsf{T} + GWG^\mathsf{T} \tag{6.53}$$

with

$$R = \operatorname{diag}(R_1 \ R_2), \quad W = \operatorname{diag}(W_1 \ W_2), \quad V = \operatorname{diag}(V_1 \ V_2) \tag{6.54}$$

and

$$D = \begin{bmatrix} D_{11} & \varepsilon D_{12} \\ \varepsilon D_{21} & D_{22} \end{bmatrix}, \quad G = \begin{bmatrix} G_{11} & \varepsilon G_{12} \\ \varepsilon G_{21} & G_{22} \end{bmatrix} \tag{6.55}$$

To be able to give the order estimate of the quantities involved in derivations that follow, we need the following assumption of the order of the quantities involved in the problem formulation.

Assumption 6.6   Matrices $A_{ij}, B_{ij}, G_{ij}, C_{ij}, D_{ij}, i, j = 1, 2$, and $R_l, W_l, V_l, l = 1, 2$ are $O(1)$.

Due to the block dominant structure of the problem matrices and under Assumption 6.6, the required solutions of the regulator and filter algebraic Riccati equations (Equations 6.52 and 6.53), that is P and Q, are compatible partitioned as

$$P = \begin{bmatrix} P_{11} & \varepsilon P_{12} \\ \varepsilon P_{12}^T & P_{22} \end{bmatrix}, \quad Q = \begin{bmatrix} Q_{11} & \varepsilon Q_{12} \\ \varepsilon Q_{12}^T & Q_{22} \end{bmatrix} \qquad (6.56)$$

In order to obtain the required solutions of Equations 6.52 and 6.53 in terms of the reduced-order problems and to overcome the complicated partitioned form of the discrete-time algebraic Riccati equation, we have used the method developed in Section 2.4.3 to transform the discrete-time algebraic Riccati equations (Equations 6.52 and 6.53) into the continuous-time algebraic Riccati equations of the form

$$A_R^T \mathbf{P} + \mathbf{P} A_R - \mathbf{P} S_R \mathbf{P} + D_R^T D_R = 0, \quad S_R = B_R R_R^{-1} B_R^T \qquad (6.57)$$

$$A_F \mathbf{Q} + \mathbf{Q} A_F^T - \mathbf{Q} S_F \mathbf{Q} + G_F W_F G_F^T = 0, \quad G_F W_F G_F^T = 0, \quad S_F = C_F^T V_F^{-1} C_F \quad (6.58)$$

such that the solutions of Equations 6.52 and 6.53 are equal to the solutions of Equations 6.57 and 6.58, that is

$$P = \mathbf{P}, \quad Q = \mathbf{Q} \qquad (6.59)$$

where

$$A_R = I - 2\left(\Delta_R^{-1}\right)^T \qquad (6.60)$$
$$B_R R_R^{-1} B_R^T = 2(I + A)^{-1} B R^{-1} B^T \Delta_R^{-1}$$

$$D_R^T D_R = 2\Delta_R^{-1} D^T D(1 + A)^{-1}$$
$$\Delta_R = (I + A^T) + D^T D(I + A)^{-1} B R^{-1} B^T$$

and

$$A_F = I - 2(\Delta_F^{-1})$$
$$C_F^T V_F^{-1} C_F = 2(I + A^T)^{-1} C^T V^{-1} C \Delta_F^{-1} \qquad (6.61)$$
$$G_F W_F G_F^T = 2\Delta_F^{-1} G W G^T (I + A^T)^{-1}$$
$$\Delta_F = (I + A) + G W G^T (I + A^T)^{-1} C^T V^{-1} C$$

It is shown in Section 2.4.3 that Equations 6.57 and 6.58 preserve the structure of weakly coupled systems. These equations can be solved in terms of the reduced-order problems very efficiently by using the recursive method developed in Chapter 2, which converges with the rate of convergence of $O(\varepsilon^2)$. Solutions of Equations 6.57

and 6.58 are found in terms of the reduced-order problems by imposing standard stabilizability–detectability assumptions on subsystems.

Getting approximate solution for P and Q in terms of the reduced-order problems will produce savings in off-line computations. However, in the case of stochastic systems, where the additional dynamical system—filter—has to be built, one is particularly interested in the reduction of online computations. In this section, the savings of online computation will be achieved by using a discrete-time equivalent of the decoupling transformation of Gajić and Shen (1989).

The Kalman filter (Equation 6.48) is viewed as a system driven by the innovation process. However, one might study the filter form when it is driven by both measurements and control. The filter form under consideration is obtained from Equation 5.124 as

$$
\begin{aligned}
\hat{x}_1(n+1) = {} & (A_{11} - B_{11}F_{11} - \varepsilon^2 B_{12}F_{12})\hat{x}_1(n) \\
& + \varepsilon(A_{12} - B_{11}F_{12} - B_{12}F_{22})\hat{x}_2(n) + K_{11}v_1(n) + \varepsilon K_{12}v_2(n) \\
\hat{x}_2(n+1) = {} & \varepsilon(A_{21} - B_{21}F_{11} - B_{22}F_{21})\hat{x}_1(n) \\
& + (A_{22} - \varepsilon^2 B_{21}F_{12} - B_{22}F_{22})\hat{x}_2(n) + \varepsilon K_{21}v_1(n) + K_{22}v_2(n)
\end{aligned}
\tag{6.62}
$$

with the innovation process

$$
\begin{aligned}
v_1(n) &= y_1(n) - C_{11}\hat{x}_1(n) - \varepsilon C_{12}\hat{x}_2(n) \\
v_2(n) &= y_2(n) - \varepsilon C_{21}\hat{x}_1(n) - C_{22}\hat{x}_2(n)
\end{aligned}
\tag{6.63}
$$

The nonsingular state transformation of Gajić and Shen (1989) will block diagonalize Equation 6.62 under the condition that the subsystem feedback matrices $(A_{11} - B_{11}F_{11} - \varepsilon^2 B_{12}F_{21})$ and $(A_{22} - B_{22}F_{22} - \varepsilon^2 B_{21}F_{12})$ have no eigenvalues in common (Section 5.1). The transformation is given by

$$
\begin{bmatrix} \hat{\eta}_1(n) \\ \hat{\eta}_2(n) \end{bmatrix} = \begin{bmatrix} I_{n_1} - \varepsilon^2 LH & -\varepsilon L \\ \varepsilon H & I_{n_2} \end{bmatrix} \begin{bmatrix} \hat{x}_1(n) \\ \hat{x}_2(n) \end{bmatrix} = T_1^{-1} \begin{bmatrix} \hat{x}_1(n) \\ \hat{x}_2(n) \end{bmatrix}
\tag{6.64}
$$

with

$$
T_1 = \begin{bmatrix} I_{n_1} & \varepsilon L \\ -\varepsilon H & I_{n_2} - \varepsilon^2 HL \end{bmatrix}
\tag{6.65}
$$

where matrices L and H satisfy equations

$$
L(a_{22} + \varepsilon H a_{12}) - (a_{11} - \varepsilon^2 a_{12}H)L + a_{12} = 0
\tag{6.66}
$$

$$
Ha_{11} - a_{22}H + a_{21} - \varepsilon^2 Ha_{12}H = 0
\tag{6.67}
$$

with

$$a_{11} = A_{11} - B_{11}F_{11} - \varepsilon^2 B_{12}F_{21}$$
$$a_{12} = A_{12} - B_{12}F_{12} - B_{12}F_{22}$$
$$a_{21} = A_{21} - B_{21}F_{11} - B_{22}F_{21}$$
$$a_{22} = A_{22} - B_{22}F_{22} - \varepsilon^2 B_{21}F_{12}$$

$$(6.68)$$

The optimal feedback control, expressed in the new coordinates, has the form

$$u_1(n) = -f_{11}\hat{\eta}_1(n) - \varepsilon f_{12}\hat{\eta}_2(n)$$
$$u_2(n) = -\varepsilon f_{21}\hat{\eta}_1(n) - f_{22}\hat{\eta}_2(n)$$

$$(6.69)$$

with

$$\hat{\eta}_1(n+1) = \alpha_1\hat{\eta}_1(n) + \beta_{11}v_1(n) + \varepsilon\beta_{12}v_2(n)$$
$$\hat{\eta}_2(n+1) = \alpha_2\hat{\eta}_2(n) + \varepsilon\beta_{21}v_1(n) + \beta_{22}v_2(n)$$

$$(6.70)$$

where

$$f_{11} = F_{11} - \varepsilon^2 F_{12}H, \quad f_{12} = F_{12} + (F_{11} - \varepsilon^2 F_{12}H)L$$
$$f_{21} = F_{21} - F_{22}H, \quad f_{22} = F_{22} + \varepsilon^2(F_{21} - F_{22}H)L$$
$$\alpha_1 = a_{11} - \varepsilon^2 a_{12}H, \quad \alpha_2 = a_{22} + \varepsilon^2 Ha_{12}$$
$$\beta_{11} = K_{11} - \varepsilon^2 L(H + K_{21}), \quad \beta_{12} = K_{12} - LK_{22} - \varepsilon^2 LHK_{12}$$
$$\beta_{21} = HK_{11} + K_{21}, \quad \beta_{22} = K_{22} + \varepsilon^2 HK_{12}$$

$$(6.71)$$

The innovation processes $v_1(n)$ and $v_2(n)$ are now given by

$$v_1(n) = y_1(n) - d_{11}\hat{\eta}_1(n) - \varepsilon d_{12}\hat{\eta}_2(n)$$
$$v_2(n) = y_2(n) - \varepsilon d_{21}\hat{\eta}_1(n) - d_{22}\hat{\eta}_2(n)$$

$$(6.72)$$

where

$$d_{11} = C_{11} - \varepsilon^2 C_{12}H, \quad d_{12} = C_{11}L + C_{12} - \varepsilon^2 C_{12}HL$$
$$d_{21} = C_{21} - C_{22}H, \quad d_{22} = C_{22} + \varepsilon^2(C_{21} - C_{22}H)L$$

$$(6.73)$$

Approximate control laws are defined by perturbing coefficients F_{ij}, K_{ij}, i, $j = 1, 2$; L and H by $O(\varepsilon^k)$, $k = 1, 2, \ldots$, in other words by using kth approximations for these coefficients, where k stands for the required order of accuracy, that is

$$u_1^{(k)}(n) = -f_{11}^{(k)}\hat{\eta}_1^{(k)}(n) - \varepsilon f_{12}^{(k)}\hat{\eta}_2^{(k)}(n)$$
$$u_2^{(k)}(n) = -\varepsilon f_{21}^{(k)}\hat{\eta}_1^{(k)}(n) - f_{22}^{(k)}\hat{\eta}_2^{(k)}(n)$$

$$(6.74)$$

with

$$\hat{\eta}_1^{(k)}(n+1) = \alpha_1^{(k)}\hat{\eta}_1^{(k)}(n) + \beta_{11}^{(k)}v_1^{(k)}(n) + \varepsilon\beta_{12}^{(k)}v_2^{(k)}(n)$$
$$\hat{\eta}_2^{(k)}(n+1) = \alpha_2^{(k)}\hat{\eta}_2^{(k)}(n) + \varepsilon\beta_{21}^{(k)}v_1^{(k)}(n) + \beta_{22}^{(k)}v_2^{(k)}(n)$$

(6.75)

where

$$v_1^{(k)}(n) = y_1(n) - d_{11}^k\hat{\eta}_1^{(k)}(n) - \varepsilon d_{12}^{(k)}\hat{\eta}_2^{(k)}(n)$$
$$v_2^{(k)}(n) = y_2(n) - \varepsilon d_{21}^k\hat{\eta}_1^{(k)}(n) - d_{22}^{(k)}\hat{\eta}_2^{(k)}(n)$$

(6.76)

and

$$f_{ij}^{(k)} = f_{ij} + O(\varepsilon^k), \quad d_{ij}^{(k)} = d_{ij} + O(\varepsilon^k)$$
$$\alpha_{ij}^{(k)} = \alpha_{ij} + O(\varepsilon^k), \quad \beta_{ij}^{(k)} = \beta_{ij} + O(\varepsilon^k)$$
$$i, j = 1, 2$$

(6.77)

The approximate values of $J^{(k)}$ are obtained from the following expression

$$J^{(k)} = \frac{1}{2}E\left\{\sum_{n=0}^{\infty}\left[x^{(k)^\mathrm{T}}(n)D^\mathrm{T}Dx^{(k)}(n) + u^{(k)^\mathrm{T}}(n)Ru^{(k)}(n)\right]\right\}$$
$$= \frac{1}{2}\mathrm{tr}\left\{D^\mathrm{T}Dq_{11}^{(k)} + f^{(k)^\mathrm{T}}Rf^{(k)}q_{22}^{(k)}\right\}$$

(6.78)

where

$$q_{11}^{(k)} = \mathrm{Var}\left\{\left(x_1^{(k)}\ x_2^{(k)}\right)^\mathrm{T}\right\} \quad \text{and} \quad q_{22}^{(k)} = \mathrm{Var}\left\{\left(\hat{\eta}_1^{(k)}\ \hat{\eta}_2^{(k)}\right)^\mathrm{T}\right\}$$
$$u^{(k)} = \begin{bmatrix} u_1^{(k)}(n) \\ u_2^{(k)}(n) \end{bmatrix}, \quad f^{(k)} = \begin{bmatrix} f_{11}^{(k)} & \varepsilon f_{12}^{(k)} \\ \varepsilon f_{21}^{(k)} & f_{22}^{(k)} \end{bmatrix}$$

(6.79)

The quantities $q_{11}^{(k)}$ and $q_{22}^{(k)}$ can be obtained by studying the variance equation of the following system driven by white noise

$$\begin{bmatrix} x^{(k)}(n+1) \\ \hat{\eta}^{(k)}(n+1) \end{bmatrix} = \begin{bmatrix} A & -Bf^{(k)} \\ \beta^{(k)}C & \alpha^{(k)} - \beta^{(k)}d^{(k)} \end{bmatrix}\begin{bmatrix} x^{(k)}(n) \\ \hat{\eta}^{(k)}(n) \end{bmatrix} + \begin{bmatrix} G & 0 \\ 0 & \beta^{(k)} \end{bmatrix}\begin{bmatrix} w(n) \\ v(n) \end{bmatrix}$$

(6.80)

where

$$\alpha^{(k)} = \begin{bmatrix} \alpha_1^{(k)} & 0 \\ 0 & \alpha_2^{(k)} \end{bmatrix}, \quad \beta^{(k)} = \begin{bmatrix} \beta_{11}^{(k)} & \varepsilon\beta_{12}^{(k)} \\ \varepsilon\beta_{21}^{(k)} & \beta_{22}^{(k)} \end{bmatrix}, \quad d^{(k)} = \begin{bmatrix} d_{11}^{(k)} & \varepsilon d_{12}^{(k)} \\ \varepsilon d_{21}^{(k)} & d_{22}^{(k)} \end{bmatrix}$$

(6.81)

Equation 6.80 can be represented in the composite form

$$\Gamma^{(k)}(n+1) = \Lambda^{(k)}\Gamma^{(k)}(n) + \Pi^{(k)}w(n) \tag{6.82}$$

with obvious definitions for $\Lambda^{(k)}$, $\Pi^{(k)}$, $\Gamma^{(k)}(n)$, and $w(n)$. The variance of $\Gamma^{(k)}(n)$ at steady state denoted by $q^{(k)}$, is given by the discrete algebraic Lyapunov equation (Gajić and Qureshi 1995)

$$q^{(k)}(n+1) = \Lambda^{(k)}q^{(k)}\Lambda^{(k)^\mathrm{T}} + \Pi^{(k)}\bar{W}\Pi^{(k)^\mathrm{T}}, \quad \bar{W} = \mathrm{diag}(W, V) \tag{6.83}$$

with $q^{(k)}$ partitioned as

$$q^{(k)} = \begin{bmatrix} q_{11}^{(k)} & q_{12}^{(k)} \\ q_{12}^{(k)^\mathrm{T}} & q_{22}^{(k)} \end{bmatrix} \tag{6.84}$$

On the other hand, the optimal value of J has the very well-known form (Kwakernaak and Sivan 1972)

$$J^{\mathrm{opt}} = \frac{1}{2}\mathrm{tr}\left[D^\mathrm{T}DQ + PK(CQC^\mathrm{T} + V)K^\mathrm{T}\right] \tag{6.85}$$

where P, Q, F, and K are obtained from Equations 6.50 through 6.53.

The near-optimality of the proposed approximate control law (Equation 6.74) is established in the following theorem.

THEOREM 6.4

Let $x_1(n)$ and $x_2(n)$ be optimal trajectories and J be the optimal value of the performance criterion. Let $x_1^{(k)}(n)$, $x_2^{(k)}(n)$, and $J^{(k)}$ be corresponding quantities under the approximate control law $u^{(k)}(n)$ given by Equation 6.74. Under the condition stated in Assumptions 6.2a and 6.6 and the stabilizability–detectability subsystem assumptions imposed on the subsystems, the following holds

$$J^{\mathrm{opt}} - J^{(k)} = O(\varepsilon^k)$$
$$\mathrm{Var}\left\{x_i(n) - x_i^{(k)}(n)\right\} = O(\varepsilon^{k2}), \quad i = 1, 2, \quad k = 0, 1, 2, \ldots \tag{6.86}$$

The proof of this theorem is rather lengthy and is therefore omitted here. It follows the ideas of Theorems 1 and 2 from Khalil and Gajić (1984) obtained for another class of small parameter problems—singularly perturbed systems. These two theorems were proved in the context of weakly coupled linear systems in Shen and Gajić (1990a). In addition, due to the discrete nature of the problem, the proof for this theorem, utilizes a bilinear transformation from Power (1967) which transforms the discrete Lyapunov equation into the continuous one and compares it with the corresponding equation under the optimal control law. More about the proof can be found in Shen (1990).

6.2.1 Case Study: Distillation Column

A real-world physical example, a fifth-order distillation column control problem (Kautsky et al. 1985), demonstrates the efficiency of the proposed method. The problem matrices A and B are

$$A = 10^{-3} \begin{bmatrix} 989.50 & 5.6382 & 0.2589 & 0.0125 & 0.0006 \\ 117.25 & 814.50 & 76.038 & 5.5526 & 0.3700 \\ 8.7680 & 123.87 & 750.20 & 107.96 & 11.245 \\ 0.9108 & 17.991 & 183.81 & 668.34 & 150.78 \\ 0.0179 & 0.3172 & 1.6974 & 13.298 & 985.19 \end{bmatrix}$$

$$B^{\mathrm{T}} = 10^{-3} \begin{bmatrix} 0.0192 & 6.0733 & 8.2911 & 9.1965 & 0.7025 \\ -0.0013 & -0.6192 & -13.339 & -18.442 & -1.4252 \end{bmatrix}$$

These matrices are obtained from Kautsky et al. (1985) by performing a discretization with the sampling rate $\Delta T = 0.1$.

Remaining matrices are chosen as

$$C = \begin{bmatrix} 1 & 1 & 0 & 0 & 0 \\ 0 & 0 & 1 & 1 & 1 \end{bmatrix}, \quad Q = I_5, \quad R = I_2$$

It is assumed that $G = B$, and that the white noise intensity matrices are given by

$$W_1 = 1, \quad W_2 = 2, \quad V_1 = 0.1, \quad V_2 = 0.1$$

The simulation results are presented in Table 6.2.

In practice, how the problem matrices are partitioned will determine the choice of the coupling parameter which in turn determines the rate of convergence and the domain of attraction of the iterative scheme to the optimal solution. It is desirable to

TABLE 6.2
Approximate Values for Criterion

k	$J^{(k)}$	$J^{(k)} - J$
0	0.80528×10^{-2}	0.6989×10^{-3}
1	0.75977×10^{-2}	0.2438×10^{-3}
2	0.74277×10^{-2}	0.7380×10^{-4}
4	0.73887×10^{-2}	0.3480×10^{-4}
6	0.73546×10^{-2}	0.5000×10^{-6}
8	0.73539×10^{-2}	$< 1.000 \times 10^{-7}$
Optimal	0.73539×10^{-2}	

get as small as possible a value of the small coupling parameter. This will speed up the convergence process. However, the small parameter is built into the problem and one cannot go beyond the physical limits. The small weak coupling parameter ε can be roughly estimated from the strongest coupled matrix—in this case matrix B. Apparently, the strongest coupling is in the third row, that is

$$\varepsilon = \frac{b_{31}}{b_{32}} = \frac{8.2911}{13.339} \approx 0.62$$

It can be seen that despite the relatively big value of the coupling parameter $\varepsilon = 0.62$, we have very rapid convergence to the optimal solution.

In summary, the near-optimum (up to any desired accuracy) steady-state regulators are obtained for stochastic linear weakly coupled discrete systems. The proposed method reduces considerably the size of required off-line and online computations since it introduces full parallelism in the design procedure.

Exercise 6.1: It is well known that the initial condition of the optimal Kalman filter has to be set to the mean value of the system initial state. Derive an expression for the optimal variance of the estimation error in the case when this condition is not satisfied. Consider both the continuous- and discrete-time domains. Hint: It is easier to solve this problem in the discrete-time domain.

6.3 STOCHASTIC OUTPUT FEEDBACK OF DISCRETE SYSTEMS

The problem of designing optimal controllers for linear systems with a limited number of output measurements available for control implementation was an area of active research for many years (see, e.g., Halyo and Broussard 1981; Qureshi et al. 1992 and references therein). The problem is defined as one in which the design engineer does not have a full set of state variables directly available for feedback purposes. The control engineers in such cases have two options: either to build the Kalman filter (or Luenberger observer) or to use the output feedback control. Very often it is not desirable to feedback all state variables in a complex system such as an aircraft. The design of the Kalman filter requires a dynamic system of the same order as the system under consideration (to be built or run on a computer as a software program). This might be costly. The output feedback control as the other alternative is more convenient from the point of view of implementation.

This chapter develops the recursive reduced-order parallel algorithm for the solution of the static output feedback control problem of quasi-weakly coupled (Skatarić et al. 1993) discrete stochastic linear systems by following the work of Hogan and Gajić (1994).

A discrete stochastic linear system is given by

$$x(k + 1) = Ax(k) + Bu(k) + Gw(k) \tag{6.87}$$

$$y(k) = Cx(k) + v(k) \tag{6.88}$$

where

$x(k) \in \Re^n$ is the state vector

$u(k) \in \Re^m$ is the control input

$y(k) \in \Re^r$ is the measured output

$w(k) \in \Re^s$ and $v(k) \in \Re^r$ are stationary uncorrelated zero-mean Gaussian white noise stochastic processes with intensities $W > 0$ and $V > 0$, respectively

The matrices A, B, G, and C are constant matrices of compatible dimensions. With Equations 6.87 and 6.88, consider the performance criterion

$$J = E\left\{ \sum_{k=0}^{\infty} \left[x(k)^{\mathrm{T}} Q x(k) + u^{\mathrm{T}}(k) R u(k) \right] \right\} \qquad (6.89)$$

with positive definite R and positive semidefinite Q, which has to be minimized. In addition, the control input is constrained to

$$u(k) = Fy(k) = FCx(k) + Fv(k) \qquad (6.90)$$

The optimal solution to this control problem has been obtained in terms of high-order nonlinear matrix algebraic equations (Halyo and Broussard 1981). The optimal feedback gain is given by

$$F = -(R + B^{\mathrm{T}} LB)^{-1} B^{\mathrm{T}} LAPC^{\mathrm{T}} (CPC^{\mathrm{T}} + V)^{-1} \qquad (6.91)$$

where P and L satisfy

$$P = (A + BFC)P(A + BFC)^{\mathrm{T}} + BFVF^{\mathrm{T}} B^{\mathrm{T}} + GWG^{\mathrm{T}} \qquad (6.92)$$

$$L = (A + BFC)^{\mathrm{T}} L(A + BFC) + C^{\mathrm{T}} F^{\mathrm{T}} RFC + Q \qquad (6.93)$$

The average value of the optimal performance criterion can be obtained from

$$J = tr\left[Q + C^{\mathrm{T}} F^{\mathrm{T}} RFC)P \right] + tr[F^{\mathrm{T}} RFV] \qquad (6.94)$$

It is shown in Halyo and Broussard (1981) that the following algorithm proposed for the numerical solution of Equations 6.91 through 6.93 converges to a local minimum under nonrestrictive conditions.

ALGORITHM 6.1

Choose F such that

$$A + B^{F(0)} C \qquad (6.95)$$

is stable and solve iteratively for $i = 0, 1, 2, \ldots$

$$P^{(i+1)} = \left(A + BF^{(i)}C\right)P^{(i+1)}\left(A + BF^{(i)}C\right)^{\mathrm{T}}$$
$$+ BF^{(i)}VF^{(i)^{\mathrm{T}}}B^{\mathrm{T}} + GWG^{\mathrm{T}} \tag{6.96}$$

$$L^{(i+1)} = \left(A + BF^{(i)}C\right)^{\mathrm{T}}L^{(i+1)}\left(A + BF^{(i)}C\right) + C^{\mathrm{T}}F^{(i)^{\mathrm{T}}}RF^{(i)}C + Q \tag{6.97}$$

$$F^{(i+1)}_{\mathrm{new}} = -\left(R + B^{\mathrm{T}}L^{(i+1)}B\right)^{-1}B^{\mathrm{T}}L^{(i+1)}AP^{(i+1)}C^{\mathrm{T}} \times \left(CP^{(i+1)}C^{\mathrm{T}} + V\right)^{-1} \tag{6.98}$$

$$F^{(i+1)} = F^{(i)} + \alpha_i\left(F^{(i+1)}_{\mathrm{new}} - F^{(i)}\right) \tag{6.99}$$

The parameter $\alpha_i \in (0, 1]$ is chosen at each iteration to ensure that the minimum is not overshot. That is

$$J_{i+1} = \mathrm{tr}\left[\left(Q + C^{\mathrm{T}}F^{(i+1)^{\mathrm{T}}}_{\mathrm{new}}RF^{(i+1)}_{\mathrm{new}}C\right)P^{(i+1)}\right]$$
$$+ \mathrm{tr}\left[F^{(i+1)^{\mathrm{T}}}_{\mathrm{new}}RF^{(i+1)}_{\mathrm{new}}V\right]$$
$$< J_i = \mathrm{tr}\left[\left(Q + C^{\mathrm{T}}F^{(i)^{\mathrm{T}}}_{\mathrm{new}}RF^{(i)}_{\mathrm{new}}C\right)P^{(i)}\right] + \mathrm{tr}\left[F^{(i)^{\mathrm{T}}}_{\mathrm{new}}RF^{(i)}_{\mathrm{new}}V\right] \tag{6.100}$$

In the following, we show that in the cases of quasi-weakly coupled linear systems, Equations 6.92 and 6.93 can be decomposed into six reduced-order Lyapunov equations to get the parallel algorithm (Bertsekas and Tsitsiklis 1991) with arbitrary order of accuracy.

6.3.1 OUTPUT FEEDBACK OF QUASI-WEAKLY COUPLED LINEAR DISCRETE SYSTEMS

A very efficient parallel reduced-order algorithm that decomposes a high-order system into a low-order system for the case of the quasi-weakly coupled discrete stochastic output feedback control problem is derived in this section. The low-order system is represented by six Lyapunov equations that may be solved in parallel to reduce computational time. The required solution can be easily obtained up to an arbitrary order of accuracy, $O(\varepsilon^{2k})$, where ε is a small weak coupling parameter and k represents the number of iterations. The efficiency of the proposed method is demonstrated on two real aircraft that possess the quasi-weakly coupled structure under the assumption of a prescribed degree of stability. The aircraft are inherently nonweakly coupled systems, but since they require a high degree of stability, we will demonstrate on two real aircraft that the prescribed degree of stability assumption makes them quasi-weakly coupled systems.

The general weakly coupled discrete stochastic control system was studied in Shen and Gajić (1990c). The system is defined as

$$x_1(k + 1) = A_1x_1(k) + \varepsilon A_2x_2(k) + B_1u_1(k) + \varepsilon B_2u_2(k)$$
$$+ G_1w_1(k) + \varepsilon G_4w_2(k)$$
$$x_2(k + 1) = \varepsilon A_3x_1(k) + A_4x_2(k) + \varepsilon B_3u_1(k) + B_4u_2(k)$$
$$+ \varepsilon G_3w_1(k) + G_2w_2(k) \tag{6.101}$$

with corresponding measurements

$$y_1(k) = C_1 x_1(k) + \varepsilon C_2 x_2(k) + v_1(k)$$
$$y_3(k) = \varepsilon C_3 x_1(k) + C_4 x_2(k) + v_2(k)$$

(6.102)

where

$x_i(k) \in \Re^{n_i}$, $i = 1, 2$, are state variables
$u_i(k) \in \Re^{m_i}$, $i = 1, 2$, are control inputs
$y_i(k) \in \Re^{r_i}$, $i = 1, 2$, are measured outputs
$w_i(k) \in \Re^{s_i}$, $i = 1, 2$, and $v_i(k) \in \Re^{r_i}$, $i = 1, 2$, are stationary uncorrelated Gaussian zero-mean white noise stochastic processes with intensities $W_i > 0$ and $V_i > 0$, respectively
ε is a small coupling parameter

The performance criterion for weakly coupled discrete systems which has to be minimized is given by

$$J = E \left\{ \sum_{k=0}^{\infty} \begin{bmatrix} x_1(k) \\ x_2(k) \end{bmatrix}^{\mathrm{T}} Q \begin{bmatrix} x_1(k) \\ x_2(k) \end{bmatrix} + u^{\mathrm{T}}(k) R u(k) \right\}$$

(6.103)

where matrices Q and R are partitioned as

$$Q = \begin{bmatrix} Q_1 & \varepsilon Q_2 \\ \varepsilon Q_2^{\mathrm{T}} & Q_3 \end{bmatrix} \geq 0, \quad R = \begin{bmatrix} R_1 & 0 \\ 0 & R_2 \end{bmatrix} > 0$$

(6.104)

All problem matrices are constant matrices of compatible dimensions.

The quasi-weakly coupled linear stochastic discrete system differs from Equations 6.101 and 6.102. It can be obtained from Equations 6.87 and 6.88 with (Skatarić et al. 1993)

$$A = \begin{bmatrix} A_1 & \varepsilon A_2 \\ \varepsilon A_3 & A_4 \end{bmatrix}, \quad B = \begin{bmatrix} 0 \\ B_2 \end{bmatrix}, \quad G = \begin{bmatrix} 0 \\ G_2 \end{bmatrix}, \quad C = [0 \quad C_2]$$

(6.105)

Many real physical systems such as aircraft, flexible space structures, power systems, and chemical reactors possess the quasi-weakly coupled structure.

It is assumed that the standard weak coupling Assumption 6.2a is satisfied. In addition, we will need an assumption about the order of matrices involved in the problem formulation.

Assumption 6.7 A_{ij}, $i, j = 1, 2$, B_2, G_2, C_2, Q_l, $l = 1, 2, 3$, and R_1, R_2 are $O(1)$.

In this case, the algebraic equations (Equations 6.92 and 6.93), comprising the solution to the stochastic output feedback control problem of discrete linear systems, can be decomposed subject to Equations 6.104 and 6.105 as follows.

Partition matrices $A + BF^{(i)}C$, $BF^{(i)}VF^{(i)^T}B^T + GWG^T$, and $P^{(i)}$ as

$$A + BF^{(i)}C = \begin{bmatrix} D_1^{(i)} & \varepsilon D_2^{(i)} \\ \varepsilon D_3^{(i)} & D_4^{(i)} \end{bmatrix} \tag{6.106}$$

$$BF^{(i)}VF^{(i)^T}B^T + GWG^T = \begin{bmatrix} 0 & 0 \\ 0 & S_3^{(i)} \end{bmatrix} \tag{6.107}$$

$$P^{(i)} = \begin{bmatrix} P_1^{(i)} & \varepsilon P_2^{(i)} \\ \varepsilon P_2^{(i)^T} & P_3^{(i)} \end{bmatrix} \tag{6.108}$$

where

$$D_1^{(i)} = A_1, \quad D_2^{(i)} = A_2, \quad D_3^{(i)} = A_3, \quad D_4^{(i)} = A_4 + B_2 F^{(i)} C_2$$
$$S_3^{(i)} = B_2 F^{(i)} VF^{(i)^T} B_2^T + G_2 WG_2^T \tag{6.109}$$

Using these partitions to expand Equation 6.96, the following equations are obtained:

$$P_1^{(i+1)} = D_1^{(i)} P_1^{(i+1)} D_1^{(i)^T}$$
$$+ \varepsilon^2 \left[D_2^{(i)} P_2^{(i+1)^T} D_1^{(i)^T} + D_1^{(i)} P_2^{(i+1)} D_2^{(i)^T} + D_2^{(i)} P_3^{(i+1)} D_2^{(i)^T} \right] \tag{6.110}$$

$$P_2^{(i+1)} = D_1^{(i)} P_2^{(i+1)} D_4^{(i)^T} + D_1^{(i)} P_1^{(i+1)} D_3^{(i)^T}$$
$$+ D_2^{(i)} P_3^{(i+1)} D_4^{(i)^T} + \varepsilon^2 D_2^{(i)} P_2^{(i+1)^T} D_3^{(i)^T} \tag{6.111}$$

$$P_3^{(i+1)} = D_4^{(i)} P_3^{(i+1)} D_4^{(i)^T} + S_3$$
$$+ \varepsilon^2 \left[D_3^{(i)} P_1^{(i+1)} D_3^{(i)^T} + D_4^{(i)} P_2^{(i+1)^T} D_3^{(i)^T} + D_3^{(i)} P_2^{(i+1)} D_4^{(i)^T} \right] \tag{6.112}$$

Let us define $O(\varepsilon^2)$ approximations of Equations 6.110 through 6.112

$$\mathbf{P}_1^{(i+1)} = D_1^{(i)} \mathbf{P}_1^{(i+1)} D_1^{(i)^T} \tag{6.113}$$

$$\mathbf{P}_2^{(i+1)} = D_1^{(i)} \mathbf{P}_2^{(i+1)} D_4^{(i)^T} + D_2^{(i)} \mathbf{P}_3^{(i+1)} D_4^{(i)^T} + D_1^{(i)} \mathbf{P}_1^{(i+1)} D_3^{(i)^T} \tag{6.114}$$

$$\mathbf{P}_3^{(i+1)} = D_4^{(i)} \mathbf{P}_3^{(i+1)} D_4^{(i)^T} + S_3 \tag{6.115}$$

Assuming that the matrix $D_1^{(i)}$ is stable (Harkara et al. 1989), Equation 6.113 has the solution, $\mathbf{P}_1^{(i+1)} = 0$. As long as the subsystem matrix $D_4^{(i)}$ is stable, Equations 6.114 and 6.115 can be solved sequentially for $\mathbf{P}_3^{(i+1)}$ and $\mathbf{P}_2^{(i+1)}$, respectively.

Defining the errors as

$$P_m^{(i+1)} = \mathbf{P_m}^{(i+1)} + \varepsilon^2 E_m, \quad m = 1, 2, 3 \tag{6.116}$$

then, subtracting Equations 6.113 through 6.115 from Equations 6.110 through 6.112 and doing some algebra, the error equations are obtained:

$$E_1 = D_1^{(i)} E_1 D_1^{(i)^T} + \left[D_2^{(i)} P_2^{(i+1)^T} D_1^{(i)^T} + D_1^{(i)} P_2^{(i+1)} D_2^{(i)^T} + D_2^{(i)} P_3^{(i+1)} D_2^{(i)^T} \right] \quad (6.117)$$

$$E_2 = D_1^{(i)} E_2 D_4^{(i)^T} + D_1^{(i)} E_1 D_3^{(i)^T} + D_2^{(i)} E_3 D_4^{(i)^T} + D_2^{(i)} P_2^{(i+1)^T} D_3^{(i)^T} \quad (6.118)$$

$$E_3 = D_4^{(i)} E_3 D_4^{(i)^T} + \left[D_3^{(i)} P_1^{(i+1)} D_3^{(i)^T} + D_4^{(i)} P_2^{(i+1)^T} D_3^{(i)^T} + D_3^{(i)} P_2^{(i+1)} D_4^{(i)^T} \right] \quad (6.119)$$

The above equations can be solved efficiently by proposing the following reduced-order parallel synchronous algorithm in the spirit of those studied in Bertsekas and Tsitsiklis (1991).

ALGORITHM 6.2

$$E_1^{(j+1)} = D_1^{(i)} E_1^{(j+1)} D_1^{(i)^T} + D_1^{(i)} \left(\mathbf{P}_1^{(i+1)} + \varepsilon^2 E_2^{(j)} \right) D_2^{(i)^T}$$
$$+ D_2^{(i)} \left(\mathbf{P}_2^{(i+1)} + \varepsilon^2 E_2^{(i)} \right)^T D_1^{(i)^T} + D_2^{(i)} \left(\mathbf{P}_3^{(i+1)} + \varepsilon^2 E_3^{(i)} \right) D_2^{(i)^T} \quad (6.120)$$

$$E_2^{(j+1)} = D_1^{(i)} E_2^{(j+1)} D_4^{(i)^T} + D_1^{(i)} E_1^{(i)} D_3^{(i)^T}$$
$$+ D_2^{(i)} E_3^{(j)} D_4^{(i)^T} + D_2^{(i)} \left(\mathbf{P}_2^{(i+1)} + \varepsilon^2 E_2^{(j)} \right)^T D_3^{(i)^T} \quad (6.121)$$

$$E_3^{(j+1)} = D_4^{(i)} E_3^{(j+1)} D_4^{(i)^T} + D_3^{(i)} \left(\mathbf{P}_1^{(i+1)} + \varepsilon^2 E_1^{(j)} \right) D_3^{(i)^T}$$
$$+ D_4^{(i)} \left(\mathbf{P}_2^{(i+1)} + \varepsilon^2 E_2^{(j)} \right)^T D_3^{(i)^T} + D_3^{(i)} \left(\mathbf{P}_2^{(i+1)} + \varepsilon^2 E_2^{(j)} \right) D_4^{(i)^T} \quad (6.122)$$

Initial conditions were chosen as $E_1^{(0)} = 0$, $E_2^{(0)} = 0$, $E_3^{(0)} = 0$.

The following theorem presents the features of the proposed algorithm.

THEOREM 6.5

Algorithm 6.2 converges to the required solutions E_1, E_2, and E_3 with the rate of convergence of $O(\varepsilon^2)$.

Proof Let $e_m^{(j)} = E_m^{(j)} - E_m^{(j-1)}$, $m = 1, 2, 3$, then from Equations 6.120 through 6.122 it follows that

$$D_1^{(i)} e_1^{(j+1)} D_1^{(i)^T} - e_1^{(j+1)} = -\varepsilon^2 \left[D_2^{(i)} e_2^{(j)} D_1^{(i)^T} + D_1^{(i)} e_2^{(j)} D_2^{(i)^T} + D_2^{(i)} e_3^{(j)} D_2^{(i)^T} \right] \quad (6.123)$$

$$D_1^{(i)} e_2^{(j+1)} D_4^{(i)^T} - e_2^{(j+1)} = -\varepsilon^2 D_2^{(i)} e_2^{(j)} D_3^{(i)^T} - D_2^{(i)} e_3^{(j)} D_4^{(i)^T} - D_1^{(i)} e_1^{(j)} D_3^{(i)^T} \quad (6.124)$$

$$D_4^{(i)} e_3^{(j+1)} D_4^{(i)^T} - e_3^{(j+1)} = -\varepsilon^2 \left[D_3^{(i)} e_1^{(j)} D_3^{(i)^T} + D_4^{(i)} e_2^{(j)^T} D_3^{(i)^T} + D_3^{(i)} e_2^{(j)} D_4^{(i)^T} \right] \quad (6.125)$$

By stability assumption imposed on $D_1^{(i)}$ and $D_4^{(i)}$, we have from Equations 6.123 and 6.125

$$\left\| e_m^{(j)} \right\| = O(\varepsilon^2), \quad m = 1, 3 \tag{6.126a}$$

and using results of Equation 6.126a in Equation 6.124 we get

$$\left\| e_2^{(j)} \right\| = O(\varepsilon^2) \tag{6.126b}$$

or

$$\left\| E_m^{(j)} - E_m^{(j-1)} \right\| = O(\varepsilon^2), \quad m = 1, 2, 3; \quad j = 1, 2, \ldots \tag{6.127}$$

Continuing the same procedure, it follows by analogy that

$$\left\| E_m^{(j)} - E_m \right\| = O(\varepsilon^{2j}), \quad m = 1, 2, 3; \quad j = 1, 2, \ldots \tag{6.128}$$

Thus, the proposed algorithm is convergent.

Using $E_m^{(\infty)}$, $m = 1, 2, 3$, in Equations 6.120 through 6.122 and comparing to Equations 6.117 through 6.119 implies that the algorithm (Equations 6.120 through 6.122) converges to the unique solution of Equations 6.117 through 6.119. There-fore, $P^{(i+1)}$ can be solved iteratively with an arbitrary order of accuracy.

Similarly, the lower order decomposition can be obtained for Equation 6.97. Partitioning matrices as follows

$$C^{\mathrm{T}} F^{(i)^{\mathrm{T}}} R F^{(i)} C + Q = \begin{bmatrix} q_1^{(i)} & 0 \\ 0 & q_3^{(i)} \end{bmatrix}, \quad L^{(i)} = \begin{bmatrix} L_1^{(i)} & \varepsilon L_2^{(i)} \\ \varepsilon L_2^{(i)^{\mathrm{T}}} & L_3^{(i)} \end{bmatrix} \tag{6.129}$$

where

$$q_1^{(i)} = Q_1, \quad q_3^{(i)} = C_2^{\mathrm{T}} F^{(i)^{\mathrm{T}}} R F^{(i)} C_2 + Q_3 \tag{6.130}$$

Using these partitions to expand Equation 6.97, the following equations are obtained:

$$\begin{aligned} L_1^{(i+1)} = {}& D_1^{(i)^{\mathrm{T}}} L_1^{(i+1)} D_1^{(i)} + q_1^{(i)} \\ & + \varepsilon^2 \left[D_3^{(i)^{\mathrm{T}}} L_2^{(i+1)^{\mathrm{T}}} D_1^{(i)} + D_1^{(i)^{\mathrm{T}}} L_2^{(i+1)} D_3^{(i)} + D_3^{(i)^{\mathrm{T}}} L_3^{(i+1)} D_3^{(i)} \right] \end{aligned} \tag{6.131}$$

$$\begin{aligned} L_2^{(i+1)} = {}& D_1^{(i)^{\mathrm{T}}} L_2^{(i+1)} D_4^{(i)^{\mathrm{T}}} + D_1^{(i)^{\mathrm{T}}} L_1^{(i+1)} D_2^{(i)} \\ & + D_3^{(i)^{\mathrm{T}}} L_3^{(i+1)} D_4^{(i)} + \varepsilon^2 D_3^{(i)^{\mathrm{T}}} L_2^{(i+1)^{\mathrm{T}}} D_2^{(i)} \end{aligned} \tag{6.132}$$

$$\begin{aligned} L_3^{(i+1)} = {}& D_4^{(i)^{\mathrm{T}}} L_3^{(i+1)} D_4^{(i)} + q_3^{(i)} \\ & + \varepsilon^2 \left[D_2^{(i)^{\mathrm{T}}} L_1^{(i+1)} D_2^{(i)} + D_4^{(i)^{\mathrm{T}}} L_2^{(i+1)^{\mathrm{T}}} D_2^{(i)} + D_2^{(i)^{\mathrm{T}}} L_2^{(i+1)} D_4^{(i)} \right] \end{aligned} \tag{6.133}$$

Let us define $O(\varepsilon^2)$ approximations of Equations 6.131 through 6.133 as

$$\mathbf{L}_1^{(i+1)} = D_1^{(i)^T} \mathbf{L}_1^{(i+1)} D_1^{(i)} + q_1^{(i)} \tag{6.134}$$

$$\mathbf{L}_2^{(i+1)} = D_1^{(i)^T} \mathbf{L}_2^{(i+1)} D_4^{(i)} + D_1^{(i)^T} \mathbf{L}_1^{(i+1)} D_2^{(i)} + D_3^{(i)^T} \mathbf{L}_3^{(i+1)} D_4^{(i)} \tag{6.135}$$

$$\mathbf{L}_3^{(i+1)} = D_4^{(i)^T} \mathbf{L}_3^{(i+1)} D_4^{(i)} + q_3^{(i)} \tag{6.136}$$

Defining the approximation errors as

$$L_m^{(i+1)} = \mathbf{L}_m^{(i+1)} + \varepsilon^2 Z_m, \quad m = 1, 2, 3 \tag{6.137}$$

the following error equations are obtained from Equations 6.131 through 6.133 and Equations 6.134 through 6.137:

$$Z_1 = D_1^{(i)^T} Z_1 D_1^{(i)} + D_3^{(i)^T} L_2^{(i+1)^T} D_1^{(i)} + D_1^{(i)^T} L_2^{(i+1)} D_3^{(i)} + D_3^{(i)^T} L_3^{(i+1)} D_3^{(i)} \tag{6.138}$$

$$Z_2 = D_1^{(i)^T} Z_2 D_4^{(i)} + D_1^{(i)^T} Z_1 D_2^{(i)} + D_3^{(i)^T} Z_3 D_4^{(i)} + D_3^{(i)^T} L_2^{(i+1)^T} D_2^{(i)} \tag{6.139}$$

$$Z_3 = D_4^{(i)^T} Z_3 D_4^{(i)} + D_2^{(i)^T} L_1^{(i+1)} D_2^{(i)} + D_4^{(i)^T} L_2^{(i+1)^T} D_2^{(i)} + D_2^{(i)^T} L_2^{(i+1)} D_4^{(i)} \tag{6.140}$$

The above equations can be solved by proposing a similar kind of algorithm as Equations 6.120 through 6.122. ■

ALGORITHM 6.3

Initialize $Z_1^{(0)} = 0$, $Z_2^{(0)} = 0$, $Z_3^{(0)} = 0$ and calculate the following equations iteratively:

$$Z_1^{(j+1)} D_1^{(i)^T} Z_1^{(j+1)} D_1^{(i)} = D_3^{(i)^T} \left(L_2^{(i+1)} + \varepsilon^2 Z_2^{(j)} \right)^T D_1^{(i)}$$
$$+ D_1^{(i)^T} \left(L_2^{(i+1)} + \varepsilon^2 Z_2^{(j)} \right) D_3^{(i)} = D_3^{(i)^T} \left(L_3^{(i+1)} + \varepsilon^2 Z_3^{(j)} \right) D_3^{(i)} \tag{6.141}$$

$$Z_2^{(j+1)} - D_1^{(i)^T} Z_2^{(j+1)} D_4^{(i)} = D_1^{(i)^T} Z_1^{(j)} D_2^{(i)}$$
$$+ D_3^{(i)^T} Z_3^{(j)} D_4^{(i)} + D_3^{(i)^T} \left(L_2^{(i+1)} + \varepsilon^2 Z_2^{(j)} \right)^T D_2^{(i)} \tag{6.142}$$

$$Z_3^{(j+1)} - D_4^{(i)^T} Z_3^{(j+1)} D_4^{(i)} = D_2^{(i)^T} \left(L_1^{(i+1)} + \varepsilon^2 Z_1^{(j)} \right) D_2^{(i)}$$
$$+ D_4^{(i)^T} \left(L_2^{(i+1)} + \varepsilon^2 Z_2^{(j)} \right)^T D_2^{(i)} + D_2^{(i)^T} \left(L_2^{(i+1)} + \varepsilon^2 Z_2^{(j)^T} \right) D_4^{(i)} \tag{6.143}$$

Algorithm 6.3 (Equations 6.141 through 6.143) has the same properties of the algorithm 6.2 (Equations 6.120 through 6.122); so we have the following theorem.

THEOREM 6.6

Algorithm 6.3 converges to the solutions of Z_m, $m = 1, 2, 3$, with the rate of convergence of $O(\varepsilon^2)$, that is

$$\begin{aligned}
\left\| Z_m^{(j)} - Z_m^{(j-1)} \right\| &= O(\varepsilon^2) \\
\left\| Z_m^{(j)} - Z_m \right\| &= O(\varepsilon^{2j})
\end{aligned} \tag{6.144}$$

The proof of Theorem 6.6 is identical to the proof of Theorem 6.5 and is thus omitted.

To summarize, P can be found from Equations 6.113 through 6.115 and Equations 6.120 through 6.122, and L can be computed from Equations 6.134 through 6.136 and Equations 6.141 through 6.143. Since these algorithms are independent of one another, the computation can be done in parallel that leads to six reduced-order Lyapunov equations.

In the remaining part of this section, we modify the quasi-weakly coupled algorithm for the case when the zero elements of matrices B, G, and C, defined in Equation 6.105, are replaced by $O(\varepsilon)$ quantities. This structure results in the process of discretization of the continuous-time quasi-weakly coupled systems. Namely, these matrices are given by

$$B = \begin{bmatrix} \varepsilon B_1 \\ B_2 \end{bmatrix}, \quad G = \begin{bmatrix} \varepsilon G_1 \\ G_2 \end{bmatrix}, \quad C = [\varepsilon C_1 \; C_2] \tag{6.145}$$

The introduced changes will produce the following modifications

$$BF^{(i)} V F^{(i)^{\mathrm{T}}} B^{\mathrm{T}} + GWG^{\mathrm{T}} = \begin{bmatrix} \varepsilon^2 S_1^{(i)} & \varepsilon S_2^{(i)} \\ \varepsilon S_2^{(i)^{\mathrm{T}}} & S_3^{(i)} \end{bmatrix} \tag{6.146}$$

with

$$\begin{aligned}
S_1^{(i)} &= B_1 F^{(i)} V F^{(i)^{\mathrm{T}}} B_1^{\mathrm{T}} + G_1 W G_1^{\mathrm{T}} \\
S_2^{(i)} &= B_1 F^{(i)} V F^{(i)^{\mathrm{T}}} B_2^{\mathrm{T}} + G_1 W G_2^{\mathrm{T}}
\end{aligned} \tag{6.147}$$

Also the matrices $D_m^{(i)}$, $1, 2, 3$, are changed into

$$\begin{aligned}
D_1^{(i)} &= A_1 + \varepsilon^2 B_1 F^{(i)} C_1, \quad D_2^{(i)} = A_2 + B_1 F^{(i)} C_2 \\
D_3^{(i)} &= A_3 + B_2 F^{(i)} C_1
\end{aligned} \tag{6.148}$$

Using these partitions to expand Equation 6.96, the following variations are obtained for Equations 6.110 and 6.111:

$$P_1^{(i+1)} = D_1^{(i)} P_1^{(i+1)} D_1^{(i)^T} + \varepsilon^2 S_1$$
$$+ \varepsilon^2 \left[D_2^{(i)} P_2^{(i+1)^T} D_1^{(i)^T} + D_1^{(i)} P_2^{(i+1)} D_2^{(i)^T} + D_2^{(i)} P_3^{(i+1)} D_2^{(i)^T} \right] \tag{6.149}$$

$$P_2^{(i+1)} = S_2 + D_1^{(i)} P_2^{(i+1)} D_4^{(i)^T} + D_1^{(i)} P_1^{(i+1)} D_3^{(i)^T}$$
$$+ D_2^{(i)} P_3^{(i+1)} D_4^{(i)^T} + \varepsilon^2 D_2^{(i)} P_2^{(i+1)^T} D_3^{(i)^T} \tag{6.150}$$

Equation 6.112 remains unchanged.

By perturbing Equations 6.149 and 6.150 and Equation 6.112 by an $O(\varepsilon^2)$, the zeroth-order approximation can be found to be the same as in Equations 6.113 and 6.115 except for Equation 6.114 which becomes

$$\mathbf{P_2^{(i+1)}} - D_1^{(i)} \mathbf{P_2^{(i+1)}} D_4^{(i)^T} = S_2 + D_2^{(i)} \mathbf{P_3^{(i+1)}} D_4^{(i)^T} + D_1^{(i)} \mathbf{P_1^{(i+1)}} D_3^{(i)^T} \tag{6.151}$$

Corresponding equations for the error terms will differ only in equation for E_1 which now has the form

$$E_1 = D_1^{(i)} E_1 D_1^{(i)^T} + D_2^{(i)} P_2^{(i+1)^T} D_1^{(i)^T}$$
$$+ D_1^{(i)} P_2^{(i+1)} D_2^{(i)^T} + D_2^{(i)} P_3^{(i+1)} D_2^{(i)^T} + S_1 \tag{6.152}$$

As a consequence, Equation 6.120 in the algorithm (Equations 6.120 through 6.122) has to be modified to

$$E_1^{(j+1)} - D_1^{(i)} E_1^{(j+1)} D_1^{(i)^T} = D_2^{(i)} \left[\mathbf{P_2^{(i+1)}} + \varepsilon^2 E_2^{(j)} \right]^T D_1^{(i)^T}$$
$$+ D_1^{(i)} \left[\mathbf{P_2^{(i+1)}} + \varepsilon^2 E_2^{(j)} \right] D_2^{(i)^T} + D_2^{(i)} \left[\mathbf{P_3^{(i+1)}} + \varepsilon^2 E_2^{(j)} \right] D_2^{(i)^T} + S_1 \tag{6.153}$$

Similarly, the lower order decomposition can be obtained for L equations. The new matrix partition structures defined in Equation 6.145 will produce

$$C^T F^{(i)^T} R F^{(i)} C + Q = \begin{bmatrix} q_1^{(i)} & \varepsilon q_2^{(i)} \\ \varepsilon q_2^{(i)^T} & q_3^{(i)} \end{bmatrix} \tag{6.154}$$

with

$$q_1^{(i)} = Q_1 + \varepsilon^2 C_1^T F^{(i)^T} R F^{(i)} C_1, \quad q_2^{(i)} = C_1^T F^{(i)^T} R F^{(i)} C_2 \tag{6.155}$$

which will change Equation 6.132 to

$$L_2^{(i+1)} - D_1^{(i)^T} L_2^{(i+1)^T} D_4^{(i)} = D_1^{(i)^T} L_1^{(i+1)} D_2^{(i)} + D_3^{(i)^T} L_3^{(i+1)} D_4^{(i)}$$
$$+ \varepsilon^2 D_3^{(i)^T} L_2^{(i+1)^T} D_2^{(i)} + q_2^{(i)} \tag{6.156}$$

so that the corresponding equation in the zeroth-order approximation (Equation 6.132) becomes

$$\mathbf{L}_2^{(i+1)} = D_1^{(i)^T} \mathbf{L}_2^{(i+1)^T} D_4^{(i)} + D_1^{(i)^T} \mathbf{L}_1^{(i+1)} D_2^{(i)} + D_3^{(i)^T} \mathbf{L}_3^{(i+1)} D_4^{(i)} + q_2^{(i)} \qquad (6.157)$$

However, the error equations are found to be the same equations as Equations 6.138 through 6.140. These equations can be solved by the same algorithm as Equations 6.141 through 6.143.

6.3.2   CASE STUDIES: FLIGHT CONTROL SYSTEMS FOR AIRCRAFT

Case Study 1

In order to demonstrate the efficiency of the proposed algorithm we ran a sixth-order example of a flight control system of an aircraft (Shapiro et al. 1981). The problem matrices are defined as follows:

$$A = \begin{bmatrix} -0.746 & 0.387 & -12.9 & 6.05 & 0.952 & 0 \\ 0.024 & -0.174 & 0.4 & -0.416 & -1.76 & 0 \\ 0.006 & -0.999 & -0.058 & -0.0012 & 0.0092 & 0.0369 \\ 0 & 0 & 0 & -5 & 0 & 0 \\ 0 & 0 & 0 & 0 & -10 & 0 \\ 1 & 0 & 0 & 0 & 0 & 0 \end{bmatrix}$$

$$B^T = \begin{bmatrix} 0 & 0 & 0 & 10 & 0 & 0 \\ 0 & 0 & 0 & 0 & 20 & 0 \end{bmatrix}$$

The remaining matrices are chosen as

$$C = \begin{bmatrix} 0 & 0 & 0 & 0 & 1 & 0 \\ 0 & 0 & 0 & 0 & 0 & 1 \end{bmatrix}$$

and

$$Q = I_6, \quad R = I_2, \quad G = B, \quad V = I_2, \quad W = 1$$

The continuous-time matrices above were multiplied by a permutation matrix to obtain the form used in this section. The system was then discretized with the sampling period of 0.1. Before the system was discretized, a prescribed degree of stability factor (Anderson and Moore 1990) of $\sigma = 10$ was introduced so that the matrix $A + \sigma I$ would become diagonally dominant and thus weakly coupled. The discretized matrices are

$$A_D = \begin{bmatrix} 0.341 & 0.036 & -0.45 & -0.001 & 0.019 & 0.168 \\ 0.001 & 0.354 & 0.155 & 0 & -0.04 & -0.012 \\ 0 & -0.036 & 0.358 & 0.001 & 0.003 & 0.001 \\ 0.035 & 0.001 & -0.023 & 0.368 & 0.001 & 0.001 \\ 0 & 0 & 0 & 0 & 0.135 & 0 \\ 0 & 0 & 0 & 0 & 0 & 0.223 \end{bmatrix}$$

$$B_D^T = \begin{bmatrix} 0.035 & 0.07 & 0.003 & 0.001 & 0.865 & 0 \\ 0.135 & -0.009 & 0 & 0.004 & 0 & 0.518 \end{bmatrix}$$

The remaining matrices are the same in discrete-time as they are in continuous-time.

The obtained results are presented in Table 6.3. The required number of iterations needed for good approximations for P and L was determined from the requirements

$$\left\| E_m^{(i+1)} \right\| - \left\| E_m^i \right\| \le 10^{-7}, \quad \left\| Z_m^{(i+1)} \right\| - \left\| Z_m^i \right\| \le 10^{-7}, \quad m = 1, 2, 3$$

For this example, the typical number of iterations needed to achieve such approximations was 6.

It can be seen from Table 6.3 that the results of the global algorithm (Equations 6.96 and 6.97) and proposed reduced-order algorithm agree up to four decimal digits at each iteration. The additional advantages (uniqueness and choice of a good initial guess) of the reduced-order algorithms for the output feedback control problem were discussed in Harkara et al. (1989). Parameters α and ε are given as $\alpha = 0.1$ and $\varepsilon = 0.3$.

TABLE 6.3

Optimal and Reduced Criteria
per Iteration

i	$J_{red}^{(i)} = J_{opt}^{(i)}$
1	1.0988
2	1.0979
3	1.0971
4	1.0964
5	1.0958
10	1.0938
20	1.0922
30	1.0917
40	1.0915
46	1.0914
	$J_{opt} = 1.0914$

Case Study 2

The second numerical example is an aircraft system considered in Anderson and Moore (1990). The problem matrices are shown below:

$$A = \begin{bmatrix} -0.28 \times 10^{-8} & 0 & 0 & 0 & 0 & 0 \\ 0 & -6.76 \times 10^{-3} & 0 & 0 & 0 & 0 \\ 0 & 0 & -0.122 & 1.57 & 0 & 0 \\ 0 & 0 & -1.57 & -0.122 & 0 & 0 \\ 0 & 0 & 0 & 0 & -0.071 & 21.3 \\ 0 & 0 & 0 & 0 & -21.3 & -0.71 \end{bmatrix}$$

$$B = G = \begin{bmatrix} -4.06 \times 10^{-4} & -1.65 \times 10^{-7} \\ -2.20 \times 10^{-1} & 7.61 \times 10^{-5} \\ 8.84 \times 10^{-2} & 3.05 \times 10^{-2} \\ -3.08 \times 10^{-1} & -1.05 \times 10^{-2} \\ 6.39 \times 10^{-2} & 4.46 \times 10^{-3} \\ -1.08 & -1.02 \times 10^{-2} \end{bmatrix}$$

$$C = 10^{-3} \begin{bmatrix} -0.35 & -157 & -156 & 4.61 & -337 & -0.24 \\ -0.205 & -154 & 281 & -98.5 & -1020 & -76.8 \end{bmatrix}$$

The remaining matrices are chosen as

$$Q = I_6, \quad R = I_2, \quad V = I_2, \quad W = 1$$

The continuous-time matrices above were multiplied by a permutation matrix to obtain the form used in Section 6.3.1. The system was then discretized with the sampling period of 0.1. Before the system was discretized, a prescribed degree of stability factor of $\sigma = I$ was introduced so that the matrix $A + \sigma I$ would become diagonally dominant and thus weakly coupled. The discretized matrices are

$$A_D = \begin{bmatrix} 0.905 & 0 & 0 & 0 & 0 & 0 \\ 0 & 0.904 & 0 & 0 & 0 & 0 \\ 0 & 0 & 0.883 & 0.14 & 0 & 0 \\ 0 & 0 & -0.14 & 0.883 & 0 & 0 \\ 0 & 0 & 0 & 0 & 0.447 & 0.714 \\ 0 & 0 & 0 & 0 & -0.714 & 0.447 \end{bmatrix}$$

$$B_D = \begin{bmatrix} -0.386 \times 10^{-4} & -0.157 \times 10^{-7} \\ -0.209 \times 10^{-1} & 0.724 \times 10^{-5} \\ 0.557 \times 10^{-2} & 0.280 \times 10^{-2} \\ -0.364 \times 10^{-1} & -0.121 \times 10^{-2} \\ -0.675 \times 10^{-1} & -0.488 \times 10^{-3} \\ -0.460 \times 10^{-1} & -0.684 \times 10^{-3} \end{bmatrix}$$

The remaining matrices are the same in both continuous and discrete time.

TABLE 6.4

Optimal and Reduced Criterion per Iteration

i	$J_{red}^{(i)}$	$J_{opt}^{(i)}$
1	1.1053	1.1127
2	0.81667	0.81802
3	0.62366	0.62313
4	0.48724	0.48624
5	0.38720	0.38613
10	0.14571	0.14533
20	0.04558	0.04555
30	0.03379	0.03348
40	0.03235	0.03235
50	0.03217	0.03218
71	0.03215	0.03215
J_{opt}	0.03215	0.03215

The obtained results are shown in Table 6.4. Parameters α and ε are given as $\alpha = 0.1$ and $\varepsilon = 0.3$. The number of iterations for approximating P and L are chosen according to the criterion given in the previous example.

6.4 OPTIMAL CONTROL OF STOCHASTIC JUMP PARAMETER LINEAR SYSTEMS

Linear systems subject to random parameter changes have been studied as a special class of hybrid stochastic systems. The linear-quadratic optimal control problem with Markovian random changes in parameters of the system was solved in Sworder (1969) and has since seen addressed by a number of researchers (see, e.g., Mariton 1990). The weakly coupled jump parameter optimal linear-quadratic control problem was considered for the first time in Borno and Gajić (1995). Our presentation in this section is based closely on this work.

Weakly coupled jump parameter linear stochastic system composed of two subsystems is defined by

$$\dot{x}_1(t) = A_1(\rho)x_1(t) + \varepsilon A_2(\rho)x_2(t) + B_1(\rho)u_1(t) + \varepsilon B_2(\rho)u_2(t)$$
$$\dot{x}_2(t) = \varepsilon A_3(\rho)x_1(t) + A_4(\rho)x_2(t) + \varepsilon B_3(\rho)u_1(t) + B_4(\rho)u_2(t)$$

(6.158)

where
  $x_i(t) \in \Re^{n_i}$, $u_i(t) \in \Re^{r_i}$, $i = 1, 2$, are the system state variables and system control inputs, respectively
  ε is a small weak coupling parameter
  ρ represents the mode of the system that takes on values in a discrete set $\Psi = \{1, 2, \ldots, N\}$

Matrices A_j, B_j, are mode dependent real matrices of appropriate dimensions.

In this section, we focus on the case where the random mode changes are Markovian. The probability distribution of the Markov chain is determined by the differential equation

$$\dot{\theta}(t) = \theta(t)\Pi \tag{6.159}$$

where
$\theta(t)$ is an N-dimensional row vector of unconditional probabilities
Π is the transition rate matrix given by

$$\Pi = \begin{bmatrix} \pi_{11} & \pi_{12} & \cdots & \pi_{1N} \\ \pi_{21} & \pi_{22} & \cdots & \pi_{2N} \\ \vdots & \vdots & \vdots & \vdots \\ \pi_{N1} & \pi_{N2} & \cdots & \pi_{NN} \end{bmatrix} \tag{6.160}$$

The matrix Π has the property that $\pi_{ij} \geq 0$, $i \neq j$, i, $j = 1, 2, \ldots, N$ and $\pi_{ii} = -\sum_{j=1, j \neq i}^{N} \pi_{ij}$, $i = 1, 2, \ldots, N$ (Kumar and Varaiya 1986).

A quadratic performance criterion in the form of the conditional expectation is associated with the system

$$V = E\left\{ \int_{t_0}^{\infty} \left[x^T(t)Q(\rho)x(t) + u^T(t)R(\rho)u(t) \right] dt \Big| x_0, j_0, t_0 \right\} \tag{6.161}$$

where $Q(\rho) \geq 0$ and $R(\rho) > 0$ are mode-dependent weighting symmetric matrices. The quantities x_0, j_0, and t_0 denote the initial state, initial mode, and initial time, respectively. The optimal control law for the above defined stochastic problem is given by (Sworder 1969)

$$u_{\text{opt}}(x(t)) = -R_j^{-1}B_j^T P_j x(t) \tag{6.162}$$

where
j indicates that the system is in mode $\rho = j$
$P_j's$, $j = 1, 2, \ldots, N$ are the positive definite stabilizing solutions of a set of coupled algebraic Riccati equations

$$\tilde{A}_j^T P_j + P_j \tilde{A}_j + Q_j - P_j S_j P_j + \sum_{k=1, k \neq j}^{N} \pi_{jk} P_k = 0, \quad j = 1, 2, \ldots, N \tag{6.163}$$

where

$$\tilde{A}_j = A_j + \frac{1}{2}\pi_{jj}I, \quad S_j = B_j R_j^{-1} B_j^T \tag{6.164}$$

The existence of positive definite stabilizing solution of the coupled algebraic Riccati equations (Equations 6.163 and 6.164) was established in Ji and Chizeck (1990) under the following assumption.

Assumption 6.8 The pairs $(A_j, \mathrm{Chol}(Q_j))$, $j = 1, 2, \ldots, N$ are observable and (A_j, B_j), $j = 1, 2, \ldots, N$ are stochastically stabilizable.

For the weakly coupled structure, matrices Q_j and R_j are partitioned as

$$Q_j = \begin{bmatrix} Q_{j1} & \varepsilon Q_{j2} \\ \varepsilon Q_{j2}^T & Q_{j3} \end{bmatrix} \geq 0, \quad R_j = \begin{bmatrix} R_{j1} & 0 \\ 0 & R_{j2} \end{bmatrix} > 0 \tag{6.165}$$

The matrix S_j for weakly coupled systems is given by

$$S_j = \begin{bmatrix} S_{j1} + \varepsilon^2 Z_{j1} & \varepsilon S_{j2} \\ \varepsilon S_{j2}^T & S_{j3} + \varepsilon^2 Z_{j3} \end{bmatrix} \tag{6.166}$$

where

$$\begin{aligned} S_{j1} &= B_{j1} R_{j1}^{-1} B_{j1}^T, \quad S_{j2} = B_{j1} R_{j1}^{-1} B_{j3}^T + B_{j2} R_{j2}^{-1} B_{j4}^T \\ S_{j3} &= B_{j4} R_{j2}^{-1} B_{j4}^T, \quad Z_{j1} = B_{j2} R_{j2}^{-1} B_{j2}^T, \quad Z_{j3} = B_{j3} R_{j1}^{-1} B_{j3}^T \end{aligned} \tag{6.167}$$

The solutions P_j, $j = 1, 2, \ldots, N$, also have the weakly coupled structure

$$P_j = \begin{bmatrix} P_{j1} & \varepsilon P_{j2} \\ \varepsilon P_{j2}^T & P_{j3} \end{bmatrix} \tag{6.168}$$

under the following assumption.

Assumption 6.9 All matrices involved in the problem formulation are $O(1)$.

Using the above partitioning consistent with the weakly coupled structure, the set of coupled algebraic Riccati equations can be written as

$$\begin{aligned} & \tilde{A}_{j1}^T P_{j1} + P_{j1} \tilde{A}_{j1} + Q_{j1} - P_{j1} S_{j1} P_{j1} + \sum_{k=1, k \neq j}^{N} \pi_{jk} P_{k_1} \\ & - \varepsilon^2 \left(P_{j2} S_{j2}^T P_{j1} + P_{j1} Z_{j1} P_{j1} + P_{j1} S_{j2} P_{j2}^T + P_{j2} S_{j3} P_{j2}^T \right) \\ & + \varepsilon^2 \left(A_{j3}^T P_{j2}^T + P_{j2} A_{j3} \right) - \varepsilon^4 P_{j2} Z_{j3} P_{j2}^T = 0, \quad j = 1, 2, \ldots, N \end{aligned} \tag{6.169}$$

$$\tilde{A}_{j4}^T P_{j3} + P_{j3}\tilde{A}_{j4} + Q_{j3} - P_{j3}S_{j3}P_{j3} + \sum_{k=1,\,k\neq j}^{N} \pi_{jk}P_{k3}$$

$$+ \varepsilon^2 \left[A_{j2}^T P_{j2} + P_{j2}A_{j2} - \left(P_{j2}^T S_{j2} + P_{j3}Z_{j3} \right) P_{j3} \right]$$

$$- \varepsilon^2 \left(P_{j3}S_{j2}^T + P_{j2}^T \left(S_{j1} + \varepsilon^2 Z_{j1} \right) \right) P_{j2} = 0, \quad j = 1, 2, \ldots, N \quad (6.170)$$

$$\tilde{A}_{j1}^T P_{j2} + P_{j2}\tilde{A}_{j4} + Q_{j2} + \sum_{k=1,\,k\neq j}^{N} \pi_{jk}P_{k2} + A_{j3}^T P_{j3} + P_{j1}A_{j2}$$

$$- P_{j2}S_{j3}P_{j3} - P_{j1}S_{j1}P_{j2} - P_{j1}S_{j2}P_{j3}$$

$$- \varepsilon^2 \left[\left(P_{j2}S_{j2}^T + P_{j1}Z_{j1} \right) P_{j2} + P_{j2}Z_{j3}P_{j3} \right] = 0, \quad j = 1, 2, \ldots, N \quad (6.171)$$

where $\tilde{A}_{jl} = A_{jl} + 0.5\pi_{jj}\, I$, $l = 1, 4$.

The zeroth-order approximation can be obtained from the above set of three matrix algebraic equations by setting $\varepsilon = 0$, which leads to

$$\tilde{A}_{j1}^T P_j^{(0)} + P_{j1}^{(0)}\tilde{A}_{j1} + Q_{j1} - P_{j1}^{(0)}S_{j1}P_{j1}^{(0)} + \sum_{k=1,\,k\neq j}^{N} \pi_{jk}P_{k_1}^{(0)} = 0, \quad j = 1, 2, \ldots, N$$

$$(6.172)$$

$$\tilde{A}_{j4}^T P_{j3}^{(0)} + P_{j3}^{(0)}\tilde{A}_{j4} + Q_{j3} - P_{j3}^{(0)}S_{j3}P_{j3}^{(0)} + \sum_{k=1,\,k\neq j}^{N} \pi_{jk}P_{k_3}^{(0)} = 0 \quad j = 1, 2, \ldots, N \quad (6.173)$$

$$\tilde{A}_{j1}^T P_{j2}^{(0)} + P_{j2}^{(0)}\tilde{A}_{j4} + Q_{j2} + \sum_{k=1,\,k\neq j}^{N} \pi_{jk}P_{k2}^{(0)} + A_{j3}^T P_{j3}^{(0)} + P_{j1}^{(0)}A_{j2}$$

$$- P_{j2}^{(0)}\,S_{j3}P_{j3}^{(0)} - P_{j1}^{(0)}S_{j1}P_{j2}^{(0)} - P_{j1}^{(0)}S_{j2}P_{j3}^{(0)} = 0, \quad j = 1, 2, \ldots, N \quad (6.174)$$

This system of algebraic equations can be solved as follows. Solve Equations 6.172 and 6.173 as independent systems of coupled algebraic Riccati equations corresponding to each subsystem. A standard algorithm for solving such equations can be found in Gajić and Borno (1995). The unique positive definite stabilizing solutions of Equations 6.172 and 6.173 exist under the following assumption.

Assumption 6.10 The subsystem pairs $(A_{j1}, \mathrm{Chol}(Q_{j1}))$ and $(A_{j4}, \mathrm{Chol}(Q_{j3}))$ $j = 1, 2, \ldots, N$, are observable and the pairs (A_{j1}, B_{j1}) and (A_{j4}, B_{j4}), $j = 1, 2, \ldots, N$, are stochastically stabilizable.

Having found solutions for $P_{j1}^{(0)}$ and $P_{j3}^{(0)}$, the solution for $P_{j2}^{(0)}$ can be obtained from Sylvester algebraic equations:

$$\tilde{A}_{j4}^T P_{j2}^{(0)} + P_{j2}^{(0)}\tilde{A}_{j4} + \hat{Q}_{j2} + \sum_{k=1,\,k\neq j}^{N} \pi_{jk}P_{k2}^{(0)} = 0 \quad (6.175)$$

$$j = 1, 2, \ldots, N$$

where

$$\tilde{A}_{j1} = \tilde{A}_{j1} - S_{j1}P_{j1}^{(0)}, \quad \tilde{A}_{j4} = \tilde{A}_{j4} - S_{j3}P_{j3}^{(0)}$$
$$\hat{Q}_{j2} = Q_{j2} + A_{j3}^{T}P_{j3}^{(0)} + P_{j1}^{(0)}A_{j2} - P_{j1}^{(0)}S_{j2}P_{j3}^{(0)}$$

(6.176)

The errors resulting from the above approximation by setting $\varepsilon = 0$ is $O(\varepsilon^2)$. Thus, we define the error equations as follows

$$P_{jl} = P_{jl}^{(0)} + \varepsilon^2 E_{jl}, \quad l = 1, 2, 3$$

(6.177)

where E_{jl} represents the error terms. Subtracting Equations 6.172 through 6.174 from Equations 6.169 through 6.171 and using Equation 6.177, after some algebra, we obtain the following error equations

$$\tilde{A}_{j1}^{T}E_{j1} + E_{j1}\tilde{A}_{j1} + \sum_{k=1,\,k=1}^{N} \pi_{jk}E_{k1} + W_{j1} = 0$$

(6.178)

$$\tilde{A}_{j4}^{T}E_{j3} + E_{j3}\tilde{A}_{j4} + \sum_{k=1,\,k\neq1}^{N} \pi_{jk}E_{k3} + W_{j3} = 0$$

(6.179)

$$\tilde{A}_{j1}^{T}E_{j2} + E_{j2}\tilde{A}_{j4} + E_{j1}\Delta_{j1} + \Delta_{j2}^{T}E_{j2} + \sum_{k=1,\,k=j}^{N} \pi_{jk}E_{k2} + W_{j2} = 0$$

(6.180)

with

$$\Delta_{j1} = A_{j2} - S_{j1}P_{j2}^{(0)} - S_{j2}P_{j3}^{(0)}$$
$$\Delta_{j2} = A_{j3} - S_{j2}^{T}P_{j1}^{(0)} - S_{j3}P_{j2}^{(0)^{T}}$$

(6.181)

$$W_{j1} = A_{j3}^{T}P_{j2}^{T} + -P_{j2}A_{j3} - P_{j2}S_{j1}^{T}P_{j1} - P_{j1}Z_{j1}P_{j1} - P_{j1}S_{j2}P_{j2}^{T}$$
$$\quad - P_{j2}S_{j3}P_{j2}^{T} - \varepsilon^2\left(P_{j2}Z_{j3}P_{j2}^{T} + E_{j1}S_{j1}E_{j1}\right)$$

$$W_{j2} = -\left(P_{j2}S_{j2}^{T} + P_{j1}Z_{j1}\right)P_{j2} - P_{j2}Z_{j3}P_{j3}$$
$$\quad - \varepsilon^2\left[E_{j1}(S_{j2}E_{j3} + S_{j1}E_{j2}) + E_{j2}S_{j3}E_{j3}\right]$$

(6.182)

$$W_{j3} = A_{j2}^{T}P_{j2} + P_{j2}^{T}A_{j2} - \left(P_{j3}S_{j2}^{T} + P_{j2}^{T}\right)(S_{j1} + \varepsilon^2 Z_{j1})P_{j2}$$
$$\quad - \left(P_{j2}^{T}S_{j2} + P_{j3}Z_{j3}\right)P_{j3} - \varepsilon^2 E_{j3}S_{j3}E_{j3}$$

We propose the following recursive reduced-order algorithm for finding the error terms.

ALGORITHM 6.4

$$\tilde{A}_{j1}^{T} E_{j1}^{(i+1)} + E_{j1}^{(i+1)} \tilde{A}_{j1} + \sum_{k=1,\,k\neq j}^{N} \pi_{jk} E_{k1}^{(i+1)} + W_{j1}^{(i)} = 0 \qquad (6.183)$$

$$\tilde{A}_{j4}^{T} E_{j3}^{(i+1)} + E_{j3}^{(i+1)} \tilde{A}_{j4} + \sum_{k=1,\,k\neq j}^{N} \pi_{jk} E_{k3}^{(i+1)} + W_{j3}^{(i)} = 0 \qquad (6.184)$$

$$\tilde{A}_{j1}^{T} E_{j2}^{(i+1)} + E_{j2}^{(i+1)} \tilde{A}_{j4} + \sum_{k=1,\,k\neq j}^{N} \pi_{jk} E_{k2}^{(i+1)} + E_{j1}^{(i+1)} \Delta_{j1} + \Delta_{j2}^{T} E_{j3}^{(i+1)} + W_{j2}^{(i)} = 0 \quad (6.185)$$

where

$$P_{jl}^{(i)} = P_{jl}^{(0)} + \varepsilon^2 E_{jl}^{(i)}, \quad l = 1, 2, 3 \qquad (6.186)$$

Quantities $W_{jl}^{(i)}$, $l = 1, 2, 3$, are obtained from Equation 6.182 by evaluating these formulas at the ith iteration. Equations 6.183 through 6.185 represent a parallel synchronous algorithm.

The many property of this algorithm is stated in Theorem 6.7.

THEOREM 6.7

Under established assumptions, Algorithm 6.4 converges with the rate of convergence of $O(\varepsilon^2)$.

Proof First, we establish the existence of a unique bounded solution for the system of algebraic Equations 6.178 through 6.180. This can be accomplished by showing that the Jacobian matrix of this system is nonsingular when $\varepsilon = 0$. It can be shown that

$$J = \begin{bmatrix} J_{11} & 0 & 0 \\ * & J_{22} & * \\ 0 & 0 & J_{33} \end{bmatrix} \qquad (6.187)$$

where * denotes terms not important for nonsingularity of the Jacobian, and

$$J_{11} = \begin{bmatrix} L_{11} & \pi_{12}I & \pi_{13}I & \cdots & \pi_{1N}I \\ \pi_{21}I & L_{21} & \pi_{23}I & \cdots & \pi_{2N}I \\ \vdots & \vdots & \vdots & \vdots & \vdots \\ \pi_{N1}I & \pi_{N2}I & \cdots & \pi_{N,N-1}I & L_{N1} \end{bmatrix} \qquad (6.188)$$

$$
J_{22} = \begin{bmatrix}
L_{12} & \pi_{12}I & \pi_{13}I & \cdots & \pi_{1N}I \\
\pi_{21}I & L_{22} & \pi_{23}I & \cdots & \pi_{2N}I \\
\vdots & \vdots & \vdots & \vdots & \vdots \\
\pi_{N1}I & \pi_{N2}I & \cdots & \pi_{N,N-1}I & L_{N2}
\end{bmatrix}
\tag{6.189}
$$

$$
J_{33} = \begin{bmatrix}
L_{13} & \pi_{12}I & \pi_{13}I & \cdots & \pi_{1N}I \\
\pi_{21}I & L_{22} & \pi_{23}I & \cdots & \pi_{2N}I \\
\vdots & \vdots & \vdots & \vdots & \vdots \\
\pi_{N1}I & \pi_{N2}I & \cdots & \pi_{N,N-1}I & L_{N3}
\end{bmatrix}
\tag{6.190}
$$

where

$$
L_{j1} = 1 \otimes \hat{A}_{j1}^{\mathrm{T}} + \hat{A}_{j1}^{\mathrm{T}} \otimes I
$$

$$
L_{j2} = 1 \otimes \hat{A}_{j1}^{\mathrm{T}} + \hat{A}_{j4}^{\mathrm{T}} \otimes I
\tag{6.191}
$$

$$
L_{j3} = 1 \otimes \hat{A}_{j4}^{\mathrm{T}} + \hat{A}_{j4}^{\mathrm{T}} \otimes I
$$

$j = 1, 2, \ldots, N$, and $\otimes$ denotes the Kronecker product. The Jacobian given by Equation 6.187 is nonsingular if its diagonal blocks J_{ll}, $l = 1, 2, 3$ are nonsingular. The matrix J_{11} is the Kronecker product representation of the coefficient matrices of a coupled set of Lyapunov equations of the form (Mariton 1990)

$$
\tilde{A}_{j1}^{\mathrm{T}} Y_j + Y_j \tilde{A}_{j1}^{\mathrm{T}} + \sum_{k=1, k \neq j}^{N} \pi_{jk} Y_k = -N_j, \quad j = 1, 2, \ldots, N
\tag{6.192}
$$

where N_j are positive definite matrices. Matrices $\hat{A}_{j1}$, $j = 1, 2, \ldots, N$, are stable by Assumption 6.10 (consequence of stochastic stabilizability [Ji and Chizeck 1990]), hence unique solutions Y_j, $j = 1, 2, \ldots, N$, exist for Equation 6.192. As a result, we conclude that J_{11} is nonsingular. Similarly, it can be shown that J_{22} and J_{33} are nonsingular. Thus, the Jacobian is nonsingular. Furthermore, by the implicit function theorem (Ortega and Rheinboldt 2000) for sufficiently small ε, there exists a unique bounded solution for the system of algebraic Equations 6.178 through 6.180.

Second, we have to prove convergence of the algorithm. We define the errors at each iteration as follows. For $i = 0$, subtracting Equation 6.183 from Equation 6.178 yields

$$
\tilde{A}_{j1}^{\mathrm{T}} \left(E_{j1} - E_{j1}^{(1)} \right) + \left(E_{j1} - E_{j1}^{(1)} \right) \tilde{A}_{j1}^{\mathrm{T}} + \sum_{k=1, k \neq j}^{N} \pi_{jk} \left(E_{k1} - E_{k1}^{(1)} \right)
$$
$$
= \varepsilon^2 g_{j1}(E_{k1}, E_{j1}, \varepsilon^2), \quad j = 1, 2, \ldots, N
\tag{6.193}
$$

This system of equations can be written as (Mariton 1990)

$$
J_{11} \left(E_1 - E_1^{(1)} \right) = \varepsilon^2 g(E_1, E_2, \varepsilon^2)
\tag{6.194}
$$

where $E_1 - E_1^{(1)}$ and g are column vectors formed by orderly stacking the columns of E_{j1} and $g_{j1}, j = 1, 2, \ldots, N$. Thus, by nonsingularity of J_{11}

$$\left\| E_{j1} - E_{j1}^{(1)} \right\| = O(\varepsilon^2) \tag{6.195}$$

Similarly, it can be shown that

$$\left\| E_{j3} - E_{j3}^{(1)} \right\| = O(\varepsilon^2) \tag{6.196}$$

$$\left\| E_{j2} - E_{j2}^{(1)} \right\| = O(\varepsilon^2) \tag{6.197}$$

Continuing the same procedure, it follows that

$$\left\| E_{j1} - E_{jl}^{(i)} \right\| = O(\varepsilon^{2i}), \quad l = 1, 2, 3 \tag{6.198}$$

Thus, for sufficiently small ε, the algorithm converges to the solutions of Equations 6.178 through 6.180, which proves Theorem 6.7. ∎

The approximative optimal control can be now obtained from

$$u_{\text{app}}^{(i)}(x(t)) = -R_j^{-1} B^{\mathrm{T}} P_j^{(i)} x(t) \tag{6.199}$$

where

$$P_j^{(i)} = \begin{bmatrix} P_{j1}^{(i)} & \varepsilon P_{j2}^{(i)} \\ \varepsilon P_{j2}^{(i)\mathrm{T}} & P_{j3}^{(i)} \end{bmatrix} \tag{6.200}$$

The next example verifies the result stated in Theorem 6.7.

Example 6.1

Consider a jump parameter linear stochastic system with two modes and the state and control matrices given by

$$A_1 = \begin{bmatrix} -2.7218 & 0.284 & 0.4865\varepsilon & 0.9092\varepsilon \\ 0.2173 & -2.7782 & 0.8977\varepsilon & 0.0606\varepsilon \\ 0.9047\varepsilon & 0.5163\varepsilon & -3.6659 & -0.6927 \\ 0.5045\varepsilon & 0.319\varepsilon & -1.2857 & -1.3341 \end{bmatrix}$$

$$B_1 = \begin{bmatrix} 3.473 & 0.6295\varepsilon \\ 2.2087 & 0.7254\varepsilon \\ 0.9995\varepsilon & 2.6657 \\ 0.2332\varepsilon & 1.5316 \end{bmatrix}$$

TABLE 6.5

Dependence of the Number of Iterations on ε

ε	Number of Iterations
0.01	1
0.1	2
0.5	5
0.9	7

$$A_2 = \begin{bmatrix} -3.0434 & 0.367 & 0.2771\varepsilon & 0.5297\varepsilon \\ 0.0541 & -3.4566 & 0.9138\varepsilon & 0.4644\varepsilon \\ 0.941\varepsilon & 0.7615\varepsilon & -2.3289 & -0.0959 \\ 0.0501\varepsilon & 0.7702\varepsilon & -0.7577 & -3.1711 \end{bmatrix}$$

$$B_1 = \begin{bmatrix} 1 & 0.351\varepsilon \\ 2.5664 & 0.5911\varepsilon \\ 0.846\varepsilon & 1.6483 \\ 0.8415\varepsilon & 1.0773 \end{bmatrix}$$

The weighted matrices are chosen as identities, that is, $R_1 = R_2 = I_2$ and $Q_1 = Q_2 = I_4$. The transition rate matrix is chosen as

$$\Pi = \begin{bmatrix} -3 & 3 \\ 2.5 & -2.5 \end{bmatrix}$$

In Table 6.5, we have presented the required number of iterations needed to achieve the accuracy $\max_j \left\| P_j - P_j^{(i)} \right\|_2 < 10^{-10}$, $j = 1, 2$, as a function of the weak coupling parameter ε.

6.5 COMMENTS

It is interesting to observe that weakly coupled linear stochastic systems are still an interesting research area (see Sagara et al. 2007), where stochastic H_∞ optimal control problem of weakly coupled linear systems with state-dependent noise has been considered. Since the initial work on linear-quadratic stochastic control problem of weakly coupled systems (Haddad and Cruz 1970), new methodologies have been developing for this class of stochastic control systems. Applications of weakly coupled systems to computer communication networks, including the Internet, seems to be an important and interesting future research area.

REFERENCES

Anderson, B. and J. Moore, *Optimal Control: Linear Quadratic Methods*, Prentice Hall, Englewood Cliffs, NJ, 1990.

Bertsekas, D. and J. Tsitsiklis, Some aspects of parallel and distributed algorithms—A survey, *Automatica*, 27, 3–21, 1991.

Borno, I. and Z. Gajić, Parallel algorithms for optimal control of weakly coupled and singularly perturbed jump linear systems, *Automatica*, 31, 85–988, 1995.

Gajić, Z. and I. Borno, Lyapunov iterations for optimal control of jump linear systems, *IEEE Transactions on Automatic Control*, AC-40, 1971–1975, 1995.

Gajić, Z. and M. Qureshi, *Lyapunov Matrix Equation in Systems Stability and Control*, Academic Press, San Diego, NJ, 1995.

Gajić, Z. and X. Shen, Decoupling transformation for weakly coupled linear systems, *International Journal of Control*, 50, 1517–1523, 1989.

Gajić, Z., D. Petkovski, and X. Shen, *Singularly Perturbed and Weakly Coupled Linear Control System—A Recursive Approach*, Springer-Verlag, New York, Lecture Notes in Control and Information Sciences, 140, 1990.

Geromel, J. and P. Pres, Decentralized load-frequency control, *Proceedings of IEE, Part D.*, 132, 225–230, 1985.

Haddad, A. and J. Cruz, ε-coupling for near-optimum design of large-scale linear systems, *Proceedings of IEE, Part D.*, 117, 223, 1970.

Halyo, N. and J. Broussard, A convergent algorithm for the stochastic infinite-time discrete optimal output feedback problem, *Proceedings of the American Control Conference*, Charlottesville, VA, WA-1E, 1981.

Harkara, N., D. Petkovski, and Z. Gajić, The recursive algorithm for the optimal static output feedback control problem of linear weakly coupled systems, *International Journal of Control*, 50, 1–11, 1989.

Hogan, S. and Z. Gajić, Stochastic output feedback control of quasi-weakly coupled linear discrete systems, *Control-Theory and Advanced Technology*, 10, 221–234, 1994.

Ji, Y. and H. Chizeck, Controllability, stabilizability, and continuous-time Markovian jump linear quadratic control, *IEEE Transactions on Automatic Control*, AC-35, 777–788, 1990.

Kautsky, J., N. Nichols, and P. Van Douren, Robust pole assignment in linear state feedback, *International Journal of Control*, 41, 1129–1155, 1985.

Khalil, H. and Z. Gajić, Near-optimum regulators for stochastic linear singularly perturbed systems, *IEEE Transactions on Automatic Control*, AC-29, 531–541, 1984.

Kokotović, P. and J. Cruz, An approximation theorem for linear optimal regulators, *Journal of Mathematical Analysis and Applications*, 27, 249–252, 1969.

Kumar, P. and P. Varaiya, *Stochastic Systems: Estimation, Identification, and Adaptive Control*, Prentice Hall, Upper Saddle River, NJ, 1986.

Kwakernaak, H. and R. Sivan, *Linear Optimal Control Systems*, Wiley-Interscience, New York, 1972.

Mariton, M., *Jump Linear Systems in Automatic Control*, Marcel-Dekker, New York, 1990.

Ortega, J. and W. Rheinboldt, *Iterative Solution of Nonlinear Equations in Several Variables*, SIAM Publishers, Philadelphia, PA, 2000.

Petrović, B. and Z. Gajić, The recursive solution of linear quadratic Nash games for weakly interconnected systems, *Journal of Optimization Theory and Applications*, 56, 463–477, 1988.

Power, H., Equivalence of Lyapunov matrix equations for continuous and discrete systems, *Electronic Letters*, 3, 83, 1967.

Qureshi, M., X. Shen, and Z. Gajić, Output feedback control of discrete linear singularly perturbed stochastic systems, *International Journal of Control*, 55, 361–371, 1992.

Sagara, M., H. Mukaidani, and T. Yamamoto, Stochastic H_∞ control problem with state-dependent noise for weakly coupled large-scale systems, *IEEJ Transactions of Electrical and Electronic Engineering*, 127, 571–878, 2007.

Shapiro, E., D. Fredericks, and R. Roony, Suboptimal constant output feed-back and its application to modern flight control system design, *International Journal of Control*, 33, 505–517, 1981.

Shen, X., *Near-Optimum Reduced-Order Stochastic Control of Linear Discrete and Continuous Systems with Small Parameters*, PhD dissertation, Rutgers University, NJ, 1990.

Shen, X. and Z. Gajić, Near-optimum steady state regulators for stochastic linear weakly coupled systems, *Automatica*, 26, 919–923, 1990a.

Shen, X. and Z. Gajić, Optimal reduced-order solution of the weakly coupled discrete Riccati equation, *IEEE Transactions on Automatic Control*, AC-35, 600–602, 1990b.

Shen, X. and Z. Gajić, Approximate parallel controllers for discrete weakly coupled linear stochastic systems, *Optimal Control Applications and Methods*, 11, 345–354, 1990c.

Skatarić, D., D. Arnautovic, and Z. Gajić, Reduced-order design of optimal controllers for quasi-weakly coupled linear control systems, *Control-Theory and Advanced Technology*, 9, 481–490, 1993.

Sworder, D., Feedback control for a class of linear systems with jump parameters, *IEEE Transactions on Automatic Control*, AC-14, 9–14, 1969.

7 Nash Differential Games

In this chapter, we present a parallel synchronous algorithm for solving the Nash differential game of weakly coupled linear systems. The results presented mostly follow the work of Petrović and Gajić (1988). In this chapter we limit our presentation to linear-quadratic Nash games (Engwerda 2005), whose optimal solution is given in terms of coupled algebraic matrix Riccati equations. The use of weakly coupled Nash differential games in stochastic set up has been recently considered in Huang et al. (2005). Interesting approaches to Nash differential games of linear weakly coupled systems have been recently developed in Mukaidani (2003, 2006a,b). Sign indefinite differential games of weakly coupled systems have been considered in Mukaidani (2007). Potential applications of the results reported in this chapter could be to mobile wireless communication networks, in which every mobile user is weakly coupled to all other mobile users that use the same communication channel (Koskie and Gajić 2005).

7.1 WEAKLY COUPLED LINEAR-QUADRATIC NASH GAMES

The linear-quadratic Nash game strategies of large-scale weakly interconnected (coupled) systems were studied for the first time in Ozguner and Perkins (1977) using a power-series expansion method with respect to a small coupling parameter ε. This approach, originated in Kokotović et al. (1969), is not recursive in its application and can be inferior compared to the hierarchical type decentralized control method (especially when ε is not very small), as was pointed out in Mahmoud (1978). In this section, we develop a recursive technique which will recover the importance of ideas presented in Kokotović et al. (1969). Motivated by previous results for singularly perturbed systems (Gajić 1986), we have shown that weak coupling produces algebraic problems similar to those of Gajić (1986) and the fixed point method used in Gajić (1986) is very efficient in this case also.

As a matter of fact, we have developed an algorithm which converges very rapidly to the exact, nonnegative definite stabilizing solution of the coupled algebraic Riccati equations of linear-quadratic Nash differential games and thus to the optimal linear Nash strategies, even in the case when ε is not small.

A controlled linear dynamic system under consideration is given by

$$\dot{x}(t) = A(\varepsilon)x(t) + B_1(\varepsilon)u_1(t) + B_2(\varepsilon)u_2(t) \tag{7.1}$$

where

  $x(t) \in R^n$ is a state vector
  $u_1(t) \in R^{m_1}$ and $u_2(t) \in R^{m_2}$ are control inputs
  $A(\varepsilon)$, $B_i(\varepsilon)$ $i = 1, 2$, are bounded matrix functions of a small parameter ε with
    compatible dimensions

Quadratic type functionals are associated with each control agent

$$J_1 = \frac{1}{2} \int_0^\infty \left[x^\mathrm{T}(t)Q_1(\varepsilon)x(t) + u_1^\mathrm{T}(t)R_1(\varepsilon)u_1(t) + u_2^\mathrm{T}(t)R_{12}(\varepsilon)u_2(t) \right] dt \qquad (7.2a)$$

$$J_2 = \int_0^\infty \left[x^\mathrm{T}(t)Q_2(\varepsilon)x(t) + u_1^\mathrm{T}(t)R_{21}(\varepsilon)u_1(t) + u_2^\mathrm{T}(t)R_2(\varepsilon)u_2(t) \right] dt \qquad (7.2b)$$

with symmetric weighting matrices satisfying

$$Q_i(\varepsilon) \geq 0, \quad R_i(\varepsilon) > 0, \quad R_{ij}(\varepsilon) \geq 0, \quad i = 1, 2; \quad j = 1, 2.$$

The optimal feedback solution to the given problems with the conflict of interest and simultaneous decision making, obtained in Star and Ho (1969), leads to the so-called Nash strategies $u_1^*(x(t))$ and $u_2^*(x(t))$ satisfying

$$J_1(u_1^*(x(t)), u_2^*(x(t))) \leq J_1(u_1(x(t)), u_2^*(x(t))) \qquad (7.3a)$$

$$J_2(u_1^*(x(t)), u_2^*(x(t))) \leq J_2(u_1^*(x(t)), u_2(x(t))) \qquad (7.3b)$$

It was shown in Starr and Ho (1969) that the optimal closed-loop Nash strategies are given by

$$u_1^*(x(t)) = R_i^{-1}(\varepsilon)B_i^\mathrm{T}(\varepsilon)K_i(\varepsilon)x(t), \quad i = 1, 2 \qquad (7.4)$$

where K_i, $i = 1, 2$, satisfy coupled algebraic Riccati equations

$$K_1(\varepsilon)A(\varepsilon) + A^\mathrm{T}(\varepsilon)K_1(\varepsilon) + Q_1(\varepsilon) - K_1(\varepsilon)S_1(\varepsilon)K_1(\varepsilon)$$
$$\quad - K_1(\varepsilon)S_2(\varepsilon)K_2(\varepsilon) - K_2(\varepsilon)S_2(\varepsilon)K_1(\varepsilon) + K_2(\varepsilon)Z_2(\varepsilon)K_2(\varepsilon)$$
$$= 0 = \mathbf{N_1}(K_1, K_2) \qquad (7.5a)$$

$$K_2(\varepsilon)A(\varepsilon) + A^\mathrm{T}(\varepsilon)K_2(\varepsilon) + Q_2(\varepsilon) - K_2(\varepsilon)S_2(\varepsilon)K_2(\varepsilon)$$
$$\quad - K_2(\varepsilon)S_1(\varepsilon)K_1(\varepsilon) - K_1(\varepsilon)S_1(\varepsilon)K_2(\varepsilon) + K_1(\varepsilon)Z_1(\varepsilon)K_1(\varepsilon)$$
$$= 0 = \mathbf{N_2}(K_1, K_2) \qquad (7.5b)$$

with

$$S_i(\varepsilon) = B_i(\varepsilon)R_i^{-1}(\varepsilon)B_i^\mathrm{T}(\varepsilon), \quad i = 1, 2$$
$$Z_i(\varepsilon) = B_i(\varepsilon)R_i^{-1}(\varepsilon)R_{ji}(\varepsilon)R_i^{-1}(\varepsilon)B_i^\mathrm{T}(\varepsilon)$$
$$i = 1, 2; \quad j = 1, 2, \quad i = j$$

The existence of the nonlinear optimal Nash strategies was established in Basar (1974), so that Equation 7.4, in fact, are the best linear optimal strategies. Since a linear control law is very desirable from a practical point of view, the linear strategies (Equation 7.4) attract the attention of many researchers.

The existence of Nash strategies (Equation 7.4) and solutions of the coupled algebraic Riccati equations (Equation 7.5) has been studied in Papavassilopoulos et al. (1979), by means of Brower's fixed point theorem and by imposing norm conditions on the given matrices. Original attempts to find solutions of Equation 7.5 can be found in Bertrand (1985) and Papavassilopoulos and Olsder (1984). In the paper of Li and Gajić (1995), under control-oriented assumptions (Kucera 1972; Wonham 1968), the algorithm for finding the nonnegative definite stabilizing solutions of equations (Equation 7.5) was derived (see Appendix 7.2). An overview of available methods for solving algebraic equations (Equation 7.5) can be found in Engwerda (2007).

In this section, the Nash game problem is considered for a special case of weakly interconnected systems characterized by

$$A(\varepsilon) = \begin{bmatrix} A_1(\varepsilon) & \varepsilon A_{12}(\varepsilon) \\ \varepsilon A_{21}(\varepsilon) & A_2(\varepsilon) \end{bmatrix}$$

$$B_1(\varepsilon) = \begin{bmatrix} B_{11}(\varepsilon) \\ \varepsilon B_{21}(\varepsilon) \end{bmatrix}, \quad B_2 = (\varepsilon) \begin{bmatrix} \varepsilon B_{21}(\varepsilon) \\ B_{22}(\varepsilon) \end{bmatrix}$$

$$Q_1(\varepsilon) = \varepsilon \begin{bmatrix} U_1(\varepsilon) & \varepsilon U_{12}(\varepsilon) \\ \varepsilon U_{12}^T(\varepsilon) & \varepsilon^2 U_2(\varepsilon) \end{bmatrix}, \quad Q_2(\varepsilon) = \begin{bmatrix} \varepsilon^2 V_1(\varepsilon) & \varepsilon V_{12}(\varepsilon) \\ \varepsilon V_{12}^T(\varepsilon) & V_2(\varepsilon) \end{bmatrix}$$

These partitions decompose the state vector $x(t)$ into two vectors $x_1(t) \in \mathbb{R}^{n_1}$ and $x_2(t) \in \mathbb{R}^{n_2}$ such that $n_1 + n_2 = n$.

The standard weak coupling assumption (Chow and Kokotović 1983) is imposed on the system matrix.

Assumption 7.1 Matrices $A_i(\varepsilon)$, $i = 1, 2, 3, 4$, are $O(1)$ and magnitudes of all system eigenvalues are $O(1)$, that is, $|\lambda_j| = O(1)$, $j = 1, 2, \ldots, n$, which implies that matrices $A_1(\varepsilon)$, $A_4(\varepsilon)$ are nonsingular with $\det\{A_1(\varepsilon)\} = O(1)$ and $\det\{A_4(\varepsilon)\} = O(1)$.

Note that when this assumption is not satisfied, the system defined in Equation 7.1 in addition of weak coupling also displays multiple timescale phenomena (singular perturbations) (Phillips and Kokotović 1981; Kokotović and Khalil 1986).

Since the small coupling parameter ε cannot change the basic structure of the subsystems by destroying their main properties (otherwise we cannot talk about the weak coupling), it is very natural to adopt the following form for the subsystem matrices.

Assumption 7.2 The problem matrices are $O(1)$ and

$$A_i(\varepsilon) = A_{i0} + \varepsilon A_{0i}(\varepsilon), \quad B_{ii}(\varepsilon) = B_{i0} + \varepsilon B_{0i}(\varepsilon)$$
$$U_1(\varepsilon) = U_{10} + \varepsilon U_{01}(\varepsilon), \quad V_2(\varepsilon) = V_{20} + \varepsilon V_{02}(\varepsilon)$$
$$R_i(\varepsilon) = R_{i0} + \varepsilon R_{0i}(\varepsilon), \quad i = 1, 2$$

where $A_{0i}(\varepsilon)$, $B_{0i}(\varepsilon)$, $R_{0i}(\varepsilon)$, $i = 1, 2$; $U_{01}(\varepsilon)$ and $V_{02}(\varepsilon)$ are continuous matrix functions of ε whose magnitudes are $O(1)$, whereas A_{i0}, B_{i0}, R_{i0}, $i = 1, 2$, and U_{10}, V_{20} are constant matrices independent of ε and also of $O(1)$.

In order to simplify the algebra, we will assume, without loss of generality, that $U_{12}(\varepsilon) = 0$, $V_{12}(\varepsilon) = 0$, $R_{12}(\varepsilon) = 0$, $R_{21}(\varepsilon) = 0$, $U_2(\varepsilon) = 0$, $V_1(\varepsilon) = 0$, $B_{12}(\varepsilon) = 0$, $B_{21}(\varepsilon) = 0$. Note that we are studying a more general case than the one studied in Ozguner and Perkins (1977) because of the ε-dependence of the problem matrices. In addition, we do not need to impose the analyticity assumption with respect to ε, which must be done for the power-series expansion method.

The following scaling of $K_1(\varepsilon)$ and $K_2(\varepsilon)$ is consistent with the nature of the solution of Equation 7.5, which can be established using information about the order of the problem matrices as stated in Assumptions 7.1 and 7.2

$$K_1(\varepsilon) = \begin{bmatrix} M_1(\varepsilon) & \varepsilon M_{12}(\varepsilon) \\ \varepsilon M_{12}^T(\varepsilon) & \varepsilon^2 M_2(\varepsilon) \end{bmatrix}, \quad K_2(\varepsilon) = \begin{bmatrix} \varepsilon^2 N_1(\varepsilon) & \varepsilon N_{12}(\varepsilon) \\ \varepsilon N_{12}^T(\varepsilon) & N_2(\varepsilon) \end{bmatrix} \tag{7.6}$$

The very well-known ε-decoupling method (Kokotović et al. 1969), based on the power-series expansion with respect to ε, will convert the given full-order problem (Equation 7.5) to a family of reduced-order problems (Ozguner and Perkins 1977). However, the power-series expansion method is not recursive in nature and in the case when we are interested in a high order of accuracy or when ε is not very small, the size of the required computations can be considerable. Moreover, when the problem matrices are functions of ε, the power-series method demands the analyticity of all matrices. On the other hand, the expansion of quadratic terms (e.g., $K_1(\varepsilon)B_1(\varepsilon)R_1^{-1}(\varepsilon)B_1^T(\varepsilon)K_1(\varepsilon)$) will produce an enormous numbers of terms, so that the reduced-order advantage of the series expansion method becomes questionable. The presence of a small parameter ε will be exploited in the next section from a different point of view, leading to the recursive scheme for the solution of Equation 7.5. Since the proposed method is of the fixed-point type, the boundness of all problem matrices over a compact set $\varepsilon \in [0, \varepsilon_1]$ has to be imposed. This is a much milder condition than the analyticity requirement of the power-series expansion method.

7.2 SOLUTION OF COUPLED ALGEBRAIC RICCATI EQUATIONS

Partitioning Equation 7.5 compatibly with Equation 7.6, we get the following set of equations

$$M_1(\varepsilon)A_1(\varepsilon) + A_1^T(\varepsilon)M_1(\varepsilon) + U_1(\varepsilon) - M_1(\varepsilon)S_{11}(\varepsilon)M_1(\varepsilon)$$
$$+ \varepsilon^2 \{ M_{12}(\varepsilon)A_{21}(\varepsilon) + A_{21}^T(\varepsilon)M_{12}^T(\varepsilon) - M_{12}(\varepsilon)S_{22}(\varepsilon)N_{12}^T(\varepsilon)$$
$$- N_{12}(\varepsilon)S_{22}(\varepsilon)M_{12}^T(\varepsilon) \} = 0 \tag{7.7a}$$

$$M_1(\varepsilon)A_{12}(\varepsilon) + M_{12}(\varepsilon)A_2(\varepsilon) - M_1(\varepsilon)S_{11}(\varepsilon)M_{12}(\varepsilon)$$
$$- M_{12}(\varepsilon)S_{22}(\varepsilon)N_2(\varepsilon) - \varepsilon^2 \{ N_{12}(\varepsilon)S_{22}(\varepsilon)M_2(\varepsilon) - A_{21}^T(\varepsilon)M_2(\varepsilon) \}$$
$$+ A_1^T(\varepsilon)M_{22}(\varepsilon) = 0 \tag{7.7b}$$

$$M_2(\varepsilon)A_2(\varepsilon) + A_2^{\mathrm{T}}(\varepsilon)M_2(\varepsilon) - M_2(\varepsilon)S_{22}(\varepsilon)N_2(\varepsilon)$$
$$- N_2(\varepsilon)S_{22}(\varepsilon)M_2(\varepsilon) + M_{12}^{\mathrm{T}}(\varepsilon)A_{21}(\varepsilon) + A_{21}^{\mathrm{T}}(\varepsilon)M_{12}(\varepsilon)$$
$$- M_{12}^{\mathrm{T}}(\varepsilon)S_{11}(\varepsilon)M_{12}(\varepsilon) = 0 \tag{7.7c}$$

$$N_1(\varepsilon)A_1(\varepsilon) + A_1^{\mathrm{T}}(\varepsilon)N_1(\varepsilon) - N_1(\varepsilon)S_{11}(\varepsilon)M_1(\varepsilon)$$
$$- M_1(\varepsilon)S_{11}(\varepsilon)N_1(\varepsilon) + N_{12}(\varepsilon)A_{21}(\varepsilon) + A_{21}^{\mathrm{T}}(\varepsilon)N_{12}^{\mathrm{T}}(\varepsilon)$$
$$- N_{12}(\varepsilon)S_{22}(\varepsilon)N_{12}^{\mathrm{T}}(\varepsilon) = 0 \tag{7.7d}$$

$$\varepsilon^2 N_1(\varepsilon)A_{12}(\varepsilon) + N_{12}(\varepsilon)A_2(\varepsilon) - N_{12}(\varepsilon)S_{22}(\varepsilon)N_2(\varepsilon)$$
$$- \varepsilon^2 N_1(\varepsilon)S_{11}(\varepsilon)M_{12}(\varepsilon) - M_1(\varepsilon)S_{11}(\varepsilon)N_{12}(\varepsilon)$$
$$+ A_{21}^{\mathrm{T}}(\varepsilon)N_2(\varepsilon) + A_1^{\mathrm{T}}(\varepsilon)N_{12}(\varepsilon) = 0 \tag{7.7e}$$

$$N_2(\varepsilon)A_2(\varepsilon) + A_2^{\mathrm{T}}(\varepsilon)N_2(\varepsilon) + V_2(\varepsilon) - N_2(\varepsilon)S_{22}(\varepsilon)N_2(\varepsilon)$$
$$+ \varepsilon^2 \left\{ N_{12}^{\mathrm{T}}(\varepsilon)A_{12}(\varepsilon) + A_{12}^{\mathrm{T}}(\varepsilon)N_{12}(\varepsilon) - N_{12}^{\mathrm{T}}(\varepsilon)S_{11}(\varepsilon)M_{12}(\varepsilon) \right.$$
$$\left. - M_{12}^{\mathrm{T}}(\varepsilon)S_{11}(\varepsilon)N_{12}(\varepsilon) \right\} = 0 \tag{7.7f}$$

where

$$S_{ii}(\varepsilon) = B_{ii}(\varepsilon)R_i^{-1}(\varepsilon)B_{ii}^{\mathrm{T}}(\varepsilon), \quad i = 1, 2$$

7.2.1 ZEROTH-ORDER APPROXIMATION

Let us define the $O(\varepsilon^2)$ perturbation of Equation 7.7 as

$$\overline{M}_1(\varepsilon)A_1(\varepsilon) + A_1^{\mathrm{T}}(\varepsilon)\overline{M}_1(\varepsilon) + U_1(\varepsilon) - \overline{M}_1(\varepsilon)S_{11}(\varepsilon)\overline{M}_1(\varepsilon) = 0 \tag{7.8a}$$

$$\overline{M}_{12}(\varepsilon)D_2(\varepsilon) + D_1^{\mathrm{T}}(\varepsilon)\overline{M}_{12}(\varepsilon) = -\overline{M}_1(\varepsilon)A_{12}(\varepsilon) \tag{7.8b}$$

$$\overline{M}_2(\varepsilon)D_2(\varepsilon) + D_2^{\mathrm{T}}(\varepsilon)\overline{M}_2(\varepsilon) = \overline{M}_{12}^{\mathrm{T}}(\varepsilon)S_{11}(\varepsilon)\overline{M}_{12}(\varepsilon)$$
$$- \overline{M}_{12}^{\mathrm{T}}(\varepsilon)A_{12}(e) - A_{12}^{\mathrm{T}}(\varepsilon)\overline{M}_{12}(\varepsilon) \tag{7.8c}$$

$$\overline{N}_1(\varepsilon)D_1(\varepsilon) + D_1^{\mathrm{T}}(\varepsilon)\overline{N}_1(\varepsilon) = \overline{N}_{12}(\varepsilon)S_{22}(\varepsilon)\overline{N}_{12}^{\mathrm{T}}(\varepsilon)$$
$$- \overline{N}_{12}(\varepsilon)A_{21}(\varepsilon) - A_{21}^{\mathrm{T}}(\varepsilon)\overline{M}_{12}^{\mathrm{T}}(\varepsilon) \tag{7.8d}$$

$$\overline{N}_{12}(\varepsilon)D_2(\varepsilon) + D_1^{\mathrm{T}}(\varepsilon)\overline{N}_{12}(\varepsilon) = -A_{21}^{\mathrm{T}}(\varepsilon)\overline{N}_2(\varepsilon) \tag{7.8e}$$

$$\overline{N}_2(\varepsilon)A_2(\varepsilon) + A_2^{\mathrm{T}}(\varepsilon)\overline{N}_2(\varepsilon) + V_2(\varepsilon) - \overline{N}_2(\varepsilon)S_{22}(\varepsilon)\overline{N}_2(\varepsilon) = 0 \tag{7.8f}$$

where

$$D_1(\varepsilon) = A_1(\varepsilon) - S_{11}(\varepsilon)\overline{M}_1(\varepsilon)$$

$$D_2(\varepsilon) = A_2(\varepsilon) - S_{22}(\varepsilon)\overline{N}_2(\varepsilon)$$

This system of equations has decoupled form and can be solved like two lower order Riccati equations (Equation 7.8a and f) and four lower order Lyapunov equations (Equation 7.8b through e). The nonnegative definite stabilizing solution of Equation 7.8a and f exists under the well-known stabilizability–detectability assumption (Kucera 1972; Wonham 1968).

Assumption 7.3   The triples $(A_1(0), B_1(0) \, \mathrm{Chol}(U_1(0)))$ and $(A_2(0), B_2(0), \mathrm{Chol} (V_2(0)))$ are stabilizable–detectable.

Under the same assumption, the unique solutions of Equation 7.8b through e exist since $D_1(\varepsilon)$ and $D_2(\varepsilon)$ are stable matrices (Kucera 1972; Wonham 1968).

7.2.2   SOLUTION OF HIGHER ORDER OF ACCURACY

The zeroth-order solutions $\overline{M}(\varepsilon)$ and $\overline{N}(\varepsilon)$ are $O(\varepsilon^2)$ close to the exact ones. The exact solutions can be sought in the form

$$K_1(\varepsilon) = \begin{bmatrix} \overline{M}_1(\varepsilon) + \varepsilon^2 E_1(\varepsilon) & \varepsilon \left[\overline{M}_{12}(\varepsilon) + \varepsilon^2 E_{12}(\varepsilon) \right] \\ \varepsilon \left[\overline{M}_{12}(\varepsilon) + \varepsilon^2 E_{12}(\varepsilon) \right]^{\mathrm{T}} & \varepsilon^2 \left[\overline{M}_2(\varepsilon) + \varepsilon^2 E_2(\varepsilon) \right] \end{bmatrix} \tag{7.9a}$$

$$K_2(\varepsilon) = \begin{bmatrix} \varepsilon^2 \left[\overline{N}_1(\varepsilon) + \varepsilon^2 G_1(\varepsilon) \right] & \varepsilon \left[\overline{N}_{12}(\varepsilon) + \varepsilon^2 G_{12}(\varepsilon) \right] \\ \varepsilon \left[\overline{N}_{12}(\varepsilon) + \varepsilon^2 G_{12}(\varepsilon) \right]^{\mathrm{T}} & \overline{N}_2(\varepsilon) + \varepsilon^2 G_2(\varepsilon) \end{bmatrix} \tag{7.9b}$$

Obviously, $O(\varepsilon^2)$ approximations of $E(\varepsilon)'s$ and $G(\varepsilon)'s$ will produce $O(\varepsilon^{k+2})$ approximations of the required solutions, which is why we are interested in finding convenient forms for these error terms and the appropriate algorithm for their solutions.

Subtracting Equation 7.8 from corresponding Equation 7.7 and after doing some algebra we get the following expressions for the error equations

$$E_1 D_1 + D_1^{\mathrm{T}} E_1 = C_1 + \varepsilon^2 F_1(E_1, E_{12}, G_{12}) \tag{7.10a}$$

$$E_1 D_{12} + E_{12} D_2 + D_1^{\mathrm{T}} E_{12} - \overline{M}_{12} S_{22} G_2 = C_2 + \varepsilon^2 F_2(E_1, E_{12}, G_{12}, E_2, G_2) \tag{7.10b}$$

$$E_{12}^{\mathrm{T}} D_{12} + D_{12}^{\mathrm{T}} E_{12} + E_2 D_2 + D_2^{\mathrm{T}} E_2 - G_2 S_{22} \overline{M}_2 - \overline{M}_2 S_{22} G_2$$
$$= \varepsilon^2 F_3(E_{12}, E_2, G_2) \tag{7.10c}$$

$$G_1 D_1 + D_1^{\mathrm{T}} G_2 + G_{12} D_{21} + D_{21}^{\mathrm{T}} G_{12}^{\mathrm{T}} - E_1 S_{11} \overline{N}_1 - \overline{N}_1 S_{11} E_1$$
$$= \varepsilon^2 F_4(E_1, G_{12}, G_2) \tag{7.10d}$$

$$G_1 D_2 + D_1^{\mathrm{T}} G_{12} + D_{21}^{\mathrm{T}} G_2 - E_1 S_{11} \overline{N}_{12} = C_5 + \varepsilon^2 F_5(E_1, E_{12}, G_1, G_{12}, G_2) \tag{7.10e}$$

$$G_2 D_2 + D_2^{\mathrm{T}} G_2 = C_6 + \varepsilon^2 F_6(E_{12}, G_{12}, G_2) \tag{7.10f}$$

where

$$D_{12} = D_{12}(\varepsilon) = A_{12}(\varepsilon) - S_{11}(\varepsilon)\overline{M}_{12}(\varepsilon)$$

$$D_{21} = D_{21}(\varepsilon) = A_{21}(\varepsilon) - S_{22}(\varepsilon)\overline{N}_{12}(\varepsilon)$$

Matrices $F_i, i = 1, 2, \ldots, 6$, and constant matrices C_j are given in Appendix 7.1. In order to simplify notation, the ε-dependence of the problem matrices in Equation 7.10 and in the remaining part of the chapter is omitted.

The weakly coupled and hierarchical structure of Equation 7.10 can be exploited by proposing the following recursive scheme, which leads, after some algebra, to the six low-order completely decoupled Lyapunov equations.

ALGORITHM 7.1

$$E_1^{(i+1)}D_1 + D_1^T E_1^{(i+1)} = \varepsilon^2 E_1^{(i)} S_{11} E_1^{(i)} - M_{12}^{(i)} D_{21}^{(i)} - D_{21}^{T^{(i)}} M_{12}^{T^{(i)}} \tag{7.11a}$$

$$E_{12}^{(i+1)}D_2 + D_1^T E_{12}^{(i+1)} = -E_1^{(i+1)}D_{12}^{(i)} + M_{12}^{(i)} S_{22} G_2^{(i+1)} - D_{21}^{T^{(i)}} M_2^{(i)} \tag{7.11b}$$

$$E_2^{(i+1)}D_2 + D_2^T E_2^{(i+1)} = M_2^{(i)} S_{22} G_2^{(i+1)} + G_2^{(i+1)} S_{22} M_2^{(i)} - E_{12}^{T^{(i+1)}} D_{12}^{(i)}$$
$$- D_{12}^{T^{(i)}} E_{12}^{(i+1)} + \varepsilon^2 E_{12}^{T^{(i+1)}} S_{11} E_{12}^{(i+1)} \tag{7.11c}$$

$$G_1^{(i+1)}D_1 + D_1^T G_1^{(i+1)} = E_1^{(i+1)} S_{11} N_1^{(i)} + N_1^{(i)} S_{11} E_1^{(i+1)}$$
$$- G_{12}^{(i+1)} D_{21}^{(i)} - D_{21}^{T^{(i)}} G_{12}^{T^{(i+1)}} + \varepsilon^2 G_{12}^{(i+1)} S_{22} G_{12}^{T^{(i+1)}} \tag{7.11d}$$

$$G_{12}^{(i+1)}D_2 + D_1^T G_{12}^{(i+1)} = -D_{21}^{T^{(i)}} G_2^{(i+1)} + E_1^{(i+1)} S_{11} N_{12}^{(i)} - N_1^{(i)} D_{12}^{(i)} \tag{7.11e}$$

$$G_2^{(i+1)}D_2 + D_2^T G_2^{(i+1)} = \varepsilon^2 G_2^{(i)} S_{22} G_2^{(i)} - N_{12}^{T^{(i)}} D_{12}^{(i)} - D_{12}^{T^{(i)}} N_{12}^{(i)} \tag{7.11f}$$

$$i = 0, 1, 2, 3, \ldots$$

with initial conditions chosen as

$$E_1^{(0)} = E_{12}^{(0)} = E_2^{(0)} = G_1^{(0)} = G_{12}^{(0)} = G_2^{(0)} = 0$$

where

$$M_{12}^{(i)} = M_{12} + \varepsilon^2 E_{12}^{(i)}, \quad N_{12}^{(i)} = \overline{N}_{12} + e^2 G_{12}^{(i)}$$

$$N_1^{(i)} = \overline{N}_1 + \varepsilon^2 G_1^{(i)}, \quad M_2^{(i)} = \overline{M}_2 + \varepsilon^2 E_2^{(i)}$$

$$D_{12}^{(i)} = A_{12} - S_{11} M_{12}^{(i)}, \quad D_{21}^{(i)} = A_{21} - S_{22} N_{12}^{T^{(i)}}$$

$$i = 1, 2, 3, \ldots$$

These Lyapunov equations have to be solved in the given order, that is, first E_1 and G_2, then E_{12} and G_{12}, and finally E_2 and G_1.

The following theorem indicates the features of the proposed recursive scheme.

THEOREM 7.1

Under imposed weak coupling and stabilizability and detectability assumptions, Assumptions 7.1 and 7.2, algorithm (Equation 7.11) converges to the exact solution of the error terms, and thus of $K_1(\varepsilon)$ and $K_2(\varepsilon)$, with the rate of convergence of $O(\varepsilon^2)$, that is

$$\left\|E_j(\varepsilon) - E_j^{(i)}(\varepsilon)\right\| = O(\varepsilon^{2i})$$

$$\left\|G_j(\varepsilon) - G_j^{(i)}(\varepsilon)\right\| = O(\varepsilon^{2i})$$

$$\left\|E_{12}(\varepsilon) - E_{12}^{(i)}(\varepsilon)\right\| = O(\varepsilon^{2i}) \qquad (7.12)$$

$$\left\|G_{12}(\varepsilon) - G_{12}^{(i)}(\varepsilon)\right\| = O(\varepsilon^{2i})$$

$$j = 1,2; \quad i = 1,2,3,\dots$$

and

$$\left\|K_j(\varepsilon) - K_j^{(i)}(\varepsilon)\right\| = O(\varepsilon^{2i+2}) \qquad (7.13)$$

$$j = 1,2; \quad i = 0,1,2,\dots$$

Proof As a starting point, we need to show the existence of a bounded solution of Equation 7.10 in the neighborhood of $\varepsilon = 0$. By the implicit function theorem it is enough to show that the corresponding Jacobian is nonsingular at $\varepsilon = 0$. The Jacobian is given by

$$J(\varepsilon)|_{\varepsilon=0} = \begin{bmatrix} \Gamma_1 & 0 & 0 & 0 & 0 & 0 \\ * & \Gamma_2 & 0 & 0 & 0 & * \\ 0 & * & \Gamma_3 & 0 & 0 & * \\ * & 0 & 0 & \Gamma_1 & * & 0 \\ * & 0 & 0 & 0 & \Gamma_2 & * \\ 0 & 0 & 0 & 0 & 0 & \Gamma_3 \end{bmatrix} \qquad (7.14)$$

where the "asterisk" denotes terms which are not important for the Jacobian nonsingularity. Γ's are given by the Kronecker product representation

$$\Gamma_i = I_{n_i} \otimes D_i^{\mathrm{T}}(0) + D_i^{\mathrm{T}}(0) \otimes I_{n_i}, \quad i = 1, 3$$

$$\Gamma_2 = I_{n_2} \otimes D_2^{\mathrm{T}}(0) + D_1^{\mathrm{T}}(0) \otimes I_{n_1}$$

where I_{n_i} and I_{n_2} are identity matrices. Under Assumptions 7.1 and 7.2, $D_1(0)$ and $D_2(0)$ are stable matrices for any sufficiently small $\varepsilon \in [0, \varepsilon_2]$ and by the well-known properties of the Kronecker product (Lancaster and Tismenetsky 1985), so are matrices Γ_1, Γ_2, and Γ_3. It is easy to see that the nonsingularity of the Jacobian is guaranteed by the nonsingularity of Γ_1, Γ_2, and Γ_3.

The second step in the proof of the given theorem is to give an estimate of the rate of convergence.

For $i = 0$, Equations 7.10a and 7.11a imply

$$\left(E_1 - E_1^{(1)}\right)D_1 + D_1^{\mathrm{T}}\left(E_1 - E_1^{(1)}\right) = \varepsilon^2 F_1(E_1, E_{12}, G_{12})$$

which by stability of D_1 and the existence of the bounded solution of Equation 7.10 gives

$$\left\|E_1 - E_1^{(1)}\right\| = O(\varepsilon^2) \tag{7.15a}$$

By the same arguments, from Equations 7.10f and 7.11f we have

$$\left\|G_2 - G_2^{(1)}\right\| = O(\varepsilon^2) \tag{7.15b}$$

Subtracting Equation 7.11b from Equation 7.10b and using Equation 7.15a and f and the expression for F_3 (from Appendix 7.1) lead to

$$\left(E_{12} - E_{12}^{(1)}\right)D_2 + D_1^{\mathrm{T}}\left(E_{12} - E_{12}^{(1)}\right) = O(\varepsilon)^2$$

which implies that

$$\left\|E_{12} - E_{12}^{(1)}\right\| = O(\varepsilon^2) \tag{7.15c}$$

By analogy (Equation 7.10b and e have similar forms), Equation 7.10e and 7.11e will produce

$$\left\|G_{12} - G_{12}^{(1)}\right\| = O(\varepsilon^2) \tag{7.15d}$$

Also, from Equation 7.10c, 7.11c, and 7.15a,b,e,f and Appendix 7.1, we have

$$\left(E_2 - E_2^{(1)}\right)D_2 + D_2^{\mathrm{T}}\left(E_2 - E_2^{(1)}\right) = O(\varepsilon^2)$$

that is

$$\left\| E_2 - E_2^{(1)} \right\| = O(\varepsilon^2) \tag{7.15e}$$

and, by analogy, from Equations 7.10d and 7.11d we get

$$\left\| G_1 - G_1^{(1)} \right\| = O(\varepsilon^2) \tag{7.15f}$$

Using these starting observations and forms of $F_j's$ and $C_j's$, it can be shown that

$$\left\| F_j - F_j^{(i)} \right\| = O(\varepsilon^{2i}), \quad j = 1, 2; \quad i = 1, 2, 3, \dots \tag{7.16}$$

For example, for $j = 1$

$$
\begin{aligned}
F_1 - F_1^{(i)} = {} & \left(E_1 - E_1^{(i)} \right) S_{11} E_1^{(i)} + E_1 S_{11} \left(E_1 - E_1^{(i)} \right) \\
& - \left(E_{12} - E_{12}^{(i)} \right) D_{21} - D_{21}^{\mathrm{T}} \left(E_{12} - E_{12}^{(i)} \right)^{\mathrm{T}} \\
& + \left(G_{12} - G_{12}^{(i)} \right) S_{22} M_{12}^{\mathrm{T}^{(i)}} + M_{12}^{(i)} S_{22} \left(G_{12} - G_{12}^{(i)} \right)^{\mathrm{T}}
\end{aligned}
$$

so that for $i = 1$, from Equation 7.15 we have $F_1 - F_1^{(1)} = O(\varepsilon^2)$, that is

$$\left(E_1 - E_1^{(2)} \right) D_1 + D_1^{\mathrm{T}} \left(E_1 - E_1^{(2)} \right) = \varepsilon^2 \left(F_1 - F_1^{(1)} \right) = O(\varepsilon^4)$$

which implies that

$$\left(E_1 - E_1^{(2)} \right) = O(\varepsilon^4)$$

Continuing the same procedure, we can verify Equation 7.15, which by the existence of the bounded solutions of $E's$ and $G's$ will imply Equation 7.12. Note that the solution of Equation 7.11 exists at each iteration since the corresponding Jacobian is always given by Equation 7.13, and thus nonsingular at $\varepsilon = 0$ in every iteration. ∎

We would like to point out that the imposed form of solution (Equation 7.9) is an additional limiting factor for a small parameter ε. Since the solution of Equation 7.10 is symmetric only (which can easily be seen from the form of corresponding equations), the small parameter ε has to be constrained to the set $\varepsilon \in [0, \varepsilon_3]$ such that $\forall \varepsilon$, $K_1(\varepsilon)$ and $K_2(\varepsilon)$ preserve the required nonnegative definiteness. Thus, the presented method is applicable for $\varepsilon \in [0, \varepsilon^*]$ where $\varepsilon^* = \min \{\varepsilon_1, \varepsilon_2, \varepsilon_3, \dots \}$. However, the limiting condition $\varepsilon^* = \min \{\varepsilon_1, \varepsilon_2, \varepsilon_3, \dots \}$ is present in the entire theory of small parameters (weak coupling and singular perturbations); it is both method-dependent and problem-dependent, and not a direct consequence of the procedure studied in this chapter.

Let us compare the proposed algorithm (Equation 7.11), based on the fixed point iteration for weakly coupled systems and the power-series expansion algorithm for the same type of systems. The comparison is done for the case when the problem matrices are not functions of ε (which is in favor of the power-series expansion algorithm). The equations corresponding to Equation 7.11 are given by Ozguner and Perkins (1977)

$$M_1^{(i+1)}D_1 + D_1^{\mathrm{T}}M_1^{(i+1)} = Z_1^{(0,1,2,...,i)} \tag{7.17a}$$

$$N_2^{(i+1)}D_2 + D_2^{\mathrm{T}}N_2^{(i+1)} = Z_6^{(0,1,2,...,i)} \tag{7.17b}$$

$$M_{12}^{(i+1)}D_2 + D_1^{\mathrm{T}}M_{12}^{(i+1)} = Z_2^{(0,1,2,...,i)} \tag{7.17c}$$

$$N_{12}^{(i+1)}D_2 + D_1^{\mathrm{T}}G_{12}^{(i+1)} = Z_5^{(0,1,2,...,i)} \tag{7.17d}$$

$$M_2^{(i+1)}D_2 + D_2^{\mathrm{T}}M_2^{(i+1)} = Z_3^{(0,1,2,...,i)} \tag{7.17e}$$

$$N_1^{(i+1)}D_1 + D_1^{\mathrm{T}}N_1^{(i+1)} = Z_4^{(0,1,2,...,i)} \tag{7.17f}$$

where $Z_j, j = 1, 2, \ldots 6$, depend on the all previously obtained terms. For example,

$$Z_1^{(0,1,2,...,i)} = -(i+1)\left(M_{12}^{(i)}A_{21} + A_{21}^{\mathrm{T}}M_{12}^{(i)^{\mathrm{T}}}\right) + \sum_{k=2(\mathrm{even})}^{i-1} \binom{i+1}{k} M_1^{(i+1-k)}S_{11}M_1^{(k)}$$

$$+ \sum_{k=1(\mathrm{odd})}^{i} \left\{ \binom{i+1}{k} M_{12}^{(i+1-k)}S_{22}N_{12}^{(k)^{\mathrm{T}}} + N_{12}^{(i+1-k)}S_{22}M_{12}^{(k)} \right\} \tag{7.18}$$

Both approaches produce the same type of equations (Lyapunov ones), but in order to form the right-hand side, for example of Equation 7.11a, we have to perform only three matrix multiplications for every i, where for corresponding equation of the power-series expansion the number of required matrix multiplications grows very quickly as i increases (Equation 7.17a). Thus, the obvious advantages of the fixed point iteration approach are

1. The size of required computation is considerably less, and since it does not grow per iteration, the proposed method is extremely efficient for obtaining the exact solution or the solution of very high accuracy.
2. The fixed point method is recursive in nature (the power-series expansion method is not), and thus much easier to implement.

The approximations of the suboptimal Nash strategies (Equation 7.4) can be defined by

$$u_j^{(i)}(x(t)) = -R_j^{-1}(\varepsilon)B_j^{\mathrm{T}}(\varepsilon)K_j^{(i)}(\varepsilon)x(t), \quad j = 1, 2; \quad i = 0, 1, 2, 3, \ldots \tag{7.19}$$

where

$$K_1^{(i)}(\varepsilon) = \begin{bmatrix} \overline{M}_1(\varepsilon) + \varepsilon^2 E_1^{(i)}(\varepsilon) & \varepsilon\left[\overline{M}_{12}(\varepsilon) + \varepsilon^2 E_{12}^{(i)}(\varepsilon)\right] \\ \varepsilon\left[\overline{M}_{12}(\varepsilon) + \varepsilon^2 E_{12}^{(i)}(\varepsilon)\right]^{\mathrm{T}} & \varepsilon^2\left[\overline{M}_2(\varepsilon) + \varepsilon^2 E_2^{(i)}(\varepsilon)\right] \end{bmatrix} \qquad (7.20a)$$

$$K_2^{(i)}(\varepsilon) = \begin{bmatrix} \varepsilon^2\left[\overline{N}_1(\varepsilon) + \varepsilon^2 G_1^{(i)}(\varepsilon)\right] & \varepsilon\left[\overline{N}_{12}(\varepsilon) + \varepsilon^2 G_{12}^{(i)}(\varepsilon)\right] \\ \varepsilon\left[\overline{N}_{12}(\varepsilon) + \varepsilon^2 G_{12}^{(i)}(\varepsilon)\right]^{\mathrm{T}} & \overline{N}_2(\varepsilon) + \varepsilon^2 G_2^{(i)}(\varepsilon) \end{bmatrix} \qquad (7.20b)$$

Then, by following the arguments of Cruz and Chen (1971), the cost approximations produce

$$J_j^{(i)}\left(u_1^{(i)}, u_2^{(i)}\right) = J_j(u_1^*, u_2^*) + O(\varepsilon^{2i+2}), \quad j = 1, 2; \quad i = 0, 1, 2, \ldots \qquad (7.21)$$

The approximate cost functions for the other cases, when the control agents uses the approximative strategies of the different order of accuracy (e.g., $u_1^{(p)}$ and $u_2^{(q)}$, $p \neq q$), can be obtained by using results of Cruz and Chen (1971) also. But, since the proposed method is recursive in its nature, and thus very easy to implement, and since the amount of required computations is constant per iteration (does not grow with i), the accuracy of very high order can be achieved at a very low cost, so that the proposed method can be efficient for finding the exact solution as well.

Since the proposed algorithm defines the error of approximation similarly to the power-series expansion, it can be easily seen that the approximate Nash strategies (Equation 7.19) are also well posed in the sense of Khalil (1980).

7.3 NUMERICAL EXAMPLE

In order to demonstrate the efficiency of the proposed algorithm, we have run a fourth-order example. Matrices A_1, A_{12}, A_{21}, A_2, B_{11}, and B_{22} have been chosen randomly (standard deviation equals to one and mean value equals to zero) and the matrices $R_1 = R_2 = U_1 = V_2 = I$ are chosen such that the required stabilizability–detectability assumptions are satisfied

$$A_1 = \begin{bmatrix} -1.035 & -0.192 \\ 1.684 & -0.421 \end{bmatrix}, \quad A_{12} = \begin{bmatrix} -1.084 & 0.579 \\ 1.327 & -0.841 \end{bmatrix}$$

$$A_{21} = \begin{bmatrix} -1.370 & -0.533 \\ 1.069 & 0.835 \end{bmatrix}, \quad A_2 = \begin{bmatrix} -1.510 & -0.139 \\ 0.410 & 1.238 \end{bmatrix}$$

$$B_{11} = \begin{bmatrix} -1.019 & 0.602 \\ -0.912 & 1.329 \end{bmatrix}, \quad B_{22} = \begin{bmatrix} -1.641 & 0.330 \\ 1.068 & 0.243 \end{bmatrix}$$

$$U_1 = V_2 = R_1 = R_2 = I_2$$

TABLE 7.1

Dependence of Number of Iterations on ε

ε	$i =$ Number of Required Iterations Such That $\varepsilon^{(i)} < 10^{-10}$
0.8	16
0.6	11
0.4	8
0.2	5
0.1	4
0.05	3
0.01	2
0.001	1

Simulation results for different values of a coupling parameter ε are given in Table 7.1. The error is defined as

$$e^{(i)} = \max\left\{\left\|\mathbf{N_1}\left(K_1^{(i)}, K_2^{(i)}\right)\right\|_\infty, \left\|\mathbf{N_2}\left(K_1^{(i)}, K_2^{(i)}\right)\right\|_\infty\right\}$$

The results from Table 7.1 strongly support the necessity of the existence of the recursive scheme for the solution of weakly coupled linear-quadratic Nash game problem, unless ε is very small, the zeroth- and first-order approximations are far from the optimal solution. Results from Table 7.2 verify, for this particular example, the conclusions of Theorem 7.1, that is, the rate of convergence of the proposed algorithm is $O(\varepsilon^2) = O(10^{-2})$.

Therefore, the solution of the Nash strategies of weakly interconnected systems can be obtained up to an arbitrary accuracy by performing iterations on the Lyapunov equations corresponding to the local subsystem problems.

TABLE 7.2

Propagation of the Error per Iteration for a Constant Value of ε

$(\varepsilon = 0.1)\ i$	Error $\varepsilon^{(i)}$
0	0.89662×10^{-2}
1	0.65481×10^{-4}
2	0.10349×10^{-6}
3	0.40663×10^{-9}
4	0.92572×10^{-11}

APPENDIX 7.1

$$F_1 = E_1 S_{11} E_1 + \overline{M}_{12} S_{22} G_{12}^T + G_{12} S_{22} \overline{M}_{12}^T$$
$$\quad - E_{12} D_{21} - D_{21}^T E_{12}^T + \varepsilon^2 \left(E_{12} S_{22} G_{12}^T + G_{12} S_{22} E_{12}^T \right)$$
$$F_2 = E_{12} S_{22} G_2 + E_1 S_{11} E_{12} + G_{12} S_{22} \overline{M}_2$$
$$\quad - D_{21}^T E_2 + \varepsilon^2 G_{12} S_{22} E_2$$
$$F_3 = E_{12} S_{11} E_{12} + E_2 S_{22} G_2 + G_2 S_{22} E_2$$
$$F_4 = G_{12} S_{22} G_{12}^T + E_1 S_{11} G_1 + G_1 S_{11} E_1$$
$$F_5 = E_1 S_{11} G_{12} + G_{12} S_{22} G_2$$
$$\quad + \overline{N}_1 S_{11} G_{12} - G_1 D_{12} + \varepsilon^2 G_1 S_{11} E_{12}$$
$$F_6 = G_2 S_{22} G_2 + E_{12}^T S_{11} \overline{N}_{12} + \overline{N}_{12}^T S_{11} E_{12}$$
$$\quad - G_{12}^T D_{12} - D_{12}^T G_{12} + \varepsilon^2 \left(E_{12}^T S_{11} E_{12} + G_{12}^T S_{11} E_{12} \right)$$
$$C_1 = -\overline{M}_{12} A_{12} - A_{21}^T \overline{M}_{12}^T + \overline{M}_{12} S_{22} \overline{N}_{12}^T + \overline{N}_{12} S_{22} \overline{M}_{12}^T$$
$$C_2 = -D_{21}^T \overline{M}_2$$
$$C_5 = -\overline{N}_1 D_{12}$$
$$C_6 = -\overline{N}_{12}^T A_{12} - A_{12}^T \overline{N}_{12} + \overline{M}_{12}^T S_{11} \overline{N}_{12} + \overline{N}_{12}^T S_{11} \overline{M}_{12}$$

APPENDIX 7.2: ALGORITHM FOR SOLVING COUPLED ALGEBRAIC RICCATI EQUATIONS OF NASH DIFFERENTIAL GAMES

The considered coupled algebraic Riccati equations have the forms

$$K_1 A + A^T K_1 + Q_1 - K_1 S_1 K_1 - K_2 S_2 K_1 - K_1 S_2 K_2 + K_2 Z_2 K_2$$
$$= \mathcal{N}_2(K_1, K_2) = 0 \tag{A.7.1}$$

$$K_2 A + A^T K_2 + Q_2 - K_2 S_2 K_2 - K_2 S_1 K_1 - K_1 S_1 K_1 + K_1 Z_1 K_1$$
$$= \mathcal{N}_2(K_1, K_2) = 0 \tag{A.7.2}$$

with

$$S_i = B_i R_{ii}^{-1} B_i^T, \quad i = 1, 2; \quad Z_i = B_i R_{ii}^{-1} R_{ji} R_{ii}^{-1} B_i^T, \quad i, j = 1, 2; \quad i = j$$

The proposed algorithm is based on the simulation results presented in Gajić and Li (1988) (see also Li and Gajić 1995). It seems, that the numerical method proposed is valid under the following assumption.

Assumption 7.4   Either the triple $(A, B_1, \mathrm{Chol}(Q_1))$ or $(A, B_2, \mathrm{Chol}(Q_2))$ is stabilizable–detectable.

These conditions are quite natural since at least one control agent has to be able to control and observe unstable modes. Because the game is a noncooperative one, the assumption that their joint effect will handle unstable modes seems to be very idealistic.

Let us suppose that $(A, B_1, \mathrm{Chol}(Q_1))$ is stabilizable–detectable. Then, a unique positive definite solution of an auxiliary algebraic Riccati equation

$$K_1^{(0)} A + A^{\mathrm{T}} K_1^{(0)} + Q_1 - K_1^{(0)} S_1 K_1^{(0)} = 0 \tag{A.7.3}$$

exists such that $\left(A - S_1 K_1^{(0)}\right)$ is stable. By plugging $K_1 = K_1^{(0)}$ in Equation A.7.2, we get the second auxiliary Riccati equation as

$$K_2^{(0)} \left(A - S_1 K_1^{(0)}\right) + \left(A - S_1 K_1^{(0)}\right)^{\mathrm{T}} K_2^{(0)} + \left(Q_2 + K_1^{(0)} Z_1 K_1^{(0)}\right) - K_2^{(0)} S_2 K_2^{(0)} = 0 \tag{A.7.4}$$

Since $\left(A - S_1 K_1^{(0)}\right)$ is stable and $Q_2 + K_1^{(0)} Z_1 K_1^{(0)}$ is a positive semidefinite matrix, the corresponding closed-loop matrix $\left(A - S_1 K_1^{(0)} - S_2 K_2^{(0)}\right)$ is a stable matrix. In fact, the triple $\left(A - S_1 K_1^{(0)}, B_2, \mathrm{Chol}\left(Q_2 + K_1^{(0)} S_1 K_1^{(0)}\right)\right)$ is stabilizable–detectable and $K_2^{(0)}$ is uniquely determined.

An iterative algorithm was proposed by Gajić and Li (1988) for solving Equations A.7.1 and A.7.2. By decoupling these equations by using appropriately one-step delay, we get the Lyapunov type iterative scheme similarly to Kleinman (1968), with $K_1^{(0)}$ and $K_2^{(0)}$ playing the role of the initial points.

ALGORITHM 7.2

$$\left(A - S_1 K_1^{(i)} - S_2 K_2^{(i)}\right)^{\mathrm{T}} K_1^{(i+1)} + K_1^{(i+1)} \left(A - S_1 K_1^{(i)} - S_2 K_2^{(i)}\right)$$
$$= \overline{Q_1^{(i)}} = -\left(Q_1 + K_1^{(i)} S_1 K_1^{(i)} + K_2^{(i)} Z_2 K_2^{(i)}\right), \quad i = 0, 1, 2, \ldots \tag{A.5}$$

$$\left(A - S_1 K_1^{(i)} - S_2 K_2^{(i)}\right)^{\mathrm{T}} K_2^{(i+1)} + K_2^{(i+1)} \left(A - S_1 K_1^{(i)} - S_2 K_2^{(i)}\right)$$
$$= \overline{Q_2^{(i)}} = -\left(Q_2 + K_1^{(i)} Z_1 K_1^{(i)} + K_2^{(i)} S_2 K_2^{(i)}\right), \quad i = 0, 1, 2, \ldots \tag{A.6}$$

The efficiency of this algorithm was demonstrated in Gajić and Li (1988) (see also Gajić and Shen 1993), where a dynamic system of order 10 was considered, which in fact produced a system of 110 nonlinear algebraic equations.

REFERENCES

Basar, T., A counter example in linear-quadratic games: Existence of nonlinear Nash strategies, *Journal of Optimization Theory and Applications*, 14, 425–430, 1974.

Bertrand, P., A homotopy algorithm for solving coupled Riccati equations, *Optimal Control Applications and Methods*, 6, 351–357, 1985.

Chow, J. and P. Kokotović, Sparsity and time scales, *Proceedings of the American Control Conference*, San Francisco, CA, pp. 656–661, 1983.

Engwerda, J., *LQ Dynamic Optimization and Differential Games*, Wiley, Chichester, 2005.

Engwerda, J., Algorithms for computing Nash equilibria in deterministic LQ games, *Computational Management Science*, 4, 113–140, 2007.

Gajić, Z., Numerical fixed point solution of linear quadratic Gaussian control problem for singularly perturbed systems, *International Journal of Control*, 43, 373–387, 1986.

Gajić, Z. and T.-Y. Li, Simulation results for two new algorithms for solving coupled algebraic Riccati equations, *Third International Symposium on Differential Games*, Sophia Antipolis, France, 1988.

Gajić, Z. and X. Shen, *Parallel Algorithms for Optimal Control of Large Scale Linear Systems*, Springer Verlag, London, 1993.

Huang, M., R. Malhame, and P. Caines, Nash equilibria for large-population linear stochastic systems of weakly coupled agents, in *Analysis, Control, and Optimization of Complex Systems*, E. Boukas and R. Malhame (Eds.), Kluwer, p. 217, May 2005.

Khaalil, H., Approximation of Nash strategies, *IEEE Transactions on Automatic Control*, AC-25, 247–250, 1980.

Kleinman, D., On an iterative technique for Riccati equation computations, *IEEE Transactions Automatic Control*, AC-13, 114–115, 1968.

Kokotović, P., W. Perkins, J. Cruz, and G. D' Ans, ε—coupling approach for near-optimum design of large scale linear systems, *Proceedings of IEE*, Part D., 116, 889–892, 1969.

Koskie, S. and Z. Gajić, A Nash game algorithm for SIR-based power control in 3G wireless CDMA networks, *IEEE/ACM Transactions on Networking*, 12, 1017–1026, 2005.

Kucera, V., A contribution to matrix quadratic equations, *IEEE Transactions on Automatic Control*, AC-17, 344–347, 1972.

Lancaster, P. and M. Tismenetsky, *The Theory of Matrices*, Academic Press, Orlando, FL 1985.

Li, T.-Y. and Z. Gajić, Lyapunov iterations for solving coupled algebraic Riccati equations of Nash differential games and algebraic Riccati equations of zero-sum games, *Annals of Dynamic Games*, 3, 333–351, 1995.

Mahmoud, M., A quantitative comparison between two decentralized control approaches, *International Journal of Control*, 28, 261–275, 1978.

Mukaidani, H., Nash strategies for multimodeling systems, *Transactions of the Society of Instrument and Control Engineers*, 39, 559–568, 2003.

Mukaidani, H., A numerical analysis of the Nash strategy for weakly coupled large-scale systems, *IEEE Transactions of Automatic Control*, 51, 1371–1377, 2006a.

Mukaidani, H., Optimal numerical strategy for Nash games of weakly coupled large-scale systems, *Dynamics of Continuous, Discrete, and Impulsive Systems*, 13, 249–268, 2006b.

Mukaidani, H., Numerical computation of sign-indefinite linear quadratic differential games for weakly coupled linear large-scale systems, *International Journal of Control*, 80, 75–86, 2007.

Philips, R. and P. Kokotović, A singular perturbation approach to modeling and control of Markov chains, *IEEE Transactions of Automatic Control*, AC-26, 1087–1094, 1981.

Ozguner, U. and W. Perkins, A series solution to the Nash strategies for large scale inter-connected systems, *Automatica*, 13, 313–315, 1979.

Papavassilopoulos, G. and P. Olsder, On the linear-quadratic closed-loop, no memory Nash games, *Journal of Optimization Theory and Applications*, 42, 551–560, 1984.

Starr, A. and Y. Ho, Nonzero-sum differential games, *Journal of Optimization Theory and Applications*, 3, 49–79, 1969.

Wonham, W., On a matrix Riccati equation of stochastic control, *SIAM Journal on Control*, 6, 681–697, 1968.

Part II

Hamiltonian Approach for Linear Weakly Coupled Control Systems

8 Finite Time Optimal Control via Hamiltonian Method

The techniques presented in this chapter are based on block diagonalization of the Hamiltonian matrix that comes from the open- and closed-loop optimal control problems of linear weakly coupled systems with quadratic performance criteria. The Hamiltonian matrix by itself has some interesting features that can be further exploited in the case of weakly coupled linear systems facilitating solutions in terms of reduced-order dynamic systems and corresponding algebraic equations. Consequently, the approach used is called the Hamiltonian method. This name for the corresponding method is coined in Gajić and Lim (2001). Similarly to the work of Gajić and Lim (2001) done for singularly perturbed systems, the Hamiltonian approach for weakly coupled systems produces very high accuracy of corresponding systems. In this chapter, our attention is on finite time horizon optimization. In Chapter 9, we will consider infinite time horizon optimization problems of linear weakly coupled systems with quadratic performance criteria and corresponding optimal filtering problems.

In Sections 8.1 and 8.2, the reduced-order methods with an arbitrary degree of accuracy are presented for solving the linear-quadratic optimal open-loop control problems of weakly coupled systems in continuous- and discrete-time domains, respectively. In Sections 8.3 and 8.4, differential and difference Riccati equations coming from finite time horizon optimization problems are considered. It is interesting to point out that simpler results are obtained for the open-loop problem since it is less computationally involved than the closed-loop problem. Several case studies and numerical examples are included in order to demonstrate the efficiency of presented procedures.

8.1 OPEN-LOOP OPTIMAL CONTROL IN CONTINUOUS-TIME

In this section, we study the open-loop control problem (linear two-point boundary value problem) of weakly coupled linear systems. The solution is obtained by exploiting the transformation presented in Chapter 5. The transformation block diagonalizes the Hamiltonian matrix of the optimal linear-quadratic control problem. Completely decoupled sets of reduced-order differential equations are obtained. The convergence to the optimal solution is pretty rapid, due to the fact that the algorithms derived in Chapter 5 have the rate of convergence of at least $O(\varepsilon^2)$. This produces a lot of savings in the size of computations required. In addition, the proposed method is very suitable for parallel and distributed computations.

Consider the linear weakly coupled system

$$
\begin{aligned}
\dot{x}_1(t) &= A_1 x_1(t) + \varepsilon A_2 x_2(t) + B_1 u_1(t) + \varepsilon B_2 u_2(t), & x_1(t_0) &= x_{10} \\
\dot{x}_2(t) &= \varepsilon A_3 x_1(t) + A_4 x_2(t) + \varepsilon B_3 u_1(t) + B_4 u_2(t), & x_2(t_0) &= x_{20}
\end{aligned}
\tag{8.1}
$$

with

$$
z(t) = \begin{bmatrix} z_1(t) \\ z_2(t) \end{bmatrix} = D \begin{bmatrix} x_1(t) \\ x_2(t) \end{bmatrix} = \begin{bmatrix} D_1 & \varepsilon D_2 \\ \varepsilon D_3 & D_4 \end{bmatrix} \begin{bmatrix} x_1(t) \\ x_2(t) \end{bmatrix}
\tag{8.2}
$$

where $x_i(t) \in \Re^{n_i}$, $u_i(t) \in \Re^{m_i}$, $z_i(t) \in \Re^{r_i}$, $i = 1, 2$, are state, control, and output variables, respectively. The system matrices are of appropriate dimensions, and in general, they are bounded functions of a small coupling parameter ε. In this section, we will assume that all given matrices are constant. Moreover, as has been the case in the previous chapters, in order to assure that weakly coupled state variables display weak coupling dynamics, the following weak coupling assumption is introduced.

Assumption 8.1 Matrices A_i, $i = 1, 2, 3, 4$, are constant and $O(1)$. In addition, magnitudes of all system eigenvalues are $O(1)$, that is, $|\lambda_j| = O(1)$, $j = 1, 2, \ldots, n$, which implies that the matrices A_1, A_4 are nonsingular with $\det\{A_1\} = O(1)$ and $\det\{A_4\} = O(1)$.

With Equations 8.1 and 8.2, consider a quadratic performance criterion in the form

$$
\begin{aligned}
J = \frac{1}{2} \int_{t_0}^{t_f} & \left\{ \begin{bmatrix} x_1(t) \\ x_2(t) \end{bmatrix}^T D^T D \begin{bmatrix} x_1(t) \\ x_2(t) \end{bmatrix} + \begin{bmatrix} u_1(t) \\ u_2(t) \end{bmatrix}^T R \begin{bmatrix} u_1(t) \\ u_2(t) \end{bmatrix} \right\} dt \\
& + \frac{1}{2} \begin{bmatrix} x_1(t_f) \\ x_2(t_f) \end{bmatrix}^T F \begin{bmatrix} x_1(t_f) \\ x_2(t_f) \end{bmatrix}
\end{aligned}
\tag{8.3}
$$

with positive definite R and positive semidefinite F, which has to be minimized. It is assumed that matrices F and R have the weakly coupled structure, that is

$$
F = \begin{bmatrix} F_1 & \varepsilon F_2 \\ \varepsilon F_2^T & F_3 \end{bmatrix}, \quad R = \begin{bmatrix} R_1 & 0 \\ 0 & R_2 \end{bmatrix}
\tag{8.4}
$$

In the follow-up derivations, in order to be able to determine the order of magnitudes of quantities appearing in the derivations, we will need the following assumption.

Assumption 8.2   All matrices and their partitions defined in Equations 8.1 through 8.4 are $O(1)$.

The open-loop optimal control problem of Equations 8.1 through 8.4 has the solution given by

$$u(t) = -R^{-1}B^{T}p(t) \tag{8.5}$$

where $p(t) \in \Re^{n_1+n_2}$ is a costate variable satisfying

$$\begin{bmatrix} \dot{p}(t) \\ \dot{x}(t) \end{bmatrix} = \begin{bmatrix} -A^{T} & -D^{T}D \\ -S & A \end{bmatrix} \begin{bmatrix} p(t) \\ x(t) \end{bmatrix} \tag{8.6}$$

with

$$A = \begin{bmatrix} A_1 & \varepsilon A_2 \\ \varepsilon A_3 & A_4 \end{bmatrix}, \quad B = \begin{bmatrix} B_1 & \varepsilon B_2 \\ \varepsilon B_3 & B_4 \end{bmatrix}$$

$$S = BR^{-1}B^{T} = \begin{bmatrix} S_1 & \varepsilon S_2 \\ \varepsilon S_2^{T} & S_3 \end{bmatrix} \tag{8.7}$$

The boundary conditions for Equations 8.6 are expressed in the standard form as

$$W \begin{bmatrix} p(t_0) \\ x(t_0) \end{bmatrix} + G \begin{bmatrix} p(t_f) \\ x(t_f) \end{bmatrix} = c \tag{8.8}$$

where

$$W = \begin{bmatrix} 0 & 0 \\ 0 & I \end{bmatrix}, \quad G = \begin{bmatrix} I & -F \\ 0 & 0 \end{bmatrix}, \quad c = \begin{bmatrix} 0 \\ x(t_0) \end{bmatrix} \tag{8.9}$$

Partitioning $p(t)$ into $p_1(t) \in \Re^{n_1}$ and $p_2(t) \in \Re^{n_2}$ such that $p(t) = \begin{bmatrix} p_1^{T}(t) & p_2^{T}(t) \end{bmatrix}^{T}$, and rearranging rows in Equation 8.6, we get

$$\begin{bmatrix} \dot{p}_1(t) \\ \dot{x}_1(t) \\ \dot{p}_2(t) \\ \dot{x}_2(t) \end{bmatrix} = \begin{bmatrix} T_1 & \varepsilon T_2 \\ \varepsilon T_3 & T_4 \end{bmatrix} \begin{bmatrix} p_1(t) \\ x_1(t) \\ p_2(t) \\ x_2(t) \end{bmatrix} \tag{8.10}$$

where $T_i's$, $i = 1, 2, 3, 4$, are given by

$$T_1 = \begin{bmatrix} -A_1^{T} & -Q_1 \\ -S_1 & A_1 \end{bmatrix}, \quad T_2 = \begin{bmatrix} -A_3^{T} & -Q_2 \\ -S_2 & A_2 \end{bmatrix}$$

$$T_3 = \begin{bmatrix} -A_2^{T} & -Q_2^{T} \\ -S_2^{T} & A_3 \end{bmatrix}, \quad T_4 = \begin{bmatrix} -A_4^{T} & -Q_3 \\ -S_3 & A_4 \end{bmatrix} \tag{8.11}$$

with

$$Q_1 = D_1^{T}D_1 + \varepsilon^2 D_3^{T}D_3, \quad Q_2 = D_1^{T}D_2 + D_3^{T}D_4, \quad Q_3 = D_4^{T}D_4 + \varepsilon^2 D_2^{T}D_2 \tag{8.12}$$

Introduce the notation

$$\begin{bmatrix} p_1(t) \\ x_1(t) \end{bmatrix} = w(t), \quad \begin{bmatrix} p_2(t) \\ x_2(t) \end{bmatrix} = \lambda(t) \tag{8.13}$$

and the corresponding weak coupling transformation presented in Chapter 5

$$\mathbf{T}_1 = \begin{bmatrix} I & -\varepsilon L \\ \varepsilon H & I - \varepsilon^2 HL \end{bmatrix}, \quad \mathbf{T}_1^{-1} = \begin{bmatrix} I - \varepsilon^2 LH & \varepsilon L \\ -\varepsilon H & I \end{bmatrix} \tag{8.14}$$

where L and H satisfy

$$T_1 L + T_2 - L T_4 - \varepsilon^2 L T_3 L = 0 \tag{8.15}$$

$$H(T_1 - \varepsilon^2 L T_3) - (T_4 + \varepsilon^2 T_3 L)H + T_3 = 0 \tag{8.16}$$

This transformation applied to Equation 8.10 produces a decoupled system of the form

$$\dot{\eta}(t) = (T_1 - \varepsilon^2 L T_3)\eta(t) \tag{8.17}$$

$$\dot{\xi}(t) = (T_4 + \varepsilon^2 T_3 L)\xi(t) \tag{8.18}$$

with

$$\begin{bmatrix} \eta(t) \\ \xi(t) \end{bmatrix} = \mathbf{T}_1^{-1} \begin{bmatrix} w(t) \\ \lambda(t) \end{bmatrix} \tag{8.19}$$

To solve the two-point boundary value problem defined in Equations 8.6 through 8.8, we need to find either initial or terminal conditions, which can be obtained as follows. The interchange of rows for $p_2(t)$ and $x_1(t)$ in Equation 8.10 modifies matrices defined in Equations 8.8 and 8.9 as follows:

$$W_1 \begin{bmatrix} w(t_0) \\ \lambda(t_0) \end{bmatrix} + G_1 \begin{bmatrix} w(t_f) \\ \lambda(t_f) \end{bmatrix} = c_1 \tag{8.20}$$

where

$$W_1 = \begin{bmatrix} 0 & 0 & 0 & 0 \\ 0 & I_{n_1} & 0 & 0 \\ 0 & 0 & 0 & 0 \\ 0 & 0 & 0 & I_{n_2} \end{bmatrix}, \quad c_1 = \begin{bmatrix} 0 \\ x_{10} \\ 0 \\ x_{20} \end{bmatrix}$$

$$G_1 = \begin{bmatrix} I_{n_1} & -F_1 & 0 & -\varepsilon F_2 \\ 0 & 0 & 0 & 0 \\ 0 & -\varepsilon F_2^T & I_{n_2} & -F_3 \\ 0 & 0 & 0 & 0 \end{bmatrix} \tag{8.21}$$

The transformation of Equation 8.19 applied to Equation 8.20 produces

$$W_2 \begin{bmatrix} \eta(t_0) \\ \xi(t_0) \end{bmatrix} + G_2 \begin{bmatrix} \eta(t_f) \\ \xi(t_f) \end{bmatrix} = c_1 \tag{8.22}$$

with

$$W_2 = W_1 \mathbf{T}_1, \quad G_2 = G_1 \mathbf{T}_1 \tag{8.23}$$

Since the solutions of Equations 8.17 and 8.18 are given by

$$\eta(t) = e^{(T_1 - \varepsilon^2 LT_3)(t - t_0)} \eta(t_0) \tag{8.24}$$

$$\xi(t) = e^{(T_4 - \varepsilon^2 T_3 L)(t - t_0)} \xi(t_0) \tag{8.25}$$

we can eliminate $\eta(t_f)$ and $\xi(t_f)$ from Equation 8.22; that is, we can obtain

$$\left(W_2 + G_2 \begin{bmatrix} e^{(T_1 - \varepsilon^2 LT_3)(t_f - t_0)} & 0 \\ 0 & e^{(T_4 + \varepsilon^2 T_3 L)(t_f - t_0)} \end{bmatrix} \right) \begin{bmatrix} \eta(t_0) \\ \xi(t_0) \end{bmatrix} = c_1 \tag{8.26}$$

Algebraic Equation 8.26 has the compact form

$$\alpha(\varepsilon) \begin{bmatrix} \eta(t_0) \\ \xi(t_0) \end{bmatrix} = c_1 \tag{8.27}$$

with obvious definition for $\alpha(\varepsilon)$. It is shown in the next lemma that this system of linear algebraic equations has a unique solution, assuming that the coupling parameter ε is sufficiently small.

LEMMA 8.1

The matrix $\alpha(\varepsilon)$ is invertible for sufficiently small values of ε.

Proof Let the transition matrices of Equations 8.17 and 8.18 be denoted by $\Phi(t - t_0)$ and $\Psi(t - t_0)$, respectively, and partitioned as follows:

$$\Phi(t - t_0) = \begin{bmatrix} \Phi_{11}(t - t_0) & \Phi_{12}(t - t_0) \\ \Phi_{21}(t - t_0) & \Phi_{22}(t - t_0) \end{bmatrix}$$

$$\Psi(t - t_0) = \begin{bmatrix} \Psi_{11}(t - t_0) & \Psi_{12}(t - t_0) \\ \Psi_{21}(t - t_0) & \Psi_{22}(t - t_0) \end{bmatrix} \tag{8.28}$$

From Equation 8.26, we have

$$\alpha(\varepsilon) = \left(W_2 + G_2 \begin{bmatrix} \Phi(t_f - t_0) & 0 \\ 0 & \psi(t_f - t_0) \end{bmatrix} \right) \tag{8.29}$$

Using expressions for W_2 and G_2, defined by Equations 8.23 and 8.21, we obtain

$$
\alpha(\varepsilon) = \begin{bmatrix}
I_{n_1} & 0 & 0 & 0 \\
\Phi_{21} - F_1\Phi_{11} & \Phi_{22} - F_1\Phi_{12} & 0 & 0 \\
0 & 0 & I_{n_2} & 0 \\
0 & 0 & \Psi_{21} - F_3\Psi_{11} & \Psi_{22} - F_3\Psi_{12}
\end{bmatrix} + O(\varepsilon)
$$

(8.30)

Since matrices $\Phi_{22}(t_f - t_0) - F_1\Phi_{12}(t_f - t_0)$ and $\Psi_{22}(t_f - t_0) - F_3\Psi_{12}(t_f - t_0)$ are invertible (Kirk 2004, p. 211), the matrix $\alpha(\varepsilon)$ is invertible for sufficiently small values of ε.   ∎

Now we are able to find $\eta(t)$ and $j(t)$ from Equations 8.24 and 8.25. Using Equation 8.19, we can find $w(t)$ and $\lambda(t)$. Partitioning $w(t)$ and $\lambda(t)$ according to Equation 8.13 we get values for $p_1(t)$ and $p_2(t)$, in other words, we find the optimal reduced-order open-loop control defined by Equation 8.5.

The transformation matrix $\mathbf{T_1}$ from Equation 8.14 can easily be obtained, with required accuracy, by using numerical algorithms developed in Chapter 5 for solving algebraic equations (Equations 8.15 and 8.16). These algorithms converge with the rate of convergence of at least $O(\varepsilon^2)$. Thus, after k iterations, one gets the approximation $\mathbf{T_1^{(k)}} = \mathbf{T_1} + O(\varepsilon^{2k})$. The use of $\mathbf{T_1^{(k)}}$ in the design procedure instead of $\mathbf{T_1}$ will perturb the coefficients of the corresponding systems of linear differential equations by $O(\varepsilon^2)$, which implies that the approximate solutions of these differential equations are $O(\varepsilon^2)$ close to the exact ones (Kato 1995). Thus, it is of interest to obtain $\mathbf{T_1^{(k)}}$ with the desired accuracy, which produces the same accuracy in the sought solution.

As a matter of fact, we have obtained the approximate expression for the optimal open-loop control in the form

$$
u^{(k)}(t) = -R^{-1}B^T p^{(k)} = u^{\text{opt}}(t) + O(\varepsilon^{2k})
$$

(8.31)

Apparently, as k increases, the approximate control defined in Equation 8.31 converges very rapidly to the optimal solution.

Simulation results for finding the optimal open-loop control in terms of the reduced-order problems are presented in Section 8.2, where a fifth-order distillation column example is solved. It is interesting to point out that the proposed method produces a simpler system of equations for the open-loop control than for the closed-loop control. This can be seen by comparing linear systems of differential equations (Equation 8.10) and the corresponding equations for the closed-loop control problem (Equations 8.76 and 8.78). The closed-loop solution is computationally much more involved since the corresponding system of differential equations is of the order of $2 \times (2n \times n)$, whereas Equation 8.10 represents the same set of equations of the order of $2n \times 1$.

The results presented in this section are mostly based on the paper by Su and Gajić (1991).

8.1.1 CASE STUDY: DISTILLATION COLUMN

The recursive reduced-order open-loop control problem is demonstrated on a real-world problem, a fifth-order distillation column (Petkov et al. 1986). The problem matrices A and B are given by

$$A = \begin{bmatrix} -0.1094 & 0.0628 & 0 & 0 & 0 \\ 1.3060 & -2.1320 & 0.9807 & 0 & 0 \\ 0 & 1.5950 & -3.1490 & 1.5470 & 0 \\ 0 & 0.0355 & 2.6320 & -4.2570 & 1.8550 \\ 0 & 0.00227 & 0 & 0.1636 & -0.1625 \end{bmatrix}$$

$$B = \begin{bmatrix} 0 & 0.0632 & 0.0838 & 0.1004 & 0.0063 \\ 0 & 0 & -0.1396 & -0.2060 & -0.0128 \end{bmatrix}^{\mathrm{T}}$$

The remaining matrices are chosen as

$$D^{\mathrm{T}}D = \begin{bmatrix} 3 & 0 & 0.7 & 0.7 & 0.7 \\ 0 & 3 & 0.7 & 0.7 & 0.7 \\ 0.7 & 0.7 & 3 & 0 & 0 \\ 0.7 & 0.7 & 0 & 3 & 0 \\ 0.7 & 0.7 & 0 & 0 & 3 \end{bmatrix}, \quad R = I_2, \quad F = I_5$$

The initial and final times are selected as $t_0 = 0$ and $t_f = 1$. The initial conditions are chosen randomly as

$$x_{10} = [-1.259 \quad 1.437]^{\mathrm{T}}, \quad x_{20} = [-0.412 \quad -0.642 \quad 0.877]^{\mathrm{T}}$$

The system is partitioned into two subsystems with $n_1 = 2$, $n_2 = 3$, and $\varepsilon = 0.6$. The small parameter ε is roughly estimated from the strongest coupled matrix—in this case matrix B—producing $|b_{31}|/|b_{32}| = 0.0838/0.1396 = 0.6$. The open-loop control is obtained with a accuracy of 10^{-5} after six iterations. The corresponding simulation results for both components of the approximate open-loop control are presented in Table 8.1.

8.2 OPEN-LOOP OPTIMAL CONTROL IN DISCRETE-TIME

In this section, we will study the open-loop optimal control problem of discrete weakly coupled linear systems in terms of reduced-order difference equations. A weakly coupled linear discrete system is represented by Gajić et al. (1990)

$$x_1(k+1) = A_1 x_1(k) + \varepsilon A_2 x_2(k) + B_1 u_1(k) + \varepsilon B_2 u_2(k)$$
$$x_2(k+1) = \varepsilon A_3 x_1(k) + A_4 x_2(k) + \varepsilon B_3 u_1(k) + B_4 u_2(k) \qquad (8.32)$$
$$x_1(0) = x_{10}, \quad x_2(0) = x_{20}$$

TABLE 8.1

Simulation Results for the Open-Loop Control

Iteration	$u(t = 0)$	$u(t = 0.25)$	$u(t = 0.5)$	$u(t = 0.75)$
6 = optimal control	−0.01033	0.00270	0.01275	0.00945
	−0.04169	−0.00916	0.00411	0.02781
5	−0.01033	0.00270	0.01274	0.00945
	−0.04169	−0.00916	0.00411	0.02781
4	−0.01039	0.00268	0.01275	0.00948
	−0.04167	−0.00917	0.00407	0.02766
3	−0.01073	−0.00262	0.01293	0.00980
	−0.04153	−0.00938	0.00360	0.02713
2	0.02831	0.03571	0.03895	0.02694
	−0.07095	−0.03485	−0.00164	0.01403
1	0.06505	−0.06219	0.05589	0.03591
	−0.09423	−0.05219	−0.02776	0.00699
0	0.69589	0.54689	0.40566	0.25491
	−0.34489	−0.25504	−0.18864	−0.11626

with state variables $x_i(k) \in \Re^{n_i}$ and control inputs $u_i(k) \in \Re^{m_i}$, $i = 1, 2$, respectively, where ε is a small coupling parameter. We will assume that all given matrices are constant. The performance criterion of the corresponding linear-quadratic control problem is defined by

$$J = \frac{1}{2} \sum_{k=0}^{k_f - 1} \left[x^T(k)Qx(k) + u^T(k)Ru(k) \right] + \frac{1}{2} x^T(k_f)Fx(k_f) \qquad (8.33)$$

with

$$x(k) = \begin{bmatrix} x_1(k) \\ x_2(k) \end{bmatrix}, \quad Q = \begin{bmatrix} Q_1 & \varepsilon Q_2 \\ \varepsilon Q_2^T & Q_3 \end{bmatrix} \geq 0$$

$$F = \begin{bmatrix} F_1 & \varepsilon F_2 \\ \varepsilon F_2^T & F_3 \end{bmatrix} \geq 0, \quad R = \begin{bmatrix} R_1 & 0 \\ 0 & R_2 \end{bmatrix} > 0 \qquad (8.34)$$

The open-loop optimal control problem has the solution

$$u(k) = -R^{-1}B^T\lambda(k + 1) \qquad (8.35)$$

where $\lambda(k)$ is the costate variable that satisfies the Hamiltonian system

$$\begin{bmatrix} x(k + 1) \\ \lambda(k + 1) \end{bmatrix} = \mathbf{H} \begin{bmatrix} x(k) \\ \lambda(k) \end{bmatrix} \qquad (8.36)$$

with boundary conditions expressed in the standard form as

$$M \begin{bmatrix} x(0) \\ \lambda(0) \end{bmatrix} + N \begin{bmatrix} x(k_f) \\ \lambda(k_f) \end{bmatrix} = c$$

$$M = \begin{bmatrix} I & 0 \\ 0 & 0 \end{bmatrix}, \quad N = \begin{bmatrix} 0 & 0 \\ -F & I \end{bmatrix}, \quad c = \begin{bmatrix} x(0) \\ 0 \end{bmatrix} \tag{8.37}$$

H is the Hamiltonian matrix and is given by

$$\mathbf{H} = \begin{bmatrix} A + BR^{-1}B^{\mathrm{T}}A^{-\mathrm{T}}Q & -BR^{-1}B^{\mathrm{T}}A^{-\mathrm{T}} \\ -A^{-\mathrm{T}}Q & A^{-\mathrm{T}} \end{bmatrix} \tag{8.38}$$

Matrices A and S have the forms

$$A = \begin{bmatrix} A_1 & \varepsilon A_2 \\ \varepsilon A_3 & A_4 \end{bmatrix}, \quad S = BR^{-1}B^{\mathrm{T}} = \begin{bmatrix} S_1 & \varepsilon Z \\ \varepsilon Z^{\mathrm{T}} & S_2 \end{bmatrix} \tag{8.39}$$

In the following, to preserve the weakly coupled structure in calculations that follow and to provide that these structures in fact imply weakly coupled dynamic connections between subsystems, we assume that Assumptions 8.1 and 8.2 hold in this section with appropriate application to the discrete-time domain matrices of this section.

Similar to Section 8.1, done for continuous-time weakly coupled systems, the Hamiltonian matrix of discrete-time systems retains the weakly coupled form by interchanging some state and costate variables so that it can be block diagonalized via the nonsingular similarity transformations presented in Chapter 5.

Partitioning vector $\lambda(k)$ as $\lambda(k) = \begin{bmatrix} \lambda_1^{\mathrm{T}}(k) & \lambda_2^{\mathrm{T}}(k) \end{bmatrix}^{\mathrm{T}}$ with $\lambda_1(k) \in \Re^{n_1}$ and $\lambda_2(k) \in \Re^{n_2}$, we get

$$\begin{bmatrix} x_1(k+1) \\ x_2(k+1) \\ \lambda_1(k+1) \\ \lambda_2(k+1) \end{bmatrix} = \begin{bmatrix} \overline{A}_1 & \varepsilon\overline{A}_2 & \overline{S}_1 & \varepsilon\overline{S}_2 \\ \varepsilon\overline{A}_3 & \overline{A}_4 & \varepsilon\overline{S}_3 & \overline{S}_4 \\ \overline{Q}_1 & \varepsilon\overline{Q}_2 & \overline{A}_{11}^{\mathrm{T}} & \varepsilon\overline{A}_{21}^{\mathrm{T}} \\ \varepsilon\overline{Q}_3 & \overline{Q}_4 & \varepsilon\overline{A}_{12}^{\mathrm{T}} & \overline{A}_{22}^{\mathrm{T}} \end{bmatrix} \begin{bmatrix} x_1(k) \\ x_2(k) \\ \lambda_1(k) \\ \lambda_2(k) \end{bmatrix} = \mathbf{H} \begin{bmatrix} x_1(k) \\ x_2(k) \\ \lambda_1(k) \\ \lambda_2(k) \end{bmatrix} \tag{8.40}$$

See Appendix 8.1 for more details about derivations of the last equation.

Interchanging the second and third rows in Equation 8.40 produces

$$\begin{bmatrix} x_1(k+1) \\ \lambda_1(k+1) \\ x_2(k+1) \\ \lambda_2(k+1) \end{bmatrix} = \begin{bmatrix} \overline{A}_1 & \overline{S}_1 & \varepsilon\overline{A}_2 & \varepsilon\overline{S}_2 \\ \overline{Q}_1 & \overline{A}_{11}^{\mathrm{T}} & \varepsilon\overline{Q}_2 & \varepsilon\overline{A}_{21}^{\mathrm{T}} \\ \varepsilon\overline{A}_3 & \varepsilon\overline{S}_3 & \overline{A}_4 & \overline{S}_4 \\ \varepsilon\overline{Q}_3 & \varepsilon\overline{A}_{12}^{\mathrm{T}} & \overline{Q}_4 & \overline{A}_{22}^{\mathrm{T}} \end{bmatrix} \begin{bmatrix} x_1(k) \\ \lambda_1(k) \\ x_2(k) \\ \lambda_2(k) \end{bmatrix}$$

$$= \begin{bmatrix} T_1 & \varepsilon T_2 \\ \varepsilon T_3 & T_4 \end{bmatrix} \begin{bmatrix} x_1(k) \\ \lambda_1(k) \\ x_2(k) \\ \lambda_2(k) \end{bmatrix} \tag{8.41}$$

where

$$T_1 = \begin{bmatrix} \overline{A}_1 & \overline{S}_1 \\ \overline{Q}_1 & A_{11}^T \end{bmatrix}, \quad T_2 = \begin{bmatrix} \overline{A}_2 & \overline{S}_2 \\ \overline{Q}_2 & A_{21}^T \end{bmatrix}$$

$$T_3 = \begin{bmatrix} \overline{A}_3 & \overline{S}_3 \\ \overline{Q}_3 & A_{12}^T \end{bmatrix}, \quad T_4 = \begin{bmatrix} \overline{A}_4 & \overline{S}_4 \\ \overline{Q}_4 & A_{22}^T \end{bmatrix} \tag{8.42}$$

Introducing the notation

$$U(k) = \begin{bmatrix} x_1(k) \\ \lambda_1(k) \end{bmatrix}, \quad V(k) = \begin{bmatrix} x_2(k) \\ \lambda_2(k) \end{bmatrix} \tag{8.43}$$

we get the weakly coupled discrete system in the new coordinates

$$U(k+1) = T_1 U(k) + \varepsilon T_2 V(k)$$
$$V(k+1) = \varepsilon T_3 U(k) + T_4 V(k) \tag{8.44}$$

Applying the nonsingular transformation from Gajić and Shen (1989)

$$\mathbf{T}_1 = \begin{bmatrix} I & -\varepsilon L \\ \varepsilon H & I - \varepsilon^2 HL \end{bmatrix}, \quad \mathbf{T}_1^{-1} = \begin{bmatrix} I - \varepsilon^2 LH & \varepsilon L \\ -\varepsilon H & I \end{bmatrix}$$

$$\begin{bmatrix} U(k) \\ V(k) \end{bmatrix} = \mathbf{T}_1 \begin{bmatrix} \overline{U}(k) \\ \overline{V}(k) \end{bmatrix} \tag{8.45}$$

to Equation 8.41 produces two completely decoupled subsystems

$$\overline{U}(k+1) = (T_1 - \varepsilon^2 L T_3)\overline{U}(k)$$
$$\overline{V}(k+1) = (T_4 + \varepsilon^2 T_3 L)\overline{V}(k) \tag{8.46}$$

where L and H satisfy

$$H(T_1 - \varepsilon^2 L T_3) - (T_4 + \varepsilon^2 T_3 L)H + T_3 = 0$$
$$T_1 L + T_2 - L T_4 - \varepsilon^2 L T_3 L = 0 \tag{8.47}$$

Matrices L and H can be obtained with the required accuracy by using numerical techniques presented in Section 5.1 that converges with the rate of convergence $O(\varepsilon^2)$. Thus, after j iterations, one gets the approximations $L^{(j)} = L + O(\varepsilon^{2j})$ and $H^{(j)} = H + O(\varepsilon^{2j})$. Using $L^{(j)}$, $H^{(j)}$ instead of L and H, will perturb the coefficients of the corresponding systems of linear differential equations by $O(\varepsilon^2)$, which implies that the same accuracy of the system solutions is obtained.

The boundary conditions are changed due to the interchange of $\lambda_1(k)$ and $x_2(k)$, which modifies matrices in Equation 8.37 as follows:

$$M_1 \begin{bmatrix} U(0) \\ V(0) \end{bmatrix} + N_1 \begin{bmatrix} U(k_f) \\ V(k_f) \end{bmatrix} = c_1 \tag{8.48}$$

where

$$M_1 = \begin{bmatrix} I_{n_1} & 0 & 0 & 0 \\ 0 & 0 & 0 & 0 \\ 0 & 0 & I_{n_2} & 0 \\ 0 & 0 & 0 & 0 \end{bmatrix}, \quad c_1 = \begin{bmatrix} x_1(0) \\ 0 \\ x_2(0) \\ 0 \end{bmatrix}$$

$$N_1 = \begin{bmatrix} 0 & 0 & 0 & 0 \\ -F_1 & I_{n_1} & -\varepsilon F_2 & 0 \\ 0 & 0 & 0 & 0 \\ -\varepsilon F_2^T & 0 & -F_3 & I_{n_2} \end{bmatrix}$$

$$\tag{8.49}$$

The nonsingular transformation Equation 8.45 applied to Equation 8.48 produces

$$M_2 \begin{bmatrix} \overline{U}(0) \\ \overline{V}(0) \end{bmatrix} + N_2 \begin{bmatrix} \overline{U}(k_f) \\ \overline{V}(k_f) \end{bmatrix} = c_1 \tag{8.50}$$

where

$$M_2 = M_1 T_1, \quad N_2 = N_1 T_1 \tag{8.51}$$

Solutions of Equation 8.46 are then given by

$$\overline{U}(k) = \left(T_1 - \varepsilon^2 L T_3\right)^k \overline{U}(0)$$
$$\overline{V}(k) = \left(T_4 + \varepsilon^2 T_3 L\right)^k \overline{V}(0) \tag{8.52}$$

We can eliminate $\overline{U}(k_f)$ and $\overline{V}(k_f)$ from Equation 8.50 such that

$$\beta(\varepsilon) \begin{bmatrix} \overline{U}(0) \\ \overline{V}(0) \end{bmatrix} = c_1 \tag{8.53}$$

where

$$\beta(\varepsilon) = \left\{ M_2 + N_2 \begin{bmatrix} \left(T_1 - \varepsilon^2 L T_3\right)^{k_f} & 0 \\ 0 & \left(T_4 + \varepsilon^2 T_3 L\right)^{k_f} \end{bmatrix} \right\} \tag{8.54}$$

It is shown in Appendix 8.2 that $\beta(\varepsilon)$ is nonsingular, that is, this system of linear algebraic equation has a unique solution, assuming that a coupling parameter ε is sufficiently small. Since $\beta(\varepsilon)$ is invertible, $\overline{U}(0)$ and $\overline{V}(0)$ can be obtained.

Now we can find $\overline{U}(k)$ and $\overline{V}(k)$ from Equation 8.52. Using Equation 8.45, we can find $U(k)$ and $V(k)$.

After getting the solutions of $U(k)$ and $V(k)$, we can use the following relations to get the values for $\lambda_1(k)$ and $\lambda_2(k)$.

$$
\begin{aligned}
\begin{bmatrix} x_1(k) \\ \lambda_1(k) \end{bmatrix} &= \begin{bmatrix} U_1(k) \\ U_2(k) \end{bmatrix} = U(k) \\
\begin{bmatrix} x_2(k) \\ \lambda_2(k) \end{bmatrix} &= \begin{bmatrix} V_1(k) \\ V_2(k) \end{bmatrix} = V(k)
\end{aligned}
\tag{8.55}
$$

The costate variable $\lambda(k)$ and the optimal open-loop control law are therefore found such that

$$
u(k) = -R^{-1}B^{\mathrm{T}}\lambda^{(j)}(k+1) + O(\varepsilon^{2j}) \tag{8.56}
$$

Apparently, as j increases, the control defined in Equation 8.56 converges very rapidly to the optimal solution.

8.2.1 NUMERICAL EXAMPLE

In order to demonstrate the proposed method, a discrete system taken from Katzberg (1977) is studied. The system matrices are

$$
A = \begin{bmatrix} 0.964 & 0.18 & 0.017 & 0.019 \\ -0.342 & 0.802 & 0.162 & 0.179 \\ 0.016 & 0.019 & 0.983 & 0.181 \\ 0.144 & 0.179 & -0.163 & 0.82 \end{bmatrix}
$$

$$
B^{\mathrm{T}} = \begin{bmatrix} 0.019 & 0.180 & 0.005 & -0.054 \\ 0.001 & 0.019 & 0.019 & 0.181 \end{bmatrix}
$$

with the system initial conditions determined by

$$
x^{\mathrm{T}}(0) = \begin{bmatrix} 1 & 1 & 1 & 1 \end{bmatrix}
$$

The weighting matrices are chosen as $R = I_2$, $Q = 0.1I_4$ with the terminal penalty matrix $F = 0.5I_4$. The small coupling parameter ε is 0.329, and the final time is $k_f = 5$.

The approximate control is defined by

$$
u^{(j)}(k) = -R^{-1}B^{\mathrm{T}}\lambda^{(j)}(k+1)
$$

where j stands for the number of iterations used to solve for L recursively in Equation 8.47. The recursive solution for L produces for $j = 0$

TABLE 8.2

Approximate and Optimal Values of $u(k)$

k	$u^{(0)}(k)$	$u^{(1)}(k)$	$u^{(2)}(k)$	$u^{(3)}(k)$	$u^{(4)}(k)$	$u^{(5)}(k)$ = Optimal
1	−0.0745	−0.1228	−0.0612	0.0397	0.0147	0.0131
	−0.3616	−0.1738	−0.0606	−0.1632	−0.1333	−0.1323
2	0.0709	−0.0674	−0.0214	0.0695	0.0465	0.0447
	−0.3208	−0.1354	−0.0430	−0.1171	−0.0926	−0.0912
3	0.1532	−0.0255	0.0123	0.0883	0.0685	0.0668
	−0.3278	−0.0898	−0.0259	−0.0759	−0.0569	−0.0556
4	0.1955	0.0074	0.0429	0.0994	0.0834	0.0819
	−0.3835	−0.0409	−0.0047	−0.0390	−0.0251	−0.0239
5	0.2178	0.0359	0.0744	0.1068	0.0946	0.0934
	−0.4884	0.0083	0.0257	−0.0056	0.0042	0.0053

$$L^{(j+1)} - L^{(j)} = \begin{bmatrix} -1.1436 & -1.3305 & 0.0562 & -0.2300 \\ 1.8946 & 0.1277 & 0.3022 & -0.2789 \\ 0.8015 & -1.7028 & -2.5535 & 4.2259 \\ 2.4466 & 0.0174 & -2.787 & 1.5939 \end{bmatrix}$$

and for $j = 6$

$$L^{(j+1)} - L^{(j)} = 10^{-5} \begin{bmatrix} -0.0009 & -0.0013 & -0.0049 & 0.0021 \\ 0.0001 & -0.0079 & -0.0091 & -0.0007 \\ -0.0391 & 0.1256 & 0.112 & -0.1037 \\ -0.0499 & 0.0656 & 0.0934 & -0.059 \end{bmatrix}$$

Table 8.2 shows the approximate and optimal values of control $u(k)$. It can be seen that the approximative optimal control approaches the optimal control and after four iterations they are in close proximity to each other. Interestingly, the zeroth approximation values are far away from the optimal open-loop control values.

8.3   DIFFERENTIAL RICCATI EQUATION

In this section, we study the finite time closed-loop optimal control problem of weakly coupled systems. The recursive reduced-order solution will be obtained by exploiting the transformation presented in Chapter 5 which will block diagonalize the Hamiltonian form of the solution for the optimal linear-quadratic control problem. Completely decoupled sets of reduced-order differential equations are obtained. The convergence to the optimal solution is pretty rapid, due to the fact that the algorithms derived in Section 5.1 have the rate of convergence of at least of $O(\varepsilon^2)$. This produces a lot of savings in the size of computations required.

Consider the linear weakly coupled system

$$\dot{x}_1(t) = A_1 x_1(t) + \varepsilon A_2 x_2(t) + B_1 u_1(t) + \varepsilon B_2 u_2(t), \quad x_1(t_0) = x_{10}$$
$$\dot{x}_2(t) = \varepsilon A_3 x_1(t) + A_4 x_2(t) + \varepsilon B_3 u_1(t) + B_4 u_2(t), \quad x_2(t_0) = x_{20}$$
(8.57)

with

$$z(t) = \begin{bmatrix} z_{1t} \\ z_{2t} \end{bmatrix} = D \begin{bmatrix} x_1(t) \\ x_2(t) \end{bmatrix} = \begin{bmatrix} D_1 & \varepsilon D_2 \\ \varepsilon D_3 & D_4 \end{bmatrix} \begin{bmatrix} x_1(t) \\ x_2(t) \end{bmatrix}$$
(8.58)

where $x_i(t) \in R^{n_i}$, $u_i(t) \in R^{m_i}$, $z_i(t) \in R^{r_i}$, $i = 1, 2$, are state, control, and output variables, respectively. The system matrices are of appropriate dimensions and, in general, they are bounded functions of a small coupling parameter ε (Gajić et al. 1990; Harkara et al. 1989; Petrović and Gajić 1988). In this section, we will assume that all given matrices are constant.

With Equations 8.57 through 8.78, we consider the performance criterion

$$J = \frac{1}{2} \int_{t_0}^{t_f} \left\{ \begin{bmatrix} x_1(t) \\ x_2(t) \end{bmatrix}^T D^T D \begin{bmatrix} x_1(t) \\ x_2(t) \end{bmatrix} + \begin{bmatrix} u_1(t) \\ u_2(t) \end{bmatrix}^T R \begin{bmatrix} u_1(t) \\ u_2(t) \end{bmatrix} \right\} dt$$
$$+ \frac{1}{2} \begin{bmatrix} x_1(t_f) \\ x_2(t_f) \end{bmatrix}^T F \begin{bmatrix} x_1(t_f) \\ x_2(t_f) \end{bmatrix}$$
(8.59)

with positive definite R and positive semidefinite F, which has to be minimized. It is assumed that matrices F and R have the weakly coupled structures, that is

$$F = \begin{bmatrix} F_1 & \varepsilon F_2 \\ \varepsilon F_2^T & F_3 \end{bmatrix}, \quad R = \begin{bmatrix} R_1 & 0 \\ 0 & R_2 \end{bmatrix}$$
(8.60)

It has been tacitly assumed that assumptions corresponding to Assumptions 8.1 and 8.2 (all matrices and their partitions are $O(1)$ and all eigenvalues of matrices A_1 and A_4 and $O(1)$) hold in this section as well for the purpose of preserving pure weak coupling in subsystem dynamics and computations that follow.

The optimal closed-loop control law has the very well-known form (Kwakernaak and Sivan 1972)

$$u(x(t)) = \begin{bmatrix} u_1(x(t)) \\ u_2(x(t)) \end{bmatrix} = -R^{-1} \begin{bmatrix} B_1 & \varepsilon B_2 \\ \varepsilon B_3 & B_4 \end{bmatrix}^T P(t) \begin{bmatrix} x_1(t) \\ x_2(t) \end{bmatrix}$$
$$= -R^{-1} B^T P(t) x(t)$$
(8.61)

where $P(t)$ satisfies the differential Riccati equation given by

$$-\dot{P}(t) = P(t)A + A^T P(t) + D^T D - P(t)SP(t), \quad P(t_f) = F$$
(8.62)

with

$$A = \begin{bmatrix} A_1 & \varepsilon A_2 \\ \varepsilon A_3 & A_4 \end{bmatrix}, \quad S = BR^{-1}B^\mathrm{T} = \begin{bmatrix} S_1 & \varepsilon S_2 \\ \varepsilon S_2^\mathrm{T} & S_3 \end{bmatrix} \tag{8.63}$$

Due to the weakly coupled structure of all coefficients in Equation 8.62, the solution of this equation has the form

$$P(t) = \begin{bmatrix} P_1(t) & \varepsilon P_2(t) \\ \varepsilon P_2^\mathrm{T}(t) & P_3(t) \end{bmatrix} \tag{8.64}$$

In this section, we will exploit the Hamiltonian form of the solution of the Riccati differential equation and the nonsingular similarity transformation presented in Section 5.1 in order to obtain an efficient recursive method for solving Equation 8.62.

The solution of Equation 8.62 can be sought in the form

$$P(t) = M(t)N^{-1}(t), \quad P(t_f) = M(t_f) = F \tag{8.65}$$

where matrices $M(t)$ and $N(t)$ satisfy the system of linear differential equations (Kwakernaak and Sivan 1972)

$$\dot{M}(t) = -A^\mathrm{T}M(t) - D^\mathrm{T}DN(t), \quad M(t_f) = F \tag{8.66}$$

$$\dot{N}(t) = -SM(t) + AN(t), \quad N(t_f) = I \tag{8.67}$$

Equation 8.65 can easily be justified starting with $P(t)N(t) = M(t)$, taking the derivative, that is, $\dot{P}(t)\,N(t) + P(t)\,\dot{N}(t) = \dot{M}(t)$ and using Equations 8.62, 8.66, and 8.67 to show that the left-hand side is equal to the right-hand side.

The next lemma guarantees the existence of the invertible solution for $N(t)$ for all t.

LEMMA 8.2

If the triple $(A, B, \mathrm{Chol}(Q))$ *is stabilizable–observable, then the matrix* $N(t)$, *with* $N(t_f) = I$ *is invertible for any* $t \in (t_0, t_f)$.

Proof By using a dichotomy transformation introduced by Wilde and Kokotović (1972)

$$\begin{bmatrix} M \\ N \end{bmatrix} = \begin{bmatrix} K & P \\ I & I \end{bmatrix} \begin{bmatrix} \widehat{M} \\ \widehat{N} \end{bmatrix} \tag{8.68}$$

$$\begin{bmatrix} \widehat{M} \\ \widehat{N} \end{bmatrix} = \begin{bmatrix} (\underline{K} - \underline{P})^{-1} & -(\underline{K} - \underline{P})^{-1}\underline{P} \\ -(\underline{K} - \underline{P})^{-1} & I + (\underline{K} - \underline{P})^{-1}\underline{P} \end{bmatrix} \begin{bmatrix} M \\ N \end{bmatrix} \tag{8.69}$$

where $\underline{P}$ and $\underline{K}$ are unique positive definite and negative definite solutions of the algebraic Riccati equation corresponding to Equation 8.62, the system (Equations 8.66 and 8.67) can be transformed into

$$\begin{bmatrix} \dot{\widehat{M}}(t) \\ \dot{\widehat{N}}(t) \end{bmatrix} = \begin{bmatrix} A - S\underline{K} & 0 \\ 0 & A - S\underline{P} \end{bmatrix} \begin{bmatrix} \widehat{M}(t) \\ \widehat{N}(t) \end{bmatrix} \tag{8.70}$$

with terminal conditions

$$\widehat{M}(t_f) = (\underline{K} - \underline{P})^{-1}(F - \underline{P})$$
$$\widehat{N}(t_f) = I + (\underline{K} - \underline{P})^{-1}(F - \underline{P}) = I + \widehat{M}(t_f)$$

It is known that $(A - S\underline{K})$ is an unstable matrix and that matrix $(A - S\underline{P})$ is stable (Wilde and Kokotović 1972). The solution of Equation 8.70 is given by

$$\widehat{M}(t) = e^{(A-S\underline{K})(t-t_f)}\widehat{M}(t_f)$$
$$\widehat{N}(t) = e^{(A-S\underline{P})(t-t_f)}\widehat{N}(t_f) \tag{8.71}$$

Using Equations 8.68 through 8.71 it can easily be shown that

$$N(t) = e^{(A-S\underline{K})(t-t_f)}\left[I + \left(I - e^{S(\underline{P}-\underline{K})(t-t_f)}(\underline{K} - \underline{P})^{-1}(\underline{P} - F)\right)\right]N(t_f)$$

that is

$$N(t) = \phi(t - t_f)N(t_f) \tag{8.72}$$

with obvious definition of $\phi(t - t_f)$. Since $\phi(t - t_f)$ plays the role of the transition matrix of $N(t)$ and by very well-known facts is nonsingular, the regularity of $N(t)$ is determined only by $N(t_f)$. Thus, having chosen $N(t_f)$ as an identity will assure the nonsingularity of $N(t)$ for any $t < t_f$, and prove the given lemma. ∎

Knowing the nature of the solution of Equation 8.62 as given by Equation 8.64, we introduce compatible partitions of $M(t)$ and $N(t)$ matrices as

$$M(t) = \begin{bmatrix} M_1(t) & \varepsilon M_2(t) \\ \varepsilon M_3(t) & M_4(t) \end{bmatrix}, \quad N(t) = \begin{bmatrix} N_1(t) & \varepsilon N_2(t) \\ \varepsilon N_3(t) & N_4(t) \end{bmatrix} \tag{8.73}$$

Partitioning Equations 8.66 and 8.67, according to Equations 8.63 and 8.73, will reveal a decoupled structure, that is, $M_1(t)$, $M_3(t)$, $N_1(t)$, and $N_3(t)$ are independent of equations for $M_2(t)$, $M_4(t)$, $N_2(t)$, and $N_4(t)$ and vice versa. Introducing the notation

$$U(t) = \begin{bmatrix} M_1(t) \\ N_1(t) \end{bmatrix}, \quad V = \begin{bmatrix} \varepsilon M_3(t) \\ \varepsilon N_3(t) \end{bmatrix}$$
$$X(t) = \begin{bmatrix} \varepsilon M_2(t) \\ \varepsilon N_2(t) \end{bmatrix}, \quad Y(t) = \begin{bmatrix} M_4(t) \\ N_4(t) \end{bmatrix} \tag{8.74}$$

and

$$T_1 = \begin{bmatrix} -A_1^T & -Q_1 \\ -S_1 & A_1 \end{bmatrix}, \quad T_2 = \begin{bmatrix} -A_3^T & -Q_2 \\ -S_2 & A_2 \end{bmatrix}$$
$$T_3 = \begin{bmatrix} -A_2^T & -Q_2^T \\ -S_2^T & A_3 \end{bmatrix}, \quad T_4 = \begin{bmatrix} -A_4^T & -Q_3 \\ -S_3 & A_4 \end{bmatrix} \tag{8.75}$$

where

$$Q_1 = D_1^T D_1 + \varepsilon^2 D_3^T D_3, \quad Q_2 = D_1^T D_2 + D_3^T D_4$$
$$Q_3 = D_4^T D_4 + \varepsilon^2 D_2^T D_2$$

and after doing the same calculation, we get two independent systems of weakly coupled matrix differential equations

$$\dot{U}(t) = T_1 U(t) + \varepsilon T_2 V(t)$$
$$\dot{V}(t) = \varepsilon T_3 U(t) + T_4 V(t) \tag{8.76}$$

with terminal conditions

$$U(t_f) = \begin{bmatrix} F_1 \\ I \end{bmatrix}, \quad V(t_f) = \begin{bmatrix} \varepsilon F_2^T \\ 0 \end{bmatrix} \tag{8.77}$$

and

$$\dot{X}(t) = T_1 X(t) + \varepsilon T_2 Y(t)$$
$$\dot{Y}(t) = \varepsilon T_3 X(t) + T_4 Y(t) \tag{8.78}$$

with terminal conditions

$$X(t_f) = \begin{bmatrix} \varepsilon F_2 \\ 0 \end{bmatrix}, \quad Y(t_f) = \begin{bmatrix} F_3 \\ I \end{bmatrix} \tag{8.79}$$

Note that these two systems have exactly the same form and they only differ in terminal conditions. From this point, we will proceed by applying the decoupling transformation from Section 5.1 This transformation is defined by

$$\mathbf{T_1} = \begin{bmatrix} I & -\varepsilon L \\ \varepsilon H & I - \varepsilon^2 HL \end{bmatrix}, \quad \mathbf{T_1^{-1}} = \begin{bmatrix} I - \varepsilon^2 LH & \varepsilon L \\ -\varepsilon H & I \end{bmatrix} \tag{8.80}$$

where L and H satisfy

$$T_1 L + T_2 - L T_4 - \varepsilon^2 L T_3 L = 0 \tag{8.81}$$

$$H(T_1 - \varepsilon^2 L T_3) - (T_4 + \varepsilon^2 T_3 L) H + T_3 = 0 \tag{8.82}$$

Applying transformation Equation 8.86 to Equations 8.76 through 8.79 produces

$$\dot{\widehat{U}}(t) = \left(T_1 - \varepsilon^2 LT_3\right)\widehat{U}(t), \quad \widehat{U}(t_f) = U(t_f) - \varepsilon LV(t_f) \tag{8.83}$$

$$\dot{\widehat{V}}(t) = \left(T_4 + \varepsilon^2 T_3 L\right)\widehat{V}(t), \quad \widehat{V}(t_f) = \varepsilon HU(t_f) + (I - \varepsilon^2 HL)V(t_f) \tag{8.84}$$

and

$$\dot{\widehat{X}}(t) = \left(T_1 - \varepsilon^2 LT_3\right)\widehat{X}(t), \quad \widehat{X}(t_f) = X(t_f) - \varepsilon LY(t_f) \tag{8.85}$$

$$\dot{\widehat{Y}}(t) = \left(T_4 + \varepsilon^2 T_3 L\right)\widehat{Y}(t), \quad \widehat{Y}(t_f) = \varepsilon HX(t_f) + (I - \varepsilon^2 HL)Y(t_f) \tag{8.86}$$

Solutions of Equations 8.83 through 8.86 are given by

$$\widehat{U}(t) = e^{(T_1 - \varepsilon^2 LT_3)(t - t_f)}\widehat{U}(t_f) \tag{8.87}$$

$$\widehat{V}(t) = e^{(T_4 + \varepsilon^2 T_3 L)(t - t_f)}\widehat{V}(t_f) \tag{8.88}$$

$$\widehat{X}(t) = e^{(T_1 - \varepsilon^2 LT_3)(t - t_f)}\widehat{X}(t_f) \tag{8.89}$$

$$\widehat{Y}(t) = e^{(T_4 + \varepsilon^2 T_3 L)(t - t_f)}\widehat{Y}(t_f) \tag{8.90}$$

so that in the original coordinates we have

$$U(t) = (I - \varepsilon^2 LH)e^{(T_1 - \varepsilon^2 LT_3)(t - t_f)}\widehat{U}(t_f) + \varepsilon Le^{(T_4 + \varepsilon^2 T_3 L)(t - t_f)}\widehat{V}(t_f) \tag{8.91}$$

$$V(t) = -\varepsilon He^{(T_1 - \varepsilon^2 LT_3)(t - t_f)}\widehat{U}(t_f) + e^{(T_4 + \varepsilon^2 T_3 L)(t - t_f)}\widehat{V}(t_f) \tag{8.92}$$

$$X(t) = (I - \varepsilon^2 LH)e^{(T_1 - \varepsilon^2 LT_3)(t - t_f)}\widehat{X}(t_f) + \varepsilon Le^{(T_4 + \varepsilon^2 T_3 L)(t - t_f)}Y(t_f) \tag{8.93}$$

$$Y(t) = -\varepsilon He^{(T_1 - \varepsilon^2 LT_3)(t - t_f)}\widehat{X}(t_f) + e^{(T_4 + \varepsilon^2 T_3 L)(t - t_f)}\widehat{Y}(t_f) \tag{8.94}$$

Partitioning $U(t)$, $V(t)$, $X(t)$, and $Y(t)$ according to Equation 8.74 will produce all components of the matrices $M(t)$ and $N(t)$; that is

$$\begin{bmatrix} M_1(t) \\ N_1(t) \end{bmatrix} = \begin{bmatrix} U_1(t) \\ U_2(t) \end{bmatrix} = U(t), \quad \begin{bmatrix} \varepsilon M_2(t) \\ \varepsilon M_2(t) \end{bmatrix} = \begin{bmatrix} X_1(t) \\ X_2(t) \end{bmatrix} = X(t)$$

$$\begin{bmatrix} \varepsilon M_3(t) \\ \varepsilon N_3(t) \end{bmatrix} = \begin{bmatrix} V_1(t) \\ V_2(t) \end{bmatrix} = V(t), \quad \begin{bmatrix} M_4(t) \\ N_4(t) \end{bmatrix} = \begin{bmatrix} Y_1(t) \\ Y_2(t) \end{bmatrix} = Y(t) \tag{8.95}$$

so that the required solution of Equation 8.62 is given by

$$P(t) = \begin{bmatrix} U_1(t) & X_1(t) \\ V_1(t) & Y_1(t) \end{bmatrix}\begin{bmatrix} U_2(t) & X_2(t) \\ V_2(t) & Y_2(t) \end{bmatrix}^{-1} \tag{8.96}$$

Thus, in order to get the solution of Equation 8.62, $P(t)$, which has dimensions $n \times n = (n_1 + n_2) \times (n_1 + n_2)$, we have to solve two simple algebraic equations (Equations 8.81 and 8.82) of dimensions $(2n_2 \times 2n_1)$ and $(2n_1 \times 2n_2)$, respectively. (Note that due to the symmetry of the matrix $P(t)$ the actual number of differential equations needed to be solved in Equation 8.62 is $0.5n(n-1)$.) The efficient numerical algorithm based on the fixed point iterations and the Newton method for solving Equations 8.81 and 8.82 were presented in Section 5.1. Then, two exponential forms, $\exp[(T_1 - \varepsilon^2 LT_3)(t - t_f)]$ and $\exp[(T_4 + \varepsilon^2 T_3 L)(t - t_f)]$ have to be transformed into the matrix forms by using some of the well-known approaches (Molen and Van Loan 1978). Finally, the inversion of the matrix $N(t)$ has to be performed.

It is well known that the Hamiltonian matrix has eigenvalues symmetrically distributed with respect to the imaginary axis, and hence the same number of stable and unstable poles. Since matrices $M(t)$ and $N(t)$ contain unstable models of the Hamiltonian we have to perform the reinitialization of the considered system of differential equations. The reinitialization technique applied to the problem under consideration will modify only the terminal conditions in Equations 8.65, 8.77, and 8.79, respectively, leading to

$$M(k \Delta t) = P(k \Delta t) \tag{8.97}$$

$$U(k \Delta t) = \begin{bmatrix} P_1(k \Delta t) \\ I \end{bmatrix}, \quad V(k \Delta t) = \begin{bmatrix} \varepsilon P_2^T(k \Delta t) \\ 0 \end{bmatrix} \tag{8.98}$$

$$X(k \Delta t) = \begin{bmatrix} \varepsilon P_2(k \Delta t) \\ 0 \end{bmatrix}, \quad Y(k \Delta t) = \begin{bmatrix} P_3(k \Delta t) \\ I \end{bmatrix} \tag{8.99}$$

where k represents the number of steps and Δt is an integration step.

The transformation matrix T_1 from Equation 8.80 can easily be obtained, with the required accuracy, by using numerical techniques developed in Section 5.1. They converge with the rate of convergence of at least $O(\varepsilon^2)$. Thus, after k iterations, one gets the approximation $T_1^{(k)} = T_1 + O(\varepsilon^{2k})$. The use of $T_1^{(k)}$ in Equations 8.83 through 8.86 instead of T_1, will perturb the coefficients of the corresponding systems of linear differential equations by $O(\varepsilon^{2k})$, which implies that the approximate solutions of these differential equations are $O(\varepsilon^{2k})$ close to the exact ones (Kato 1995). Thus, it is of interest to obtain $T_1^{(k)}$ with the desired accuracy, which produces the same accuracy in the sought solution.

The recursive reduced-order solution of the differential Riccati equation of weakly coupled systems is demonstrated in Section 8.3.1 where a real-world example is considered.

8.3.1 Case Study: Gas Absorber

A real-world example, a six-plate gas absorber (De Vlieger et al. 1982) is considered to demonstrate the proposed method.

The problem matrices A and B are given by

$$A = \begin{bmatrix} -1.173 & 0.6341 & 0 & 0 & 0 & 0 \\ 0.5390 & -1.173 & 0.6341 & 0 & 0 & 0 \\ 0 & 0.5390 & -1.173 & 0.6341 & 0 & 0 \\ 0 & 0 & 0.5390 & -1.173 & 0.6341 & 0 \\ 0 & 0 & 0 & 0.5390 & -1.173 & 0.6341 \\ 0 & 0 & 0 & 0 & 0.5390 & -1.173 \end{bmatrix}$$

$$B^T = \begin{bmatrix} 0.5390 & 0 & 0 & 0 & 0 & 0 \\ 0 & 0 & 0 & 0 & 0 & 0.6341 \end{bmatrix}$$

Remaining matrices are chosen as

$$D^T D = \begin{bmatrix} 1 & 0 & 0 & 0 & 0 & 0 \\ 0 & 1 & 0 & 0 & 0 & 0 \\ 0 & 0 & 1 & 0 & 0 & 0 \\ 0 & 0 & 0 & 2 & 0 & 0 \\ 0 & 0 & 0 & 0 & 2 & 0 \\ 0 & 0 & 0 & 0 & 0 & 2 \end{bmatrix}, \quad R = 0.1 I_2, \quad F = I_6$$

The initial and final times are selected as $t_0 = 0$ and $t_f = 1$. The initial conditions are

$$x_{10} = [-0.0306 - 0.0568 - 0.0788]^T$$
$$x_{20} = [-0.0977 - 0.1138 - 0.1273]^T$$

The system is partitioned into two subsystems with $n_1 = 3$, $n_2 = 3$, and $\varepsilon = 0.37$. The small parameter ε is built into the problem. It can be roughly estimated from the strongest coupled matrix—in this case matrix A—producing $|a_{34}|/(|a_{32}| + |a_{33}|) = 0.6341/1.7120 = 0.37$. The simulation results for the differential Riccati equation are presented in Table 8.3. After performing four iterations, we have obtained an accuracy of 10^{-5}.

TABLE 8.3

Simulation Result for the Element $P_{11}(t)$ of the Riccati Differential Equation

Iteration	$t = 0.25$	$t = 0.5$	$t = 1$
4 = optimal	0.51024	0.39942	0.35868
3	0.51024	0.39942	0.35867
2	0.51031	0.40003	0.36180
1	0.51022	0.39922	0.35808
0	0.51023	0.39939	0.36066

TABLE 8.4
Simulation Results for the Open-Loop Control

Iteration	$u(t=0)$	$u(t=0.25)$	$u(t=0.5)$	$u(t=0.75)$
Optimal 5	0.17112	0.12257	0.08678	0.05618
	0.64956	0.33987	0.18081	0.08738
4	0.17106	0.12252	0.08674	0.05615
	0.64956	0.33987	0.18081	0.08738
3	0.16264	0.11538	0.08088	0.05123
	0.64959	0.33988	0.18081	0.08738
2	0.30392	0.23349	0.17643	0.12555
	0.65187	0.34082	0.18105	0.08723
1	0.19365	0.12912	0.08322	0.04314
	0.66051	0.34421	0.18181	0.08667
0	0.56931	0.37896	0.24325	0.12551
	0.70882	0.36203	0.18507	0.08298

We have also solved the open-loop optimal control for the same example by using the corresponding recursive reduced-order method presented in Chapter 6. Corresponding simulation results for both components of the approximate open-loop control are presented in Table 8.4.

By comparing linear systems of differential Equations 8.76 through 8.79 and 8.10, apparently the closed-loop solution is computationally much more involved since Equations 8.76 and 8.78 are of the order of $2(2n \times n)$, whereas Equation 8.10 represents the same set of equations of order $2n$.

8.4 DIFFERENCE RICCATI EQUATION

In this section, we use the approach developed in Section 8.3 to get the solution of the weakly coupled difference Riccati equation, up to any order of accuracy, by solving the reduced-order linear difference equations.

The weakly coupled linear discrete system is represented by

$$x_1(k+1) = A_1 x_1(k) + \varepsilon A_2 x_2(k) + B_1 u_1(k) + \varepsilon B_2 u_2(k)$$
$$x_2(k+1) = \varepsilon A_3 x_1(k) + A_4 x_2(k) + \varepsilon B_3 u_1(k) + B_4 u_2(k)$$

$$(8.100)$$

with states $x_i(k) \in \Re^{n_i}$, and control inputs $u_i(k) \in \Re^{m_i}$, $i=1,2$, where ε is a small coupling parameter. The performance criterion of the corresponding linear-quadratic discrete control problem is defined as in Equation 8.79, taking into account the presence of two control agents, that is

$$u(k) = \begin{bmatrix} u_1(k) \\ u_2(k) \end{bmatrix}, \quad R = R^{\mathrm{T}} = \begin{bmatrix} R_1 & 0 \\ 0 & R_2 \end{bmatrix} > 0 \qquad (8.101)$$

Introducing the notation

$$A = \begin{bmatrix} A_1 & \varepsilon A_2 \\ \varepsilon A_3 & A_4 \end{bmatrix}, \quad B = \begin{bmatrix} B_1 & \varepsilon B_2 \\ \varepsilon B_3 & B_4 \end{bmatrix} \tag{8.102}$$

the Hamiltonian forms of this optimal control problem can be written as the back recursion

$$\begin{bmatrix} x(k) \\ \lambda(k) \end{bmatrix} = \begin{bmatrix} A^{-1} & A^{-1}BR^{-1}B^{\mathrm{T}} \\ QA^{-1} & A^{\mathrm{T}} + QA^{-1}BR^{-1}B^{\mathrm{T}} \end{bmatrix} \begin{bmatrix} x(k+1) \\ \lambda(k+1) \end{bmatrix}$$

$$= \mathbf{H} \begin{bmatrix} x(k+1) \\ \lambda(k+1) \end{bmatrix}$$

The optimal control law has the very well-known form given by

$$u(k) = -R^{-1}B^{\mathrm{T}}\lambda(k+1) = -R^{-1}B^{\mathrm{T}}P(k+1)x(k+1)$$

where $P(k)$ satisfies the difference Riccati equation

$$P(k) = Q + A^{\mathrm{T}}P(k+1)[I + SP(k+1)]^{-1}A$$
$$= Q + A^{\mathrm{T}}P(k+1)A - A^{\mathrm{T}}P(k+1)B\left[R + B^{\mathrm{T}}P(k+1)B\right]^{-1}B^{\mathrm{T}}P(k+1)A \tag{8.103}$$

with

$$S = BR^{-1}B^{\mathrm{T}} = \begin{bmatrix} S_1 & \varepsilon S_2 \\ \varepsilon S_2^{\mathrm{T}} & S_3 \end{bmatrix}$$
$$S_1 = B_1 R_1^{-1}B_1^{\mathrm{T}} + \varepsilon^2 B_2 R_2^{-1}B_2^{\mathrm{T}}, \quad S_2 = \left(B_1 R_1^{-1}B_3^{\mathrm{T}} + B_2 R_2^{-1}B_4^{\mathrm{T}}\right) \tag{8.104}$$
$$S_3 = B_4 R_2^{-1}B_2^{\mathrm{T}} + \varepsilon^2 B_3 R_1^{-1}B_3^{\mathrm{T}}$$

In order to obtain an efficient numerical method for solving Equation 8.103 in terms of reduced-order problems, we will utilize the known Hamiltonian form of the solution of the difference Riccati equation and a nonsingular decoupling transformation from Section 5.1.

The presented method for solving difference Riccati equation of weakly coupled discrete systems is dual to the one developed in Section 8.3 for the reduced-order solution of the differential Riccati equation of weakly coupled continuous systems.

The solution of Equation 8.103 can be sought in the form

$$P(k) = M(k)N^{-1}(k), \quad P(k_f) = M(k_f) = F \tag{8.105}$$

where matrices $M(k)$ and $N(k)$ satisfy a system of linear difference equations

$$N(k) = A^{-1}N(k+1) + A^{-1}BR^{-1}B^{\mathrm{T}}M(k+1), \quad N(k_f) = I$$
$$M(k) = QA^{-1}N(k+1) + (A^{\mathrm{T}} + QA^{-1}BR^{-1}B^{\mathrm{T}})M(k+1), \quad M(k_f) = F \tag{8.106}$$

The next lemma guarantees the existence of the invertible solution for $N(k)$, for all values of k.

LEMMA 8.3

If the triple $(A, B, \mathrm{Chol}(Q))$ is stabilizable–observable then the matrix $N(k)$, with $N(k_f) = I$ is invertible for any $k = 0, 1, 2, \ldots, k_f$.

Proof Using the discrete version of the dichotomy transformation of Wilde and Kokotović (1972) we have

$$\begin{bmatrix} N(k) \\ M(k) \end{bmatrix} = \begin{bmatrix} I & I \\ \overline{P} & \overline{K} \end{bmatrix} \begin{bmatrix} \widehat{N}(k) \\ \widehat{M}(k) \end{bmatrix}, \quad N(k_f) = I, \quad M(k_f) = F$$

and

$$\begin{bmatrix} \widehat{N}(k) \\ \widehat{M}(k) \end{bmatrix} = \begin{bmatrix} I + (\overline{K} - \overline{P})^{-1}\overline{P} & -(\overline{K} - \overline{P})^{-1} \\ -(\overline{K} - \overline{P})^{-1}\overline{P} & (\overline{K} - \overline{P})^{-1} \end{bmatrix} \begin{bmatrix} N(k) \\ M(k) \end{bmatrix}$$

where $\overline{P}$ and $\overline{K}$ are unique positive definite and negative definite solutions of the discrete-time algebraic Riccati equation corresponding to Equation 8.103. These two solutions exist under the conditions stated in Lemma 8.3.

The system (Equation 8.106) can be transformed into

$$\begin{bmatrix} \widehat{N}(k) \\ \widehat{M}(k) \end{bmatrix} = \begin{bmatrix} A^{-1}(1 + BR^{-1}B^{\mathrm{T}}P) & 0 \\ 0 & A^{-1}(1 + BR^{-1}B^{\mathrm{T}}K) \end{bmatrix} \begin{bmatrix} \widehat{N}(k+1) \\ \widehat{M}(k+1) \end{bmatrix}$$

with terminal conditions

$$\widehat{N}(k_f) = I + (\overline{K} - \overline{P})^{-1}(\overline{P} - F)$$
$$\widehat{M}(k_f) = (\overline{K} - \overline{P})^{-1}(F - \overline{P})$$

The solution of this system is given by

$$\widehat{N}(k) = \left[A^{-1}(I + BR^{-1}B^{\mathrm{T}}\overline{P})\right]^{k_f - k}\widehat{N}(k_f)$$
$$\widehat{M}(k) = \left[A^{-1}(I + BR^{-1}B^{\mathrm{T}}\overline{K})\right]^{k_f - k}\widehat{M}(k_f)$$

It can be shown from the above formulas that

$$N(k) = \left[A^{-1}(I + BR^{-1}B^{\mathrm{T}}\overline{P})\right]^{k_f - k}\left[I + (\overline{K} - \overline{P})^{-1}(\overline{P} - F)\right]$$
$$+ \left[A^{-1}(I + BR^{-1}B^{\mathrm{T}}\overline{K})\right]^{k_f - k}(\overline{K} - \overline{P})^{-1}(F - \overline{P})$$

that is

$$N(k) = \phi(k_f - k)N(k_f), \quad N(k_f) = I$$

with obvious definition of $\phi(k_f - k)$. Since $\phi(k_f - k)$ plays the role of the transition matrix, it is nonsingular by the fact that the matrix A is nonsingular. Note that the matrix $(I + BR^{-1}B^{\mathrm{T}}\overline{P})$ is also nonsingular. The regularity of $N(k)$ is determined by $N(k_f)$ only. Thus, having chosen $N(k_f)$ as an identity will assure the nonsingularity of $N(k)$ for any $k < k_f$, and prove the given lemma. ∎

It has been assumed in this section that duals of Assumptions 8.1 and 8.2 hold (all matrices and their partitions are $O(1)$ and all eigenvalues of matrices A_1 and A_4 are $O(1)$). This is needed for the purpose of preserving pure weak coupling in subsystem dynamics and computations that follow.

Due to the weakly coupled structure of all coefficients in Equation 8.103, the solution of that equation has the form

$$P(k) = \begin{bmatrix} P_1(k) & \varepsilon P_2(k) \\ \varepsilon P_2^{\mathrm{T}}(k) & P_3(k) \end{bmatrix}, \quad P(k_f) = F = \begin{bmatrix} F_1 & \varepsilon F_2 \\ \varepsilon F_2^{\mathrm{T}} & F_3 \end{bmatrix} \tag{8.107}$$

where
$$\dim\{P_1\} = n_1 \times n_1$$
$$\dim\{P_3\} = n_2 \times n_2$$

Let compatible partitions of matrices $M(k)$ and $N(k)$ be

$$M(k) = \begin{bmatrix} M_1(k) & \varepsilon M_2(k) \\ \varepsilon M_3(k) & M_4(k) \end{bmatrix}, \quad N(k) = \begin{bmatrix} N_1(k) & \varepsilon N_2(k) \\ \varepsilon N_3(k) & N_4(k) \end{bmatrix} \tag{8.108}$$

Partitioning Equation 8.106, according to Equation 8.108, will reveal a decoupled structure, that is, equation for $M_1(k)$, $M_3(k)$, $N_1(k)$, and $N_3(k)$ are independent of equations for $M_2(k)$, $M_4(k)$, $N_2(k)$, and $N_4(k)$ and vice versa

$$\begin{bmatrix} N_1(k) \\ \varepsilon N_3(k) \\ M_1(k) \\ \varepsilon M_3(k) \end{bmatrix} = \begin{bmatrix} \overline{A_1} & \varepsilon \overline{A_2} & \overline{S_1} & \varepsilon \overline{S_2} \\ \varepsilon \overline{A_3} & \overline{A_4} & \varepsilon \overline{S_3} & \overline{S_4} \\ \overline{Q_1} & \varepsilon \overline{Q_2} & \overline{A_{11}^{\mathrm{T}}} & \varepsilon \overline{A_{21}^{\mathrm{T}}} \\ \varepsilon \overline{Q_3} & \overline{Q_4} & \varepsilon \overline{A_{12}^{\mathrm{T}}} & \overline{A_{22}^{\mathrm{T}}} \end{bmatrix} \begin{bmatrix} N_1(k+1) \\ \varepsilon N_3(k+1) \\ M_1(k+1) \\ \varepsilon M_3(k+1) \end{bmatrix}$$

$$= H \begin{bmatrix} N_1(k+1) \\ \varepsilon N_3(k+1) \\ M_1(k+1) \\ \varepsilon M_3(k+1) \end{bmatrix} \tag{8.109}$$

$$
\begin{bmatrix} \varepsilon N_2(k) \\ N_4(k) \\ \varepsilon M_2(k) \\ M_4(k) \end{bmatrix} = \begin{bmatrix} \overline{A_1} & \varepsilon\overline{A_2} & \overline{S_1} & \varepsilon\overline{S_2} \\ \varepsilon\overline{A_3} & \overline{A_4} & \varepsilon\overline{S_3} & \overline{S_4} \\ \overline{Q_1} & \varepsilon\overline{Q_2} & \overline{A_{11}^T} & \varepsilon\overline{A_{21}^T} \\ \varepsilon\overline{Q_3} & \overline{Q_4} & \varepsilon\overline{A_{12}^T} & \overline{A_{22}^T} \end{bmatrix} \begin{bmatrix} \varepsilon N_2(k+1) \\ N_4(k+1) \\ \varepsilon M_2(k+1) \\ M_4(k+1) \end{bmatrix}
$$

$$
= H \begin{bmatrix} \varepsilon N_2(k+1) \\ N_4(k+1) \\ \varepsilon M_2(k+1) \\ M_4(k+1) \end{bmatrix} \tag{8.110}
$$

Interchanging the second and third rows in Equations 8.109 and 8.110, respectively, produces

$$
\begin{bmatrix} N_1(k) \\ M_1(k) \\ \varepsilon N_3(k) \\ \varepsilon M_3(k) \end{bmatrix} = \begin{bmatrix} \overline{A_1} & \overline{S_1} & \varepsilon\overline{A_2} & \varepsilon\overline{S_2} \\ \overline{Q_1} & \overline{A_{11}^T} & \varepsilon\overline{Q_2} & \varepsilon\overline{A_{21}^T} \\ \varepsilon\overline{A_3} & \varepsilon\overline{S_3} & \overline{A_4} & \overline{S_4} \\ \varepsilon\overline{Q_3} & \varepsilon\overline{A_{12}^T} & \overline{Q_4} & \overline{A_{22}^T} \end{bmatrix} \begin{bmatrix} N_1(k+1) \\ M_1(k+1) \\ \varepsilon N_3(k+1) \\ \varepsilon M_3(k+1) \end{bmatrix}
$$

$$
= \begin{bmatrix} T_1 & \varepsilon T_2 \\ \varepsilon T_3 & T_4 \end{bmatrix} \begin{bmatrix} N_1(k+1) \\ M_1(k+1) \\ \varepsilon N_3(k+1) \\ \varepsilon M_3(k+1) \end{bmatrix} \tag{8.111}
$$

$$
\begin{bmatrix} \varepsilon N_2(k) \\ \varepsilon M_2(k) \\ N_4(k) \\ M_4(k) \end{bmatrix} = \begin{bmatrix} \overline{A_1} & \overline{S_1} & \varepsilon\overline{A_2} & \varepsilon\overline{S_2} \\ \overline{Q_1} & \overline{A_{11}^T} & \varepsilon\overline{Q_2} & \varepsilon\overline{A_{21}^T} \\ \varepsilon\overline{A_3} & \varepsilon\overline{S_3} & \overline{A_4} & \overline{S_4} \\ \varepsilon\overline{Q_3} & \varepsilon\overline{A_{12}^T} & \overline{Q_4} & \overline{A_{22}^T} \end{bmatrix} \begin{bmatrix} \varepsilon N_2(k+1) \\ \varepsilon M_2(k+1) \\ N_4(k+1) \\ M_4(k+1) \end{bmatrix}
$$

$$
= \begin{bmatrix} T_1 & \varepsilon T_2 \\ \varepsilon T_3 & T_4 \end{bmatrix} \begin{bmatrix} \varepsilon N_2(k+1) \\ \varepsilon M_2(k+1) \\ N_4(k+1) \\ M_4(k+1) \end{bmatrix} \tag{8.112}
$$

where

$$
T_1 = \begin{bmatrix} \overline{A_1} & \overline{S_1} \\ \overline{Q_1} & \overline{A_{11}^T} \end{bmatrix}, \quad T_2 = \begin{bmatrix} \overline{A_2} & \overline{S_2} \\ \overline{Q_2} & \overline{A_{21}^T} \end{bmatrix}
$$

$$
T_3 = \begin{bmatrix} \overline{A_3} & \overline{S_3} \\ \overline{Q_3} & \overline{A_{12}^T} \end{bmatrix}, \quad T_4 = \begin{bmatrix} \overline{A_4} & \overline{S_4} \\ \overline{Q_4} & \overline{A_{22}^T} \end{bmatrix} \tag{8.113}
$$

Introducing the notation

$$U(k) = \begin{bmatrix} N_1(k) \\ M_1(k) \end{bmatrix}, \quad V(k) = \begin{bmatrix} \varepsilon N_3(k) \\ \varepsilon M_3(k) \end{bmatrix}$$

$$X(k) = \begin{bmatrix} \varepsilon N_2(k) \\ \varepsilon M_2(k) \end{bmatrix}, \quad Y(k) = \begin{bmatrix} N_4(k) \\ M_4(k) \end{bmatrix} \tag{8.114}$$

we get two independent systems of weakly coupled difference equations

$$U(k) = T_1 U(k+1) + \varepsilon T_2 V(k+1)$$
$$V(k) = \varepsilon T_3 U(k+1) + T_4 V(k+1) \tag{8.115}$$

$$X(k) = T_1 X(k+1) + \varepsilon T_2 Y(k+1)$$
$$Y(k) = \varepsilon T_3 X(k+1) + T_4 Y(k+1) \tag{8.116}$$

with terminal conditions

$$U(k_f) = \begin{bmatrix} I \\ F_1 \end{bmatrix}, \quad V(k_f) = \begin{bmatrix} 0 \\ \varepsilon F_2^{\mathrm{T}} \end{bmatrix}, \quad X(k_f) = \begin{bmatrix} 0 \\ \varepsilon F_2 \end{bmatrix}, \quad Y(k_f) = \begin{bmatrix} I \\ F_3 \end{bmatrix} \tag{8.117}$$

Note that both systems (Equations 8.115 and 8.116) have exactly the same form and the only difference is in the terminal conditions.

Applying the decoupling transformation from Section 5.1

$$\mathbf{T_1} = \begin{bmatrix} I & -\varepsilon L \\ \varepsilon H & I - \varepsilon^2 HL \end{bmatrix}, \quad \mathbf{T_1^{-1}} = \begin{bmatrix} I - \varepsilon^2 LH & \varepsilon L \\ -\varepsilon H & I \end{bmatrix} \tag{8.118}$$

where L and H satisfy

$$T_1 L + T_2 - LT_4 - \varepsilon^2 LT_3 L = 0$$
$$H(T_1 - \varepsilon^2 LT_3) - (T_4 + \varepsilon^2 T_3 L)H + T_3 = 0 \tag{8.119}$$

Equations 8.115 and 8.116 produce

$$\overline{U}(k) = (T_1 - \varepsilon^2 LT_3)\overline{U}(k+1)$$
$$\overline{V}(k) = (T_4 + \varepsilon^2 T_3 L)\overline{V}(k+1) \tag{8.120}$$

$$\overline{X}(k) = (T_1 - \varepsilon^2 LT_3)\overline{X}(k+1)$$
$$\overline{Y}(k) = (T_4 + \varepsilon^2 T_3 L)\overline{Y}(k+1) \tag{8.121}$$

with terminal conditions

$$\overline{U}(k_f) = U(k_f) - \varepsilon LV(k_f)$$
$$\overline{V}(k_f) = \varepsilon HU(k_f) + (I - \varepsilon^2 HL)V(k_f)$$
$$\overline{X}(k_f) = X(k_f) - \varepsilon LY(k_f)$$
$$\overline{Y}(k_f) = \varepsilon HX(k_f) + (I - \varepsilon^2 HL)Y(k_f) \tag{8.122}$$

Matrices L an H can easily be obtained, at a very low cost, by using the recursive algorithm from Section 5.1. Solutions of Equations 8.120 and 8.121 are given by

$$\begin{aligned}
\overline{U}(k) &= (T_1 - \varepsilon^2 LT_3)^{k_f - k}\overline{U}(k_f) \\
\overline{V}(k) &= (T_4 - \varepsilon^2 T_3 L)^{k_f - k}\overline{V}(k_f) \\
\overline{X}(k) &= (T_1 - \varepsilon^2 LT_3)^{k_f - k}\overline{X}(k_f) \\
\overline{Y}(k) &= (T_4 + \varepsilon^2 T_3 L)^{k_f - k}\overline{Y}(k_f)
\end{aligned}$$
(8.123)

Corresponding solutions in the original coordinates are

$$\begin{aligned}
U(k) &= (I - \varepsilon^2 LH)(T_1 - \varepsilon^2 LT_3)^{k_f - k}\overline{U}(k_f) \\
&\quad + \varepsilon L(T_4 + \varepsilon^2 T_3 L)^{k_f - k}\overline{V}(k_f) \\
V(k) &= -\varepsilon K(T_1 - \varepsilon^2 LT_3)^{k_f - k}\overline{U}(k_f) + (T_4 + \varepsilon^2 T_3 L)^{k_f - k}\overline{V}(k_f) \\
X(k) &= (I - \varepsilon^2 LH)(T_1 - \varepsilon^2 LT_3)^{k_f - k}\overline{X}(k_f) \\
&\quad + \varepsilon L(T_4 + \varepsilon^2 T_3 L)^{k_f - k}\overline{Y}(k_f) \\
Y(k) &= -\varepsilon H(T_1 - \varepsilon^2 LT_3)^{k_f - k}\overline{X}(k_f) + (T_4 + \varepsilon^2 T_3 L)^{k_f - k}\overline{Y}(k_f)
\end{aligned}$$
(8.124)

Partitioning Equation 8.124, according to Equation 8.114, will produce all components of matrices $M(k)$ and $N(k)$, that is

$$\begin{aligned}
\begin{bmatrix} N_1(k) \\ M_1(k) \end{bmatrix} &= \begin{bmatrix} U_1(k) \\ U_2(k) \end{bmatrix} = U(k), & \begin{bmatrix} \varepsilon N_3(k) \\ \varepsilon M_3(k) \end{bmatrix} &= \begin{bmatrix} V_1(k) \\ V_2(k) \end{bmatrix} = V(k) \\
\begin{bmatrix} \varepsilon N_2(k) \\ \varepsilon M_2(k) \end{bmatrix} &= \begin{bmatrix} X_1(k) \\ X_2(k) \end{bmatrix} = X(k), & \begin{bmatrix} N_4(k) \\ M_4(k) \end{bmatrix} &= \begin{bmatrix} Y_1(k) \\ Y_2(k) \end{bmatrix} = Y(k)
\end{aligned}$$
(8.125)

The required solution of Equation 8.103 is given by

$$P(k) = \begin{bmatrix} U_2(k) & X_2(k) \\ V_2(k) & Y_2(k) \end{bmatrix} \begin{bmatrix} U_1(k) & X_1(k) \\ V_1(k) & Y_1(k) \end{bmatrix}^{-1}$$
(8.126)

Thus, in order to get the solution of Equation 8.103, $P(k)$, which has $\dim\{P(k)\} = n \times n = (n_1 + n_2) \times (n_1 + n_2)$, we solve two simple algebraic equations (Equation 8.119) of dimensions of $(2n_2 \times 2n_1)$ and $(2n_1 \times 2n_2)$, respectively. In addition, the $(k_f - k)$th powers of the matrices $T_1 - \varepsilon^2 LT$ and $T_4 + \varepsilon^2 T_3 L$ have to be found.

8.4.1 NUMERICAL EXAMPLE

In order to demonstrate the proposed method, the discrete system from Section 8.2.1 is studied. The problem matrices A and B are given in Section 8.2.1. The remaining matrices are chosen as $R = 0.5I_2$, $Q = 0.1I_4$, $k_f = 8$, and the terminal condition is given by

$$P(k_f) = F = \begin{bmatrix} 0.9 & 0 & 0.3 & 0 \\ 0 & 0.9 & 0 & 0.3 \\ 0.3 & 0 & 0.9 & 0 \\ 0 & 0.3 & 0 & 0.9 \end{bmatrix}$$

The small weak coupling parameter ε is built in the problem and can be roughly estimated from the strongest coupled matrix (matrix A). The strongest coupling is in the fourth row, where

$$\varepsilon = \frac{0.323}{0.983} \approx 0.329$$

With the proposed method, the simulation results for Equation 8.125 and the solution of the weakly coupled matrix difference Riccati Equation 8.126 at $k = 4$ are

$$\begin{bmatrix} U_1(4) & X_1(4) \\ V_1(4) & Y_1(4) \end{bmatrix} = \begin{bmatrix} 0.063 & -1.182 & 0.497 & 0.394 \\ 2.698 & 1.828 & -1.475 & -1.049 \\ 0.632 & 0.388 & 0.446 & -1.326 \\ -2.002 & -0.973 & 1.724 & 2.447 \end{bmatrix}$$

$$\begin{bmatrix} U_2(4) & X_2(4) \\ V_2(4) & Y_2(4) \end{bmatrix} = \begin{bmatrix} 0.569 & -1.190 & 0.495 & 0.077 \\ 1.050 & 0.809 & 0.163 & 0.430 \\ 0.639 & 0.349 & 0.996 & -0.651 \\ 0.125 & 0.477 & 0.933 & 1.086 \end{bmatrix}$$

$$P(4) = \begin{bmatrix} 1.273 & 0.121 & 0.181 & -0.023 \\ 0.121 & 0.814 & 0.314 & 0.675 \\ 0.181 & 0.314 & 1.192 & 0.485 \\ -0.023 & 0.675 & 0.485 & 1.000 \end{bmatrix}$$

The obtained solution $P(4)$, is identical to the solution of the global Riccati difference equation (Equation 8.103).

8.5 CONCLUDING REMARKS

It is interesting to observe that the study of finite horizon optimization problems for linear weakly coupled systems has not been an active research area since the work of Gajić and Shen (1993) in the mid-1990s. The information presented in this chapter is based heavily on this work. In contrast, many new and diverse results have recently been obtained on infinite time horizon optimization problems for weakly coupled systems (see, e.g., the results of Mukaidani [2005, 2006a,b, 2007a,b,c, 2008], Kecman [2006], and Kim and Lim [2006, 2007]). We hope this book and particularly this chapter will motivate further research in the direction of weakly coupled finite time optimization problems.

APPENDIX 8.1

From Equation 8.38 we have

$$H = \begin{bmatrix} A + BR^{-1}B^T A^{-T}Q & -BR^{-1}B^T A^{-T} \\ -A^{-T}Q & A^{-T} \end{bmatrix} \tag{A.8.1}$$

Since A^{-T} has the same structure as A^T, that is

$$A^{-T} = \begin{bmatrix} O(1) & O(\varepsilon) \\ O(\varepsilon) & O(1) \end{bmatrix} \tag{A.8.2}$$

then

$$A^{-T}Q = \begin{bmatrix} O(1) & O(\varepsilon) \\ O(\varepsilon) & O(1) \end{bmatrix}\begin{bmatrix} O(1) & O(\varepsilon) \\ O(\varepsilon) & O(1) \end{bmatrix} = \begin{bmatrix} O(1) & O(\varepsilon) \\ O(\varepsilon) & O(1) \end{bmatrix} \tag{A.8.3}$$

$$BR^{-1}B^T A^{-T} = \begin{bmatrix} O(1) & O(\varepsilon) \\ O(\varepsilon) & O(1) \end{bmatrix}\begin{bmatrix} O(1) & O(\varepsilon) \\ O(\varepsilon) & O(1) \end{bmatrix} = \begin{bmatrix} O(1) & O(\varepsilon) \\ O(\varepsilon) & O(1) \end{bmatrix}$$

$$A + BR^{-1}B^T A^{-T}Q = \begin{bmatrix} O(1) & O(\varepsilon) \\ O(\varepsilon) & O(1) \end{bmatrix} \tag{A.8.4}$$

$$H = \begin{bmatrix} \overline{A_1} & \varepsilon\overline{A_2} & \overline{S_1} & \varepsilon\overline{S_2} \\ \varepsilon\overline{A_3} & \overline{A_4} & \varepsilon\overline{S_3} & \overline{S_4} \\ \overline{Q_1} & \varepsilon\overline{Q_2} & \overline{A_{11}^T} & \varepsilon\overline{A_{21}^T} \\ \varepsilon\overline{Q_3} & \overline{Q_4} & \varepsilon\overline{A_{12}^T} & \overline{A_{22}^T} \end{bmatrix}$$

Note again that it is easy to obtain the matrices with bars in the process of programming and it is of no interest to obtain the corresponding analytical expressions.

APPENDIX 8.2

Let the transition matrices of the difference equations given in Equation 8.46 be denoted as $\phi(k)$ and $\psi(k)$, respectively, and let us partition them as follows:

$$\phi(k) = \begin{bmatrix} \phi_1(k) & \phi_2(k) \\ \phi_3(k) & \phi_4(k) \end{bmatrix}$$

$$\psi(k) = \begin{bmatrix} \psi_1(k) & \psi_2(k) \\ \psi_3(k) & \psi_4(k) \end{bmatrix} \tag{A.8.5}$$

From Equation 8.54 we have

$$\beta(\varepsilon) = \left\{ M_2 + N_2 \begin{bmatrix} \phi(k) & 0 \\ 0 & \psi(k) \end{bmatrix} \right\} \tag{A.8.6}$$

Using expressions for M_2 and N_2 defined by Equation 8.51 we get

$$\beta(\varepsilon) = \begin{bmatrix} I_{n_1} & 0 & 0 & 0 \\ \phi_3(k) - F\phi_1(k) & \phi_4(k) - F_1\phi_2(k) & 0 & 0 \\ 0 & 0 & I_{n_2} & 0 \\ 0 & 0 & \psi_3(k) - F_3\psi_1(k) & \psi_4(k) - F_3\psi_2(k) \end{bmatrix} + O(\varepsilon) \tag{A.8.7}$$

It is left as an exercise to the reader to show that under stabilizability–detectability conditions imposed on the subsystems, the matrices $\phi_4(k_f) - F_1\phi_2(k_f)$ and $\psi_4(k_f) - F_3\psi_2(k_f)$ are invertible. Thus, the matrix $\beta(\varepsilon)$ is invertible for sufficiently small values of ε.

Exercise 8.1: Show that the matrices $\phi_4(k_f) - F_1\phi_2(k_f)$ and $\psi_4(k_f) - F_3\psi_2(k_f)$ are invertible under stabilizability–detectability conditions imposed on the subsystems.

REFERENCES

De Vlieger, J., H. Verbruggen, and P. Bruijn, A time-optimal control algorithm for digital computer control, *Automatica*, 18, 239–244, 1982.

Gajić, Z. and M. Lim, *Optimal Control of Singularly Perturbed Linear Systems and Applications: High Accuracy Techniques,* Marcel Dekker, New York, 2001.

Gajić, Z., D. Petkovski, and X. Shen, *Singularly Perturbed and Weakly Coupled Linear Control Systems—A Recursive Approach*, Springer-Verlag, Lecture Notes in Control and Information Sciences, 140, New York, 1990.

Gajić, Z. and X. Shen, Decoupling transformation for weakly coupled linear systems, *International Journal of Control*, 50, 1517–1523, 1989.

Gajić, Z. and X. Shen, Parallel Algorithm for Optimal Control of Large Scale Linear Systems, Springer Verlag, London, 1993.

Harkara, N., D. Petkovski, and Z. Gajić, The recursive algorithm for the optimal static output feedback control problem of linear weakly coupled systems, *International Journal of Control*, 50, 1–11, 1989.

Kato, T., *Perturbation Theory of Linear Operators*, Springer-Verlag, New York, 1995.

Katzberg, J., Structured feedback control of discrete linear stochastic systems with quadratic cost, *IEEE Transactions on Automatic Control*, AC-22, 232–236, 1977.

Kecman, V., Eigenvector approach for reduced-order optimal control problems of weakly coupled systems, *Dynamics of Continuous Discrete and Impulsive Systems*, 13, 569–588, 2006.

Kim, Y.-J. and M.-T. Lim, Parallel robust H_∞ control for weakly coupled bilinear systems with parameter uncertainties using successive Galerkin approximation, *International Journal of Control, Automation, and Systems*, 4, 689–696, 2006.

Kim, Y.-J. and M.-T. Lim, Parallel optimal control for weakly coupled bilinear systems using successive Galerkin approximation, *Proceedings of IET—Control Theory and Applications*, 1, 909–914, 2007.

Kirk, K., *Optimal Control Theory*, Dover Publications, New York, 2004.

Kwakernaak, H. and R. Sivan, *Linear Optimal Control Systems*, Wiley-Interscience, New York, 1972.

Molen, C. and C. Van Loan, Nineteen dubious ways to compute the exponential of a matrix, *SIAM Review*, 20, 801–836, 1978.

Mukaidani, H., Numerical computation for H_2 state feedback control of large-scale systems, *Dynamics of Continuous, Discrete, and Impulsive Systems Series B: Applications and Algorithms*, 12, 281–296, 2005.

Mukaidani, H., A numerical analysis of the Nash strategy for weakly coupled large-scale systems, *IEEE Transactions on Automatic Control*, 56, 1371–1377, 2006a.

Mukaidani, H., Optimal numerical strategy for Nash games of weakly coupled large-scale systems, *Dynamics of Continuous, Discrete, and Impulsive Systems*, 13, 249–268, 2006b.

Mukaidani, H., Numerical computation of sign indefinite linear quadratic differential games for weakly coupled linear large scale systems, *International Journal of Control*, 80, 75–86, 2007a.

Mukaidani, H., Newton method for solving cross-coupled sign-indefinite algebraic Riccati equations of weakly coupled large-scale systems, *Applied Mathematics and Computation*, 188, 103–115, 2007b.

Mukaidani, H., Numerical computation for solving algebraic Riccati equation of weakly coupled systems, *Electrical Engineering of Japan*, 160, 39–48, 2007c.

Mukaidani, H., Numerical computation for H_∞ output feedback control for strongly coupled large-scale systems, *Applied Mathematics and Computation*, 197, 212–227, 2008.

Petkov, P., N. Christov, and M. Konstantinov, A computational algorithm for pole placement assignment of linear multi-input systems, *IEEE Transactions on Automatic Control*, AC-31, 1044–1047, 1986.

Petrović, B. and Z. Gajić, The recursive solution of linear quadratic Nash games for weakly interconnected systems, *Journal of Optimization Theory and Applications*, 56, 463–477, 1988.

Su, W. and Z. Gajić, Reduced-order solution to the finite time optimal control problems of linear weakly coupled systems, *IEEE Transactions on Automatic Control*, AC-36, 498–501, 1991.

Wilde, R. and P. Kokotović, A dichotomy in linear control theory, *IEEE Transactions on Automatic Control*, AC-17, 382–383, 1972.

9 Hamiltonian Method for Steady State Optimal Control and Filtering

In this chapter, we first show how the algebraic Riccati equations of weakly coupled control continuous- and discrete-time systems composed of two subsystems can be completely and exactly decomposed into two reduced-order algebraic Riccati equations corresponding to local subsystems. The decomposed algebraic Riccati equations are nonsymmetric. The Newton method is very efficient for solving these equations since good initial guesses can easily be obtained from symmetric reduced-order local algebraic Riccati equations. The initial guesses are $O(\varepsilon^2)$ close to the exact solution, where $\int$ is a small weak coupling parameter.

Having obtained the solutions of the global algebraic Riccati equations in terms of the local algebraic Riccati equations we are in the position to solve the global linear-quadratic optimal control and Kalman filtering problems using only the local (subsystem level) information. The obtained local subsystem optimal Kalman filters are driven by the system measurements. Due to complete and exact decomposition of the algebraic Riccati equations, we have obtained parallel algorithms for solving these equations. The presented procedure produces a new insight into optimal filtering and control of weakly coupled systems since the corresponding reduced-order optimal filters and controllers are completely decoupled.

At the end of this chapter, we show how the considered procedures can be extended to weakly coupled continuous-time stochastic systems composed of N subsystems.

Several real world examples are used to demonstrate considered methodologies, including a satellite, helicopter, and distillation column.

9.1 EXACT DECOMPOSITION OF THE WEAKLY COUPLED CONTINUOUS-TIME ALGEBRAIC RICCATI EQUATION

Consider the linear weakly coupled system

$$\dot{x}_1(t) = A_1 x_1(t) + \varepsilon A_2 x_2(t) + B_1 u_1(t) + \varepsilon B_2 u_2(t)$$
$$\dot{x}_2(t) = \varepsilon A_3 x_1(t) + A_4 x_2(t) + \varepsilon B_3 u_1(t) + B_4 u_2(t)$$

(9.1)

with

$$z(t) = \begin{bmatrix} z_1(t) \\ z_2(t) \end{bmatrix} = D \begin{bmatrix} x_1(t) \\ x_2(t) \end{bmatrix} = \begin{bmatrix} D_1 & \varepsilon D_2 \\ \varepsilon D_3 & D_4 \end{bmatrix} \begin{bmatrix} x_1(t) \\ x_2(t) \end{bmatrix} \tag{9.2}$$

where $x_i(t) \in \Re^{n_i}$, $u_i(t) \in \Re^{m_i}$, $z_i(t) \in \Re^{r_i}$, $i = 1, 2$, are state, control, and output variables, respectively. The system matrices are constant and of appropriate dimensions and, in general, they are bounded functions of a small coupling parameter ε (Gajić et al. 1990). It is assumed that magnitudes of all the system eigenvalues are $O(1)$, that is, $|\lambda_j| = O(1)$, $j = 1, 2, \ldots, n$ implying that matrices A_1 and A_4 are nonsingular with $\det\{A_1\} = O(1)$ and $\det\{A_4\} = O(1)$. Hence, weakly coupled linear control systems are considered in this chapter under the following assumption (Chow and Kokotović 1983).

Assumption 9.1 (Weak Coupling Assumption) The magnitudes of all system eigenvalues are $O(1)$, $|\lambda_j| = O(1)$, $j = 1, 2, \ldots, n$, which implies $\det\{A_1(\varepsilon)\} = O(1)$ and $\det\{A_4(\varepsilon)\} = O(1)$.

With Equations 9.1 and 9.2, consider the performance criterion

$$J = \frac{1}{2} \int_{t_0}^{\infty} \left\{ \begin{bmatrix} x_1(t) \\ x_2(t) \end{bmatrix}^T D^T D \begin{bmatrix} x_1(t) \\ x_2(t) \end{bmatrix} + \begin{bmatrix} u_1(t) \\ u_2(t) \end{bmatrix}^T R \begin{bmatrix} u_1(t) \\ u_2(t) \end{bmatrix} \right\} dt \tag{9.3}$$

with positive definite R, which has to be minimized. It is assumed that the matrix R has the weakly coupled structure, which for the reason of simplicity is represented as

$$R = \begin{bmatrix} R_1 & 0 \\ 0 & R_2 \end{bmatrix} \tag{9.4}$$

In this chapter, and in general in this book, we will require that all partitioned matrices involved in computations of such solutions are $O(1)$, which will provide that solutions of the corresponding algebraic equations to be introduced also preserve the weakly coupled structure. Hence, we will need the following assumption.

Assumption 9.2 Problem matrices A_i, B_i, D_i, $i = 1, 2, 3, 4$, and R_j, $j = 1, 2$, are $O(1)$.

The optimal *closed-loop* control law has the very well-known form

$$u(x(t)) = \begin{bmatrix} u_1(x(t)) \\ u_2(x(t)) \end{bmatrix} = -R^{-1} \begin{bmatrix} B_1 & \varepsilon B_2 \\ \varepsilon B_3 & B_4 \end{bmatrix}^T P \begin{bmatrix} x_1(t) \\ x_2(t) \end{bmatrix}$$

$$= -R^{-1} B^T P x(t) \tag{9.5}$$

where P satisfies the algebraic Riccati equation given by

$$0 = PA + A^T P + Q - PSP \tag{9.6}$$

with

$$A = \begin{bmatrix} A_1 & \varepsilon A_2 \\ \varepsilon A_3 & A_4 \end{bmatrix}, \quad S = BR^{-1}B^{\mathrm{T}} = \begin{bmatrix} S_1 & \varepsilon S_2 \\ \varepsilon S_2^{\mathrm{T}} & S_3 \end{bmatrix}, \tag{9.7}$$

and

$$Q = D^{\mathrm{T}}D = \begin{bmatrix} Q_1 & \varepsilon Q_2 \\ \varepsilon Q_2^{\mathrm{T}} & Q_3 \end{bmatrix} \tag{9.8}$$

The *open-loop* optimal control problem of Equations 9.1 through 9.4 has the solution given by

$$u(t) = -R^{-1}B^{\mathrm{T}}p(t) \tag{9.9}$$

where $p(t) \in \Re^{n_1 + n_2}$ is the costate variable satisfying

$$\begin{bmatrix} \dot{x}(t) \\ \dot{p}(t) \end{bmatrix} = \begin{bmatrix} A & -S \\ -Q & -A^{\mathrm{T}} \end{bmatrix} \begin{bmatrix} x(t) \\ p(t) \end{bmatrix} \tag{9.10}$$

Partitioning $p(t)$ into $p_1(t) \in \Re^{n_1}$ and $p_2(t) \in \Re^{n_2}$ and rearranging rows in Equation 9.10, we obtain

$$\begin{bmatrix} \dot{x}_1(t) \\ \dot{p}_1(t) \\ \dot{x}_2(t) \\ \dot{p}_2(t) \end{bmatrix} = \begin{bmatrix} T_1 & \varepsilon T_2 \\ \varepsilon T_3 & T_4 \end{bmatrix} \begin{bmatrix} x_1(t) \\ p_1(t) \\ x_2(t) \\ p_2(t) \end{bmatrix} \tag{9.11}$$

where $T_i's$, $i = 1, 2, 3, 4$, are given by

$$T_1 = \begin{bmatrix} A_1 & -S_1 \\ -Q_1 & -A_1^{\mathrm{T}} \end{bmatrix}, \quad T_2 = \begin{bmatrix} A_2 & -S_2 \\ -Q_2 & -A_3^{\mathrm{T}} \end{bmatrix}$$
$$T_3 = \begin{bmatrix} A_3 & -S_2^{\mathrm{T}} \\ -Q_2^{\mathrm{T}} & -A_2^{\mathrm{T}} \end{bmatrix}, \quad T_4 = \begin{bmatrix} A_4 & -S_3 \\ -Q_3 & -A_4^{\mathrm{T}} \end{bmatrix} \tag{9.12}$$

Introducing the notation

$$\begin{bmatrix} x_1(t) \\ p_1(t) \end{bmatrix} = w(t), \quad \begin{bmatrix} x_2(t) \\ p_2(t) \end{bmatrix} = \lambda(t) \tag{9.13}$$

and applying the decoupling transformation

$$\begin{bmatrix} \eta(t) \\ \xi(t) \end{bmatrix} = \mathbf{T}_{\mathbf{I}}^{-1} \begin{bmatrix} w(t) \\ \lambda(t) \end{bmatrix} \tag{9.14}$$

$$\mathbf{T}_1 = \begin{bmatrix} I & -\varepsilon L \\ \varepsilon H & I - \varepsilon^2 HL \end{bmatrix}, \quad \mathbf{T}_1^{-1} = \begin{bmatrix} I - \varepsilon^2 LH & \varepsilon L \\ -\varepsilon H & I \end{bmatrix} \tag{9.15}$$

where L and H satisfy

$$T_1 L + T_2 - L T_4 - \varepsilon^2 L T_3 L = 0 \tag{9.16}$$

$$H(T_1 - \varepsilon^2 L T_3) - (T_4 + \varepsilon^2 T_3 L)H + T_3 = 0 \tag{9.17}$$

will produce decoupled dynamic subsystems in the new coordinates

$$\dot{\eta}(t) = (T_1 - \varepsilon^2 L T_3)\eta(t) \tag{9.18}$$

$$\dot{\xi}(t) = (T_4 + \varepsilon^2 T_3 L)\xi(t) \tag{9.19}$$

The rearrangement of states in Equation 9.11 is done by using a permutation matrix E of the form

$$\begin{bmatrix} x_1(t) \\ p_1(t) \\ x_2(t) \\ p_2(t) \end{bmatrix} = \begin{bmatrix} I_{n_1} & 0 & 0 & 0 \\ 0 & 0 & I_{n_1} & 0 \\ 0 & I_{n_2} & 0 & 0 \\ 0 & 0 & 0 & I_{n_2} \end{bmatrix} \begin{bmatrix} x_1(t) \\ x_2(t) \\ p_1(t) \\ p_2(t) \end{bmatrix} = E \begin{bmatrix} x(t) \\ p(t) \end{bmatrix} \tag{9.20}$$

Combining Equations 9.14 and 9.20, we obtain the relationship between the original coordinates and the new ones

$$\begin{bmatrix} \eta_1(t) \\ \xi_1(t) \\ \eta_2(t) \\ \xi_2(t) \end{bmatrix} = E^\mathsf{T} \mathbf{T}_1^{-1} E \begin{bmatrix} x(t) \\ p(t) \end{bmatrix} = \Pi \begin{bmatrix} x(t) \\ p(t) \end{bmatrix} = \begin{bmatrix} \Pi_1 & \Pi_2 \\ \Pi_3 & \Pi_4 \end{bmatrix} \begin{bmatrix} x(t) \\ p(t) \end{bmatrix} \tag{9.21}$$

Since $p(t) = P_x(t)$, where P satisfies the algebraic Riccati equation (Equation 9.6), it follows that

$$\begin{bmatrix} \eta_1(t) \\ \xi_1(t) \end{bmatrix} = (\Pi_1 + \Pi_2 P)x(t), \quad \begin{bmatrix} \eta_2(t) \\ \xi_2(t) \end{bmatrix} = (\Pi_3 + \Pi_4 P)x(t) \tag{9.22}$$

In the original coordinates, the required optional solution has a closed-loop nature. We have the same attribute for the new systems (Equations 9.18 and 9.19), that is

$$\begin{bmatrix} \eta_2(t) \\ \xi_2(t) \end{bmatrix} = \begin{bmatrix} P_1 & 0 \\ 0 & P_2 \end{bmatrix} \begin{bmatrix} \eta_1(t) \\ \xi_1(t) \end{bmatrix} \tag{9.23}$$

Then Equations 9.22 and 9.23 yield

$$\begin{bmatrix} P_1 & 0 \\ 0 & P_2 \end{bmatrix} = (\Pi_3 + \Pi_4 P)(\Pi_1 + \Pi_2 P)^{-1} \tag{9.24}$$

Following the same logic, we can find P reversely by introducing

$$E^T T_1 E = \Omega = \begin{bmatrix} \Omega_1 & \Omega_2 \\ \Omega_3 & \Omega_4 \end{bmatrix} \tag{9.25}$$

which yields

$$P = \left(\Omega_3 + \Omega_4 \begin{bmatrix} P_1 & 0 \\ 0 & P_2 \end{bmatrix} \right) \left(\Omega_1 + \Omega_2 \begin{bmatrix} P_1 & 0 \\ 0 & P_2 \end{bmatrix} \right)^{-1} \tag{9.26}$$

The invertibility of the matrices defined in Equations 9.24 and 9.26 is proved for sufficiently small values of ε in Appendix 9.1.

Partitioning Equations 9.18 and 9.19 as

$$\begin{bmatrix} \dot{\eta}_1(t) \\ \eta_2(t) \end{bmatrix} = \begin{bmatrix} a_1 & a_2 \\ a_3 & a_4 \end{bmatrix} \begin{bmatrix} \eta_1(t) \\ \eta_2(t) \end{bmatrix} \tag{9.27}$$

$$\begin{bmatrix} \dot{\xi}_1(t) \\ \xi_2(t) \end{bmatrix} = \begin{bmatrix} b_1 & b_2 \\ b_3 & b_4 \end{bmatrix} \begin{bmatrix} \xi_1(t) \\ \xi_2(t) \end{bmatrix} \tag{9.28}$$

where

$$\begin{aligned} a_1 &= A_1 + O(\varepsilon^2), & a_2 &= -S_1 + O(\varepsilon^2) \\ a_3 &= -Q_1 + O(\varepsilon^2), & a_4 &= -A^T + O(\varepsilon^2) \end{aligned} \tag{9.29}$$

$$\begin{aligned} b_1 &= A_4 + O(\varepsilon^2), & b_2 &= -S_3 + O(\varepsilon^2) \\ b_3 &= -Q_3 + O(\varepsilon^2), & b_4 &= -A^T + O(\varepsilon^2) \end{aligned} \tag{9.30}$$

and using Equation 9.23 yield two reduced-order nonsymmetric algebraic Riccati equations corresponding to local subsystems

$$0 = P_1 a_1 - a_4 P_1 - a_3 + P_1 a_2 P_1 \tag{9.31}$$

$$0 = P_2 b_1 - b_4 P_2 - b_3 + P_2 b_2 P_2 \tag{9.32}$$

It is important to notice that the total number of scalar quadratic algebraic equations in Equations 9.31 and 9.32 is $n_1^2 + n_2^2$. On the other hand, the global algebraic Riccati equation (Equation 9.6) contains $1/2(n_1 + n_2)(n_1 + n_2 + 1)$ scalar algebraic equations. Thus, the proposed method can also reduce the number of algebraic equations if

$$n_1^2 + n_2^2 < 1/2(n_1 + n_2)(n_1 + n_2 + 1) \tag{9.33}$$

or

$$(n_1 - n_2)^2 < n_1 + n_2 \tag{9.34}$$

which will definitely be the case when n_1 is close to n_2. Furthermore, due to the split into two independent subsystems, the advantage of parallel computations becomes significant in this case.

From Equations 9.29 and 9.30, it follows that the $O(\varepsilon^2)$ perturbations of the nonsymmetric algebraic Riccati equation (Equations 9.31 and 9.32) are symmetric, namely,

$$P_1 A_1 + A_1^T P_1 + D_1^T D_1 - P_1 B_1 R_1^{-1} B_1^T P_1 + O(\varepsilon^2) = 0 \tag{9.35}$$

$$P_2 A_4 + A_4^T P_2 + D_4^T D_4 - P_2 B_4 R_2^{-1} B_4^T P_2 + O(\varepsilon^2) = 0 \tag{9.36}$$

Using these facts and the implicit function theorem (Ortega and Rheinboldt 1970), the existence of the unique solutions of Equations 9.31 and 9.32 is guaranteed by the following lemma.

LEMMA 9.1

If both the triples (A_1, B_1, D_1) and (A_4, B_4, D_4) are stablizable–detectable, then $\exists \varepsilon_0 > 0$ such that $\forall \varepsilon \le \varepsilon_0$ the solutions of Equations 9.31 and 9.32 exist.

Two numerical methods can be proposed for solving Equations 9.31 and 9.32, namely, the fixed point iterations and the Newton method. The Newton method of Su and Gajić (1992) leads to the following recursive scheme

$$P_1^{(i+1)}\left(a_1 + a_2 P_1^{(i)}\right) - \left(a_4 - P_1^{(i)} a_2\right) P_1^{(i+1)} = a_3 + P_1^{(i)} a_2 P_1^{(i)} \tag{9.37}$$

$$P_2^{(i+1)}\left(b_1 + b_2 P_2^{(i)}\right) - \left(b_4 - P_2^{(i)} b_2\right) P_2^{(i+1)} = b_3 + P_2^{(i)} b_2 P_2^{(i)} \tag{9.38}$$

with the initial conditions obtained from the symmetric reduced-order algebraic Riccati equations

$$P_1^{(0)} A_1 + A_1^T P_1^{(0)} + Q_1 - P_1^{(0)} S_1 P_1^{(0)} = 0 \tag{9.39}$$

$$P_2^{(0)} A_4 + A_4^T P_2^{(0)} + Q_3 - P_2^{(0)} S_2 P_2^{(0)} = 0 \tag{9.40}$$

Note that Equations 9.39 and 9.40 are $O(\varepsilon^2)$ perturbations of the original equations (Equations 9.31 and 9.32) which implies that the required solutions of Equations 9.31 and 9.32, P_1 and P_2, are $O(\varepsilon^2)$ close to $P_1^{(0)}$ and $P_2^{(0)}$ so that the proposed Newton algorithm (Equations 9.37 and 9.38) is very efficient for solving Equations 9.31 and 9.32 since the excellent initial guesses are available. In addition, under conditions established in Lemma 9.1, the unique positive semidefinite stabilizing solutions of Equations 9.39 and 9.40 exist.

9.1.1   CASE STUDY: A SATELLITE CONTROL PROBLEM

To demonstrate the presented method, we have solved a fourth-order example, a satellite control problem considered in Ackerson and Fu (1970). Problem matrices are given by

$$A = \begin{bmatrix} 0 & 0.667 & 0 & 0 \\ -0.667 & 0 & 0 & 0 \\ 0 & 0 & 0 & 1.53 \\ 0 & 0 & 1.53 & 0 \end{bmatrix}, \quad B = \begin{bmatrix} 0 & 0.2 \\ 1 & 0 \\ 0.4 & 0 \\ 0 & 1 \end{bmatrix}$$

Penalty matrices Q and R are chosen as identities.
Results obtained from Equations 9.31 and 9.32 are

$$P_1 = \begin{bmatrix} 2.2201 & 0.45889 \\ 0.4410 & 1.2749 \end{bmatrix}, \quad P_2 = \begin{bmatrix} 1.5056 & 0.1947 \\ 0.22817 & 1.2782 \end{bmatrix}$$

which by the use formula (Equation 9.26) produce

$$P = \begin{bmatrix} 2.2437 & 0.46218 & 0.13613 & -0.10735 \\ 0.46218 & 1.3456 & -0.2091 & -0.24753 \\ 0.13613 & -0.2091 & 1.5375 & 0.24817 \\ -0.10735 & -0.24753 & 0.24817 & 1.3396 \end{bmatrix}$$

Exactly the same result has been obtained by using the classical global method for solving the algebraic Riccati equation (Equation 9.6).

9.2   OPTIMAL FILTERING IN CONTINUOUS-TIME

It has been shown in Chapter 6 using the recursive approach that the Kalman filtering problem of linear weakly coupled systems can be facilitated in terms of reduced-order optimal local Kalman filters corresponding to subsystems with the filter coefficients obtained with an arbitrary order of accuracy, that is $O(\varepsilon^{2k})$, where a small parameter ε represents the measure of coupling between subsystems, and k represents the number of the fixed-point iterations used to calculate coefficients of the corresponding filters.

It is important to point out that the local filters from Chapter 6 are driven by the innovation process so that the additional communication channels are required to form the innovation process. In the Hamiltonian approach, the corresponding filters will be driven by the system measurements only. In addition, the optional filter gains will be completely determined in terms of the exact reduced-order local algebraic Riccati equations.

Consider the linear continuous-time invariant weakly coupled stochastic system

$$\begin{aligned} x_1(t) &= A_1 x_1(t) + \varepsilon A_2 x_2(t) + G_1 w_1(t) + \varepsilon G_2 w_2(t) \\ x_2(t) &= \varepsilon A_3 x_1(t) + A_4 x_2(t) + \varepsilon G_3 w_1(t) + G_4 w_2(t) \end{aligned} \tag{9.41}$$

with the corresponding measurements

$$y_1(t) = C_1 x_1(t) + \varepsilon C_2 x_2(t) + v_1(t)$$
$$y_2(t) = \varepsilon C_3 x_1 + C_4 x_2(t) + v_2(t)$$

(9.42)

where $x_1(t) \in R^{n_1}$ and $x_2(t) \in R^{n_2}$ are state vectors, $w_i(t) \in R^{r_i}$, $i = 1, 2$, and $v_i(t) \in R^{l_i}$, $i = 1, 2$, are zero-mean stationary, white Gaussian noise stochastic processes with intensities $W_i > 0$ and $V_i > 0$, respectively, and $y_i(t) \in R^{l_i}$ are the system measurements. In the following A_i, G_i, C_i, $i = 1, 2, 3, 4$, are constant matrices.

We assume that the system matrix satisfies the weak coupling Assumption 9.1 and that all matrices are $O(1)$, that is, the following assumption is needed.

Assumption 9.3 Problem matrices $A_i, D_i, G_i, i = 1, 2, 3, 4$, and W_j, V_j, $j = 1, 2$, are $O(1)$.

The optimal Kalman filter, corresponding to Equations 9.41 and 9.42, driven by the innovation process is given by (Gajić and Shen 1993)

$$\begin{bmatrix} \dot{\hat{x}}_1(t) \\ \dot{\hat{x}}_2(t) \end{bmatrix} = \begin{bmatrix} A_1 & \varepsilon A_2 \\ \varepsilon A_3 & A_4 \end{bmatrix} \begin{bmatrix} \hat{x}_1(t) \\ \hat{x}_2(t) \end{bmatrix}$$
$$+ \begin{bmatrix} K_1 & \varepsilon K_2 \\ \varepsilon K_3 & K_4 \end{bmatrix} \begin{bmatrix} y_1(t) - C_1 \hat{x}_1(t) - \varepsilon C_2 \hat{x}_2(t) \\ y_2(t) - \varepsilon C_3 \hat{x}_1(t) - C_4 \hat{x}_2(t) \end{bmatrix}$$

(9.43)

Introducing the notation

$$A = \begin{bmatrix} A_1 & \varepsilon A_2 \\ \varepsilon A_3 & A_4 \end{bmatrix}, \quad G = \begin{bmatrix} G_1 & \varepsilon G_2 \\ \varepsilon G_3 & G_4 \end{bmatrix}, \quad C = \begin{bmatrix} C_1 & \varepsilon C_2 \\ \varepsilon C_3 & C_4 \end{bmatrix}$$
$$W = \begin{bmatrix} W_1 & 0 \\ 0 & W_2 \end{bmatrix}, \quad V = \begin{bmatrix} V_1 & 0 \\ 0 & V_2 \end{bmatrix}, \quad Z = C^T V^{-1} C$$
$$K = \begin{bmatrix} K_1 & \varepsilon K_2 \\ \varepsilon K_3 & K_4 \end{bmatrix}, \quad P = \begin{bmatrix} P_1 & \varepsilon P_2 \\ \varepsilon P_2^T & P_3 \end{bmatrix}$$

(9.44)

the optimal filter gain K is obtained from

$$K = PC^T V^{-1}$$

(9.45)

where the matrix P represents the positive semidefinite stabilizing solution to the algebraic filter Riccati equation

$$AP + PA^T - PZP + GWG^T = 0$$

(9.46)

For the recursive approach decomposition and approximation of the linear weakly coupled Kalman filter (Equation 9.43), the transformation of Gajić and Shen (1993), Shen and Gajić (1990a,b)

$$\begin{bmatrix} \hat{\eta}_1(t) \\ \hat{\eta}_2(t) \end{bmatrix} = \begin{bmatrix} I - \varepsilon^2 LH & \varepsilon L \\ -\varepsilon H & I \end{bmatrix} \begin{bmatrix} \hat{x}_1(t) \\ \hat{x}_2(t) \end{bmatrix} \tag{9.47}$$

where L and H satisfy algebraic equations

$$\begin{aligned} A_1 L + A_2 - LA_4 - \varepsilon^2 LA_3 L &= 0 \\ H(A_1 - \varepsilon^2 LA_3) - (A_4 + \varepsilon^2 A_3 L)H + A_3 &= 0 \end{aligned} \tag{9.48}$$

The decoupling transformation (Equation 9.47) applied to Equation 9.43 produces two independent reduced-order filters driven by the innovation processes

$$\begin{aligned} \dot{\hat{\eta}}_1(t) &= \alpha_1 \hat{\eta}_1(t) + \beta_1 v_1(t) + \varepsilon \beta_2 v_2(t) \\ \dot{\hat{\eta}}_2(t) &= \alpha_2 \hat{\eta}_2 + \varepsilon \beta_3 v_1(t) + \varepsilon \beta_4 v_2(t) \end{aligned} \tag{9.49}$$

In the new coordinates the innovation processes are given by

$$\begin{aligned} v_1(t) &= y(t) - d_1 \hat{\eta}_1(t) - \varepsilon d_2 \hat{\eta}(t) \\ v_2(t) &= y(t) - \varepsilon d_3 \hat{\eta}_1(t) - d_4 \hat{\eta}_2(t) \end{aligned} \tag{9.50}$$

The newly defined matrices in Equations 9.49 and 9.50 can be found in Section 6.2.

Equations 9.46 and 9.48 are solvable and produce the unique solutions under the following assumption.

Assumption 9.4   The matrices A_1 and A_4 have no eigenvalues in common and the matrices in the Riccati equation (Equation 9.46) satisfy the standard stabilizability–detectability conditions.

In the decomposition procedure from Section 6.2, the local filters (Equation 9.49) require the additional communication channels necessary to form the innovation process (Equation 9.50). Here, we present a decomposition scheme such that the local filters are completely decoupled and both of them are driven by the system measurements. The method is based on the exact decomposition technique for solving the regulator algebraic Riccati equation of weakly coupled developed in Section 9.1, see also Su and Gajić (1992). We will give an additional interpretation of the results from Su and Gajić (1992); which will be used in this section.

Consider the linear-quadratic optimal control problem corresponding to Equation 9.1, that is

$$\begin{aligned} \dot{x}_1(t) &= A_1 x_1(t) + \varepsilon A_2 x_2(t) + B_1 u_1(t) + \varepsilon B_2 u_2(t) \\ \dot{x}_2(t) &= \varepsilon A_3 x_1(t) + A_4 x_2(t) + \varepsilon B_3 u_1(t) + B_4 u_2(t) \end{aligned}$$

$$J = \frac{1}{2} \int_0^\infty \left[\begin{pmatrix} x_1(t) \\ x_2(t) \end{pmatrix}^T Q \begin{pmatrix} x_1(t) \\ x_2(t) \end{pmatrix} + u^T(t) R u(t) \right] dt, \quad Q \geq 0, \quad R > 0 \tag{9.51}$$

where the control vector with components, $u_i(t) \in R^{m_i}$ $i = 1, 2$, has to be chosen such that the performance criterion, J, is minimized. The very well-known solution to this problem is given by

$$u(x(t)) = -R^{-1}B^T P_r x(t) = -Fx(t) \tag{9.52}$$

where P_r is the positive semidefinite stabilizing solution of the regulator algebraic Riccati equation

$$A^T P_r + P_r A + Q - P_r S P_r = 0 \tag{9.53}$$

with

$$Q = \begin{bmatrix} Q_1 & Q_2 \\ Q_2^T & Q_3 \end{bmatrix}, \quad R = \begin{bmatrix} R_1 & 0 \\ 0 & R_2 \end{bmatrix}, \quad P_r = \begin{bmatrix} P_{1r} & eP_{2r} \\ \varepsilon P_{2r}^T & P_{3r} \end{bmatrix}$$
$$S = BR^{-1}B^T = \begin{bmatrix} S_1 & \varepsilon S_2 \\ \varepsilon S_3 & S_4 \end{bmatrix}, \quad B = \begin{bmatrix} B_1 & \varepsilon B_2 \\ \varepsilon B_3 & B_4 \end{bmatrix} \tag{9.54}$$

The optimal regulator gain F is given by

$$F = \begin{bmatrix} F_1 & \varepsilon F_2 \\ \varepsilon F_3 & F_4 \end{bmatrix} = -R^{-1}B^T P_r \tag{9.55}$$

The results of interest that we need, which can be deduced from Section 9.1 (see also Su and Gajić 1992) are given in the form of the following lemma.

LEMMA 9.2

Consider the optimal closed-loop linear system

$$\dot{x}_1(t) = (A_1 - B_1 F_1 - \varepsilon^2 B_2 F_3)x_1(t) + \varepsilon(A_2 - B_1 F_2 - B_2 F_4)x_2(t)$$
$$\dot{x}_2(t) = \varepsilon(A_3 - B_2 F_1 - B_4 F_3)x_1(t) + (A_4 - B_4 F_4 - \varepsilon^2 B_3 F_2)x_2(t) \tag{9.56}$$

then there exists a nonsingular transformation **T**

$$\begin{bmatrix} \xi_1(t) \\ \xi_2(t) \end{bmatrix} = \mathbf{T} \begin{bmatrix} x_1(t) \\ x_2(t) \end{bmatrix} \tag{9.57}$$

such that

$$\dot{\xi}_1(t) = (a_1 + a_2 P_{r1})\xi(t)$$
$$\dot{\xi}_2(t) = (b_1 + b_2 P_{r2})\xi_2(t) \tag{9.58}$$

where P_{r1} and P_{r2} are the unique solutions of the exact reduced-order completely decoupled local algebraic regulator Riccati equations

$$0 = P_{r1}a_1 - a_4P_{r1} - a_3 + P_{r1}a_2P_{r1}$$
$$0 = P_{r2}b_1 - b_4P_{r2} - b_3 + P_{r2}b_2P_{r2}$$

(9.59)

Matrices a_i, b_i, $i = 1, 2, 3, 4$ are defined in Equations 9.27 and 9.28. In this section we will give their expressions for the corresponding local filters and local algebraic filter Riccati equations to be defined later. The nonsingular transformation $\mathbf{T}$ is given by

$$\mathbf{T} = (\Pi_1 + \Pi_2 P_r)$$

(9.60)

Even more, the global solution P_r can be obtained from the reduced-order exact local algebraic regulator Riccati equations, that is

$$P_r = \left(\Omega_3 + \Omega_4 \begin{bmatrix} P_{r1} & 0 \\ 0 & P_{r2} \end{bmatrix}\right)\left(\Omega_1 + \Omega_2 \begin{bmatrix} P_{r1} & 0 \\ 0 & P_{r2} \end{bmatrix}\right)^{-1}$$

(9.61)

Known matrices Ω_i, $i = 1, 2, 3, 4$, and Π_1, Π_2, are given in terms of solutions of the decoupling algebraic equations.

The desired decomposition of the Kalman filter (Equation 9.43) will be obtained by producing a lemma dual to Lemma 9.2. Consider the optimal closed-loop Kalman filter (Equation 9.43) driven by the system measurements, that is

$$\dot{\hat{x}}_1(t) = \left(A_1 - K_1C_1 - \varepsilon^2 K_2C_3\right)\hat{x}_1(t) + \varepsilon(A_2 - K_1C_2 - K_2C_4)\hat{x}_2(t) + [K_1 \ \varepsilon K_2]y(t)$$
$$\dot{\hat{x}}_2(t) = \varepsilon(A_3 - K_3C_1 - K_4C_3)\hat{x}_1(t) + \left(A_4 - K_4C_4 - \varepsilon^2 K_3C_2\right)\hat{x}_2(t) + [\varepsilon K_3 \ K_4]y(t)$$

(9.62)

with the optimal filter gains K_i, $i = 1, 2$, calculated from Equations 9.44 through 9.46. By duality between the optimal filter and regulator, the filter Riccati equation (Equation 9.46) can be solved by using the same decomposition method for solving Equation 9.53 with

$$A \to A^T, \quad Q \to GWG^T, \quad F^T = K$$
$$S = BR^{-1}B^T \to Z = C^TV^{-1}C$$

(9.63)

such that

$$\begin{bmatrix} \dot{\hat{x}}_1(t) \\ \dot{\hat{x}}_2(t) \end{bmatrix} = \begin{bmatrix} A_1 - C_1^TF_1 - \varepsilon^2 C_3^TF_3 & \varepsilon(A_2 - C_1^TF_2 - C_3^TF_4) \\ \varepsilon(A_3 - C_2^TF_1 - C_4^TF_3) & A_4 - C_4^TF_4 - \varepsilon^2 C_2^TF_2 \end{bmatrix}^T \begin{bmatrix} \hat{x}_1(t) \\ \hat{x}_2(t) \end{bmatrix}$$
$$+ \begin{bmatrix} F_1 & \varepsilon F_2 \\ \varepsilon F_3 & F_4 \end{bmatrix}^T y(t)$$

(9.64)

By invoking results from Su and Gajić (1992), and using duality, the following matrices have to be formed (see also Section 9.1):

$$
T_1 = \begin{bmatrix} A_1^T & -(C_1^T V_1^{-1} C_1 + \varepsilon^2 C_3^T V_2^{-1} C_3) \\ -(G_1 W_1 G_1^T + \varepsilon^2 G_2 W_2 G_2^T) & -A_1 \end{bmatrix}
$$

$$
T_2 = \begin{bmatrix} A_3^T & -(C_1^T V_1^{-1} C_2 + C_3^T V_2^{-1} C_4) \\ -(G_1 W_1 G_3^T + G_2 W_2 G_4^T) & -A_2 \end{bmatrix}
$$

$$
T_3 = \begin{bmatrix} A_2^T & -(C_1^T V_1^{-1} C_1 + C_4^T V_2^{-1} C_3) \\ -(G_3 W_1 G_1^T + G_4 W_2 G_2^T) & -A_3 \end{bmatrix}
$$

$$
T_4 = \begin{bmatrix} A_4^T & -(C_4^T V_2^{-1} C_4 + \varepsilon^2 C_2^T V_1^{-1} C_2) \\ -(G_4 W_2 G_4^T + \varepsilon^2 G_3 W_1 G_3^T) & -A_4 \end{bmatrix}
$$

(9.65)

These matrices (T_1, T_2, T_3, and T_4) comprise the system matrix of a standard weakly coupled system, namely

$$
\begin{bmatrix} T_1 & \varepsilon T_2 \\ \varepsilon T_3 & T_4 \end{bmatrix}
$$

The decomposition of this system is achieved by using the decoupling transformation of Gajić and Shen (1989), which is determined by the solutions of the following algebraic equations:

$$
T_1 M + T_2 - M T_4 - \varepsilon^2 M T_3 M = 0
$$
$$
N(T_1 - \varepsilon^2 M T_3) - (T_4 + \varepsilon^2 T_3 M) N + T_3 = 0
$$

(9.66)

Algebraic equations (Equation 9.66) can be efficiently solved by using either the fixed-point iterations or the Newton method (Gajić and Shen 1989).

By using the permutation matrix

$$
E = \begin{bmatrix} I_{n1} & 0 & 0 & 0 \\ 0 & 0 & I_{n1} & 0 \\ 0 & I_{n2} & 0 & 0 \\ 0 & 0 & 0 & I_{n2} \end{bmatrix}
$$

(9.67)

we can define

$$
\Pi = \begin{bmatrix} \Pi_1 & \Pi_2 \\ \Pi_3 & \Pi_4 \end{bmatrix} = E^T \begin{bmatrix} I & -\varepsilon M \\ \varepsilon N & I - \varepsilon^2 NM \end{bmatrix} E
$$

(9.68)

Then, the desired transformation is given by

$$\mathbf{T_2} = (\Pi_1 + \Pi_2 P) \tag{9.69}$$

Note that Π_i, $i = 1, 2, 3, 4$, are square matrices of dimension n.

The transformation $\mathbf{T_2}$ applied to the filter variables as

$$\begin{bmatrix} \xi_1(t) \\ \xi_2(t) \end{bmatrix} = \mathbf{T_2^{-T}} \begin{bmatrix} \hat{x}_1(t) \\ \hat{x}_2(t) \end{bmatrix} \tag{9.70}$$

produces

$$\begin{bmatrix} \dot{\xi}_1(t) \\ \dot{\xi}_2(t) \end{bmatrix} = \mathbf{T_2^{-T}} \begin{bmatrix} A_1 - K_1 C_1 - \varepsilon^2 K_2 C_3 & \varepsilon(A_2 - K_1 C_2 - K_2 C_4) \\ \varepsilon(A_3 - K_3 C_1 - K_4 C_3) & A_4 - K_4 C_4 - \varepsilon^2 K_3 C_2 \end{bmatrix}^{\mathrm{T}}$$
$$\times \mathbf{T_2^T} \begin{bmatrix} \dot{\xi}_1(t) \\ \hat{\xi}_2(t) \end{bmatrix} + \mathbf{T_2^{-T}} \begin{bmatrix} K_1 & \varepsilon K_2 \\ \varepsilon K_3 & K_4 \end{bmatrix} y(t) \tag{9.71}$$

such that the complete closed-up filter decomposition is achieved, leading to

$$\dot{\xi}_1(t) = (a_1 + a_2 P_{f1}) \hat{\xi}_1(t) + [k_1 \; k_2] y(t)$$
$$\dot{\xi}_1(t) = (b_1 + b_2 P_{f2}) \hat{\xi}_2(t) + [k_3 \; k_4] y(t) \tag{9.72}$$

The matrices in Equation 9.72 are given by

$$\begin{bmatrix} a_1 & a_2 \\ a_3 & a_4 \end{bmatrix} = (T_1 - \varepsilon^2 M T_3), \quad \begin{bmatrix} b_1 & b_2 \\ b_3 & b_4 \end{bmatrix} = (T_4 + \varepsilon^2 T_3 M) \tag{9.73a}$$

$$\begin{bmatrix} k_1 & k_2 \\ k_3 & k_4 \end{bmatrix} = \mathbf{T_2^{-T}} \begin{bmatrix} K_1 & \varepsilon K_2 \\ \varepsilon K_3 & K_4 \end{bmatrix} \tag{9.73b}$$

with the reduced-order subsystem nonsymmetric algebraic Riccati equations

$$0 = P_{f1} a_1 - a_4 P_{f1} - a_3 + P_{f1} a_2 P_{f1}$$
$$0 = P_{f2} b_1 - b_4 P_{f2} - b_3 + P_{f2} b_2 P_{f2} \tag{9.74}$$

The Newton method for solving nonsymmetric Riccati equation (Equation 9.74) and the choice of excellent initial guesses for the Newton method via the solution of symmetric algebraic Riccati equations that are $O(\varepsilon^2)$ perturbations of Equation 9.74, are discussed in the previous section. It is important to point out that the matrix P can be obtained in terms of P_{f1} and P_{f2} by using Equation 9.61 with

$$P_{r1} = P_{f1}, \quad P_{r2} = P_{f2} \tag{9.75}$$

and Ω_1, Ω_2, Ω_3, Ω_4 obtained from

$$\Omega = \begin{bmatrix} \Omega_1 & \Omega_2 \\ \Omega_3 & \Omega_4 \end{bmatrix} = E^{\mathrm{T}} \begin{bmatrix} I - \varepsilon^2 MN & \varepsilon M \\ -\varepsilon N & I \end{bmatrix} E \qquad (9.76)$$

Ω_i, $i = 1, 2, 3, 4$, are square matrices of dimension n.

A lemma dual to Lemma 9.2, which in fact produces theoretical fundamentals for the presented Kalman filtering decomposition of linear weakly coupled stochastic continuous-time systems, can be now formulated as follows.

LEMMA 9.3

Given the closed-loop optimal Kalman filter (Equation 9.62) of a linear weakly coupled system, there exists a nonsingular transformation matrix (Equation 9.69), which completely decouples (Equation 9.62) into reduced-order local filters (Equation 9.72) both driven by the system measurements. Moreover, the decoupling transformation (Equation 9.70) and the filter coefficients given in Equation 9.72 can be obtained in terms of exact reduced-order completely decoupled algebraic Riccati equation (Equation 9.74).

Note that the actual procedure for computing the new decomposition technique can be completely performed at the subsystem levels. Namely, in the proposed method we have complete separation for both off-line (calculation of filters coefficients) and online (filtering process itself) computations.

The presentation of this section is mostly based on the work of Aganović and Gajić (1995). The presented procedure for the reduced-order Kalman of continuous-time weakly coupled linear systems is summarized in the following algorithm.

ALGORITHM 9.1

---

1. Solve decoupling transformation (Equation 9.66).
2. Solve for coefficients a_i, b_i, $i = 1, 2, 3, 4$, from Equation 9.73a.
3. Solve reduced-order Riccati equation (Equation 9.74).
4. Form transformation (Equation 9.69) by using Equations 9.74 through 9.76.
5. Find the reduced-filter (local) optimal gains from Equation 9.73b.

9.2.1 A HELICOPTER FILTERING PROBLEM

In order to demonstrate the proposed method we solve the filtering problem for a helicopter in low-speed flight condition considered in Grimbble (1993). The problem matrices are given by

$$A = \begin{bmatrix} -0.0098 & 0.0305 & -1.1774 & -32.1188 & -0.0118 & -0.4597 & 0 & 0 & 0 \\ -0.1632 & -0.5333 & 51.2195 & -2.0858 & -0.0172 & -0.4941 & 1.3028 & 0 & 0 \\ 0.0110 & 0.0058 & -2.0510 & 0 & 0.0114 & 0.4320 & 0 & 0 & 0 \\ 0 & 0 & 0.9992 & 0 & 0 & 0 & 0 & 0.0406 & 0 \\ 0.0143 & 0.0189 & -0.4348 & 0.0847 & -0.0816 & 1.0326 & 32.02923 & -49.7509 & 0 \\ 0.0099 & 0.0953 & -2.2384 & 0 & -0.0658 & -10.8462 & 0 & -0.1203 & 0 \\ 0 & 0 & -0.0026 & 0 & 0 & 1 & 0 & -0.0649 & 0 \\ -0.0097 & 0.0129 & -0.3133 & 0 & 0.0101 & -1.8855 & 0 & -0.6791 & 0 \\ 0 & 0 & -0.0406 & 0 & 0 & 0 & 0 & 1.0013 & 0 \end{bmatrix}$$

$$C = \begin{bmatrix} 0.0065 & -0.0997 & 0 & 5.0666 & 0.004 & 0 & -0.0133 & 0 & 0 \\ 0 & 0 & 0 & 10 & 0 & 0 & 0 & 0 & 0 \\ 0 & 0 & 0 & 0 & 0 & 0 & 5 & 0 & 0 \\ 0 & 0 & 0 & 0 & 0 & 0 & 0 & 0 & 5 \end{bmatrix}$$

The noise input matrix has been chosen to be identical to the control input matrix. This matrix according to Grimbble (1993) is given by

$$C = \begin{bmatrix} 16.7883 & -28.6070 & 6.6517 & 0 \\ -298.3142 & -30.1066 & 0.0023 & 0 \\ 5.5417 & 26.8395 & -5.7691 & 0 \\ 0 & 0 & 0 & 0 \\ 2.2574 & -6.3561 & -30.5836 & 15.9209 \\ 21.2054 & -31.6321 & -153.5206 & -0.9705 \\ 0 & 0 & 0 & 0 \\ 16.3444 & -5.8468 & -27.3854 & -13.0709 \\ 0 & 0 & 0 & 0 \end{bmatrix}$$

Since in Grimble (1993), the linear-quadratic stochastic optimal control problem is considered for the colored noise, we have assumed in this paper that white noise intensity matrices are identities.

Equation 9.66, whose solutions comprise the desired transformation, is solved by the Newton method. Also, the reduced-order nonsymmetric algebraic Riccati equations corresponding to subsystems are solved by using Newton's algorithm from Su and Gajić (1992). We have obtained completely decoupled reduced-order filters driven by the system measurements $y(t)$ as follows:

$$\dot{\hat{\xi}}_1(t) = \begin{bmatrix} -0.0371 & 0.5975 & -0.0193 & -0.0009 \\ 1.5521 & -29.1253 & 0.6457 & 0.0127 \\ -0.9927 & 48.3968 & -2.0219 & 0.9967 \\ 245.977 & 533.0225 & -260.1043 & -21.9362 \end{bmatrix} \hat{\xi}_1(t)$$

$$+ \begin{bmatrix} 11.1 & -32.8 & -3.8 & 34.2 \\ -285.4 & 90.6 & -9.2 & -14.9 \\ -9.7 & 28.8 & -19.4 & 37.1 \\ -211.6 & 61.7 & 317.6 & -579.5 \end{bmatrix} y(t)$$

$$\dot{\hat{\xi}}_2(t) = \begin{bmatrix} 0.0245 & -0.0924 & -0.0031 & 0.004 & -0.0024 \\ 1.0290 & -10.7737 & 1.0033 & -1.8718 & 0.005 \\ -6.2605 & -455.5855 & -30.2372 & -73.9713 & -3.7847 \\ -49.9652 & -0.2289 & 0.0706 & -0.7823 & 1.0071 \\ 615.4538 & -73.461 & -3.8943 & -67.9179 & -11.4456 \end{bmatrix} \hat{\xi}_2(t)$$

$$+ \begin{bmatrix} -26.3 & 3.6 & 7.8 & -124.1 \\ 46.2 & -11.4 & 91.9 & 18.3 \\ -360.6 & 135.3 & -11 & -17.6 \\ 70 & -17 & 17.4 & 15 \\ -1524.7 & 564.6 & -32.6 & -172 \end{bmatrix} y(t)$$

All simulation results in this paper are obtained by using MATLAB package for computer-aided control system design.

9.3   OPTIMAL CONTROL AND FILTERING IN DISCRETE-TIME

In this section, the algebraic regulator and filter Riccati equations of weakly coupled discrete-time stochastic linear control systems are completely and exactly decomposed into reduced-order continuous-time algebraic Riccati equations corresponding to subsystems. That is, the exact solution of the global discrete algebraic Riccati equation is found in terms of the reduced-order subsystem nonsymmetric continuous-time algebraic Riccati equations. In addition, the optimal global Kalman filter is decomposed into local optimal filters both driven by the system measurements and the system optimal control inputs. As a result, the optimal linear-quadratic Gaussian control problem for weakly coupled linear discrete systems takes the complete decomposition and parallelism between subsystem filters and controllers. The presented approach is based on a closed-loop decomposition technique that guarantees complete decomposition of the optimal filters and regulators and distribution of all required off-line and online computations.

In the regulation problem (optimal linear-quadratic control problem), we show how to decompose exactly the weakly coupled discrete algebraic Riccati equation into two reduced-order continuous-time algebraic Riccati equations. The reduced-order continuous-time algebraic Riccati equations are nonsymmetric, but their $O(\varepsilon^2)$ approximations are symmetric. The Newton method is very efficient for solving these nonsymmetric Riccati equations since the initial guesses close $O(\varepsilon^2)$ to the exact solutions can easily be obtained. This decomposition allows us to design linear controllers for subsystems completely independently of each other and thus to achieve the complete and exact separation for the linear-quadratic regulator problem. It is important to notice that it is, in general, easier to solve the continuous-time algebraic Riccati equation than the discrete-time algebraic Riccati equation.

In the filtering problem, in addition of using duality between filter and regulator to solve the discrete-time filter algebraic Riccati equation in terms of reduced-order continuous-time algebraic Riccati equations, we have obtained completely independent reduced-order Kalman filters both driven by system measurements and system optimal control inputs. In the last part of this section, we use the separation principle to solve the linear-quadratic Gaussian control problem of weakly coupled discrete stochastic systems. Two real control system examples are solved in order to demonstrate the proposed methods.

9.3.1   LINEAR-QUADRATIC OPTIMAL CONTROL

Consider the weakly coupled linear time-invariant discrete system described by (Gajić et al. 1990; Gajić and Shen 1993):

$$x_1(k+1) = A_1 x_1(k) + \varepsilon A_2 x_2(k) + B_1 u_1(k) + \varepsilon B_2 u_2(k) x_1(0) = x_{10} \qquad (9.77)$$
$$x_2(k+1) = \varepsilon A_3 x_1(k) + A_4 x_2(k) + \varepsilon B_3 u_1(k) + B_4 u_2(k) x_2(0) = x_{20} \qquad (9.78)$$

with state variables $x_1(k) \in R^{n_1}$, $x_2(k) \in R^{n_2}$, and control inputs $u_i(k) \in R^{m_i}$, $i = 1, 2$, where ε is a small coupling parameter. The performance criterion of the corresponding linear-quadratic control problem is represented by

$$J = \frac{1}{2} \sum_{k=0}^{\infty} \left[x(k)^{\mathrm{T}} Q x(k) + u(k)^{\mathrm{T}} R u(k) \right] \qquad (9.79)$$

where

$$x(k) = \begin{bmatrix} x_1(k) \\ x_2(k) \end{bmatrix}, \quad u(k) = \begin{bmatrix} u_1(k) \\ u_2(k) \end{bmatrix} \qquad (9.80)$$

$$Q = \begin{bmatrix} Q_1 & \varepsilon Q_2 \\ \varepsilon Q_2^{\mathrm{T}} & Q_3 \end{bmatrix} \geq 0, \quad R = \begin{bmatrix} R_1 & 0 \\ 0 & R_2 \end{bmatrix} > 0 \qquad (9.81)$$

It is well known that the solution of the above optimal regulation problem is given by

$$u(k) = -R^{-1} B^{\mathrm{T}} \lambda(k+1) = -(R + B^{\mathrm{T}} P_r B)^{-1} B^{\mathrm{T}} P_r A x(k)$$
$$= -Fx(k) = -\begin{bmatrix} F_1 & \varepsilon F_2 \\ \varepsilon F_3 & F_4 \end{bmatrix} x(k) \qquad (9.82)$$

where $\lambda(k)$ is a costate variable and P_r is the positive semidefinite stabilizing solution of the discrete algebraic Riccati equation given by (Dorato and Levis 1971; Lewis 1986):

$$P_r = Q + A^{\mathrm{T}} P_r [I + S P_r]^{-1} A$$
$$= Q + A^{\mathrm{T}} P_r A - A^{\mathrm{T}} P_r B [R + B^{\mathrm{T}} P_r B]^{-1} B^{\mathrm{T}} P_r A \qquad (9.83)$$

The Hamiltonian form of Equations 9.77 through 9.79 can be written as the forward recursion (Lewis 1986):

$$\begin{bmatrix} x(k+1) \\ \lambda(k+1) \end{bmatrix} = \mathbf{H_r} \begin{bmatrix} x(k) \\ \lambda(k) \end{bmatrix} \qquad (9.84)$$

with

$$\mathbf{H_r} = \begin{bmatrix} A + BR^{-1}B^{\mathsf{T}}A^{-\mathsf{T}}Q & -BR^{-1}B^{\mathsf{T}}A^{-\mathsf{T}} \\ -A^{-\mathsf{T}}Q & A^{-\mathsf{T}} \end{bmatrix} \qquad (9.85)$$

where $\mathbf{H_r}$ is the symplectic matrix which has the property that the eigenvalues of $\mathbf{H_r}$ can be grouped into two disjoint subsets Γ_1 and Γ_2, such that for every $\lambda_c \in \Gamma_1$ there exists $\lambda_d \in \Gamma_2$, which satisfies $\lambda_c \times \lambda_d = 1$, and we can choose either Γ_1 or Γ_2 to contain only the stable eigenvalues (Salgado et al. 1988). For the weakly coupled discrete systems, corresponding matrices in Equation 9.82 are given by

$$A = \begin{bmatrix} A_1 & \varepsilon A_2 \\ \varepsilon A_3 & A_4 \end{bmatrix}, \quad B = \begin{bmatrix} B_1 & \varepsilon A_2 \\ \varepsilon B_3 & B_4 \end{bmatrix}, \quad S = BR^{-1}B^{\mathsf{T}} = \begin{bmatrix} S_1 & \varepsilon S_2 \\ \varepsilon S_2^{\mathsf{T}} & S_3 \end{bmatrix} \quad (9.86)$$

In this section, we assume that the matrix A satisfies the weak coupling Assumption 9.1. In addition, the problem matrices are assumed to be $O(1)$, that is, we impose the following assumption specific to this subsection.

Assumption 9.5 The problem matrices A_i, B_i, $i = 1, 2, 3, 4$, Q_j, $j = 1, 2, 3$, and R_j, $j = 1, 2$, are $O(1)$.

The optimal open-loop control problem is a two-point boundary value problem with the associated state-costate equations forming the Hamiltonian matrix. For weakly coupled discrete systems, the Hamiltonian matrix retains the weakly coupled form by interchanging some state and costate variables so that it can be block diagonalized via the nonsingular transformation introduced by Gajić and Shen (1989). In the following, we show how to get the solution of the discrete-time algebraic Riccati equation of weakly coupled systems exactly in terms of the solutions of two reduced-order continuous-time algebraic Riccati equations.

Partitioning vector $\lambda(k)$ such that $\lambda(k) = \begin{bmatrix} \lambda_1^{\mathsf{T}}(k) & \lambda_2^{\mathsf{T}}(k) \end{bmatrix}^{\mathsf{T}}$ with $\lambda_1(k) \in R^{n_1}$ and $\lambda_2(k) \in R^{n_2}$, we obtain

$$\begin{bmatrix} x_1(k+1) \\ x_2(k+1) \\ \lambda_1(k+1) \\ \lambda_2(k+1) \end{bmatrix} = \mathbf{H_r} \begin{bmatrix} x_1(k) \\ x_2(k) \\ \lambda_1(k) \\ \lambda_2(k) \end{bmatrix} \qquad (9.87)$$

It has been shown in Gajić and Shen (1993) (in the chapter on the open-loop control, p. 181) that the Hamiltonian matrix (Equation 9.85) has the following form:

$$\mathbf{H_r} = \begin{bmatrix} \bar{A}_{1r} & \varepsilon \bar{A}_{2r} & \bar{S}_{1r} & \varepsilon \bar{S}_{2r} \\ \varepsilon \bar{A}_{3r} & \bar{A}_{4r} & \varepsilon \bar{S}_{3r} & \bar{S}_{4r} \\ \bar{Q}_{1r} & \varepsilon \bar{Q}_{2r} & \bar{A}_{11r}^{\mathsf{T}} & \varepsilon \bar{A}_{21r}^{\mathsf{T}} \\ \varepsilon \bar{Q}_{3r} & \bar{Q}_{4r} & \varepsilon \bar{A}_{12r}^{\mathsf{T}} & \bar{A}_{22r}^{\mathsf{T}} \end{bmatrix} \qquad (9.88)$$

Note that in the following there is no need for the analytical expressions for matrices with a bar. These matrices have to be formed by the computer in the process of calculations, which can be done easily. Interchanging second and third rows in Equation 9.88 yields

$$
\begin{bmatrix} x_1(k+1) \\ \lambda_1(k+1) \\ x_2(k+1) \\ \lambda_2(k+1) \end{bmatrix} = \begin{bmatrix} \overline{A}_{1r} & \overline{S}_{1r} & \varepsilon\overline{A}_{2r} & \varepsilon\overline{S}_{2r} \\ \overline{Q}_{1r} & \overline{A}_{11r}^{-T} & \varepsilon\overline{Q}_{2r} & \varepsilon\overline{A}_{21r}^{-T} \\ \varepsilon\overline{A}_{3r} & \varepsilon\overline{S}_{3r} & \overline{A}_{4r} & \overline{S}_{4r} \\ \varepsilon\overline{Q}_{3r} & \varepsilon\overline{A}_{12r}^{-T} & \overline{Q}_{4r} & \overline{A}_{22r}^{-T} \end{bmatrix} \begin{bmatrix} x_1(k) \\ \lambda_1(k) \\ x_2(k) \\ \lambda_2(k) \end{bmatrix}
$$

$$
= \begin{bmatrix} T_{1r} & \varepsilon T_{2r} \\ \varepsilon T_{3r} & T_{4r} \end{bmatrix} \begin{bmatrix} x_1(k) \\ \lambda_1(k) \\ x_2(k) \\ \lambda_2(k) \end{bmatrix} \tag{9.89}
$$

where

$$
T_{1r} = \begin{bmatrix} \overline{A}_{1r} & \overline{S}_{1r} \\ \overline{Q}_{1r} & \overline{A}_{11r}^{-T} \end{bmatrix}, \quad T_{2r} = \begin{bmatrix} \overline{A}_{2r} & \overline{S}_{2r} \\ \overline{Q}_{2r} & \overline{A}_{21r}^{-T} \end{bmatrix}
$$

$$
T_{3r} = \begin{bmatrix} \overline{A}_{3r} & \overline{S}_{3r} \\ \overline{Q}_{3r} & \overline{A}_{12r}^{-T} \end{bmatrix}, \quad T_{4r} = \begin{bmatrix} \overline{A}_{4r} & \overline{S}_{4r} \\ \overline{Q}_{4r} & \overline{A}_{22r}^{-T} \end{bmatrix} \tag{9.90}
$$

Introducing the notation

$$
U(k) = \begin{bmatrix} x_1(k) \\ \lambda_1(k) \end{bmatrix}, \quad V(k) = \begin{bmatrix} x_2(k) \\ \lambda_2(k) \end{bmatrix} \tag{9.91}
$$

we have the weakly coupled discrete system under new notation

$$
\begin{aligned} U(k+1) &= T_{1r}U(k) + \varepsilon T_{2r}V(k) \\ V(k+1) &= \varepsilon T_{3r}U(k) + T_{4r}V(k) \end{aligned} \tag{9.92}
$$

Applying the transformation

$$
\mathbf{T_r} = \begin{bmatrix} I - \varepsilon^2 L_r H_r & -\varepsilon L_r \\ \varepsilon H_r & I \end{bmatrix}, \quad \mathbf{T_r^{-1}} = \begin{bmatrix} I & -\varepsilon L_r \\ -\varepsilon H_r & I - \varepsilon^2 H_r L_r \end{bmatrix}
$$

$$
\begin{bmatrix} \eta(k) \\ \xi(k) \end{bmatrix} = \mathbf{T_r} \begin{bmatrix} U(k) \\ V(k) \end{bmatrix} \tag{9.93}
$$

to Equation 9.92, produces two completely decoupled subsystems

$$
\begin{bmatrix} \eta_1(k+1) \\ \eta_2(k+1) \end{bmatrix} = \eta(k+1) = (T_{1r} - \varepsilon^2 T_{2r} H_r)\eta(k) \tag{9.94}
$$

$$\begin{bmatrix} \xi_1(k+1) \\ \xi_2(k+1) \end{bmatrix} = \xi(k+1) = (T_{4r} + \varepsilon^2 H_r T_{2r})\xi(k) \tag{9.95}$$

where L_r and H_r satisfy

$$H_r T_{1r} - T_{4r} H_r + T_{3r} - \varepsilon^2 H_r T_{2r} H_r = 0 \tag{9.96}$$

$$L_r(T_{4r} + \varepsilon^2 H_r T_{2r}) - (T_{1r} - \varepsilon^2 T_{2r} H_r)L_r - T_{2r} = 0 \tag{9.97}$$

The unique solutions of Equations 9.96 and 9.97 exist under condition that matrices T_{1r} and $-T_{4r}$ have no eigenvalues in common (Gajić and Shen 1989). The algebraic equations (Equations 9.96 and 9.97) can be solved by using the Newton method (Gajić and Shen 1989), which converges quadratically in the neighborhood of the sought solution. The good initial guess required in the Newton recursive scheme is easily obtained, with the accuracy of $O(\varepsilon^2)$, by setting $\varepsilon = 0$ in those equations, which requires the solution of linear algebraic Lyapunov (Sylvester) equations.

The rearrangement and modification of variables in Equation 9.89 is done by using the permutation matrix E of the form:

$$\begin{bmatrix} x_1(k) \\ \lambda_1(k) \\ x_2(k) \\ \lambda_2(k) \end{bmatrix} = \begin{bmatrix} I_{n_1} & 0 & 0 & 0 \\ 0 & 0 & I_{n_1} & 0 \\ 0 & I_{n_2} & 0 & 0 \\ 0 & 0 & 0 & I_{n_2} \end{bmatrix} \begin{bmatrix} x_1(k) \\ x_2(k) \\ \lambda_1(k) \\ \lambda_2(k) \end{bmatrix} = E \begin{bmatrix} x(k) \\ \lambda(k) \end{bmatrix} \tag{9.98}$$

From Equations 9.91, 9.93 through 9.95, and 9.98, we obtain the relationship between the original coordinates and the new ones

$$\begin{bmatrix} \eta_1(k) \\ \xi_1(k) \\ \eta_2(k) \\ \xi_2(k) \end{bmatrix} = E^T T_r E \begin{bmatrix} x(k) \\ \lambda(k) \end{bmatrix} = \Pi_r \begin{bmatrix} x(k) \\ \lambda(k) \end{bmatrix} = \begin{bmatrix} \Pi_{1r} & \Pi_{2r} \\ \Pi_{3r} & \Pi_{4r} \end{bmatrix} \begin{bmatrix} x(k) \\ \lambda(k) \end{bmatrix} \tag{9.99}$$

Since $\lambda(k) = P_r x(k)$, where P_r satisfies the discrete algebraic Riccati equation (Equation 9.83), it follows from Equation 9.99 that

$$\begin{bmatrix} \eta_1(k) \\ \xi_1(k) \end{bmatrix} = (\Pi_{1r} + \Pi_{2r} P_r)x(k), \quad \begin{bmatrix} \eta_2(k) \\ \xi_2(k) \end{bmatrix} = (\Pi_{3r} + \Pi_{4r} P_r)x(k) \tag{9.100}$$

In the original coordinates, the required optimal solution has a closed-loop nature. We have the same characteristic for the new systems (Equations 9.94 and 9.95), that is,

$$\begin{bmatrix} \eta_2(k) \\ \xi_2(k) \end{bmatrix} = \begin{bmatrix} P_{ra} & 0 \\ 0 & P_{rb} \end{bmatrix} \begin{bmatrix} \eta_1(k) \\ \xi_1(k) \end{bmatrix} \tag{9.101}$$

Then Equations 9.100 and 9.101 yield

$$\begin{bmatrix} P_{ra} & 0 \\ 0 & P_{rb} \end{bmatrix} = (\Pi_{3r} + \Pi_{4r}P_r)(\Pi_{1r} + \Pi_{2r}P_r)^{-1} \tag{9.102}$$

It can be shown that $\Pi_r = I + O(\varepsilon) \Rightarrow \Pi_{1r} = I + O(\varepsilon)$, $\Pi_{2r} = O(\varepsilon)$, which implies that the matrix inversion defined in Equation 9.102 exists for sufficiently shall ε.

Following the same logic, we can find P_r reversely by introducing

$$E^T T_r^{-1} E = \Omega_r = \begin{bmatrix} \Omega_{1r} & \Omega_{2r} \\ \Omega_{3r} & \Omega_{4r} \end{bmatrix} \tag{9.103}$$

and it yields

$$P_r \left(\Omega_{3r} + \Omega_{4r} \begin{bmatrix} P_{ra} & 0 \\ 0 & P_{rb} \end{bmatrix} \right) \left(\Omega_{1r} + \Omega_{2r} \begin{bmatrix} P_{ra} & 0 \\ 0 & P_{rb} \end{bmatrix} \right)^{-1} \tag{9.104}$$

The required matrix in Equation 9.104 is invertible for small values of ε since from Equation 9.103 we have $\Omega_r = I + O(\varepsilon) \Rightarrow \Omega_{1r} = I + O(\varepsilon)$, $\Omega_{2r} = O(\varepsilon)$. Partitioning Equations 9.94 and 9.95 as

$$\begin{bmatrix} \eta_1(k+1) \\ \eta_2(k+1) \end{bmatrix} = \begin{bmatrix} a_{1r} & a_{2r} \\ a_{3r} & b_{4r} \end{bmatrix} \begin{bmatrix} \eta_1(k) \\ \eta_2(k) \end{bmatrix} = (T_{1r} - \varepsilon^2 T_{2r} H_r) \begin{bmatrix} \eta_1(k) \\ \eta_2(k) \end{bmatrix} \tag{9.105}$$

$$\begin{bmatrix} \xi_1(k+1) \\ \xi_2(k+1) \end{bmatrix} = \begin{bmatrix} b_{1r} & b_{2r} \\ b_{3r} & b_{4r} \end{bmatrix} \begin{bmatrix} \xi_1(k) \\ \xi_2(k) \end{bmatrix} = (T_{4r} + \varepsilon^2 H_r T_{2r}) \begin{bmatrix} \xi_1(k) \\ \xi_2(k) \end{bmatrix} \tag{9.106}$$

and using Equation 9.101 yield two reduced-order nonsymmetric algebraic Riccati equations

$$P_{ra} a_{1r} - a_{4r} P_{ra} - a_{3r} + P_{ra} a_{2r} P_{ra} = 0 \tag{9.107}$$

$$P_{ra} b_{1r} - b_{4r} P_{rb} - b_{3r} + P_{rb} b_{2r} P_{rb} = 0 \tag{9.108}$$

It is very interesting that the algebraic Riccati equation of weakly coupled discrete-time control systems is completely and exactly decomposed into two reduced-order nonsymmetric continuous-time algebraic Riccati equations (Equations 9.107 and 9.108). The latter ones are much easier to solve.

It can be shown that $O(\varepsilon^2)$ perturbations of Equations 9.107 and 9.108 lead to the symmetric reduced-order *discrete-time* algebraic Riccati equations, as obtained in Shen and Gajić (1990b). The solutions of these equations can be used as very good initial guesses for the Newton method for solving the obtained nonsymmetric algebraic Riccati equations (Equations 9.107 and 9.108). Another way to find the initial guesses which are $O(\varepsilon^2)$ close to the exact solutions is simply to perturb the coefficients in Equations 9.107 and 9.108 by $O(\varepsilon^2)$, which leads to the reduced-order nonsymmetric algebraic Riccati equations

$$
\begin{aligned}
P_{\text{ra}}^{(0)}\bar{A}_{1\text{r}} - \bar{A}_{11\text{r}}^{\text{T}}P_{\text{ra}}^{(0)} - \bar{Q}_{1\text{r}} + P_{\text{ra}}^{(0)}\bar{S}_{1\text{r}}P_{\text{ra}}^{(0)} &= 0 \\
P_{\text{rb}}^{(0)}\bar{A}_{4\text{r}} - \bar{A}_{22\text{r}}^{\text{T}}P_{\text{rb}}^{(0)} - \bar{Q}_{4\text{r}} + P_{\text{rb}}^{(0)}\bar{S}_{4\text{r}}P_{\text{rb}}^{(0)} &= 0
\end{aligned}
\tag{9.109}
$$

Note that the nonsymmetric algebraic Riccati equations have been studied by several researchers (see, e.g., Medanić [1982]). An efficient algorithm for solving the general nonsymmetric algebraic Riccati equation based on subspace iterations is derived in Avramović et al. (1980). Another approach is to use the eigenvector method (Bunse-Gerstner et al. 1992).

The Newton algorithm for Equation 9.107 is given by

$$
P_{\text{ra}}^{(i+1)}\left(a_{1\text{r}} + a_{2\text{r}}P_{\text{ra}}^{(i)}\right) - \left(a_{4\text{r}} - P_{\text{ra}}^{(i)}a_{2\text{r}}\right)P_{\text{ra}}^{(i+1)} = a_{3\text{r}} + P_{\text{ra}}^{(i)}a_{2\text{r}}P_{\text{ra}}^{(i)}
$$

$$
i = 0, 1, 2, \ldots
\tag{9.110}
$$

The Newton algorithm for Equation 9.108 is similarly obtained as

$$
P_{\text{rb}}^{(i+1)}\left(b_{1\text{r}} + b_{2\text{r}}P_{\text{rb}}^{(i)}\right) - \left(b_{4\text{r}} - P_{\text{rb}}^{(i)}b_{2\text{r}}\right)P_{\text{rb}}^{(i+1)} = b_{3\text{r}} + P_{\text{rb}}^{(i)}b_{2\text{r}}P_{\text{rb}}^{(i)}
$$

$$
i = 0, 1, 2, \ldots
\tag{9.111}
$$

The proposed method is very suitable for parallel computations since it allows complete parallelism. In addition, due to complete and exact decomposition of the discrete algebraic Riccati equation, the optimal control at steady state can be performed independently and in parallel. The reduced-order subsystems in the new coordinates are given by

$$
\eta_1(k + 1) = (a_{1\text{r}} + a_{2\text{r}}P_{\text{ra}})\eta_1(k)
\tag{9.112}
$$

$$
\xi_1(k + 1) = (b_{1\text{r}} + b_{2\text{r}}P_{\text{rb}})\xi_1(k)
\tag{9.113}
$$

In summary, the optimal strategy and the optimal performance value are obtained by using the following algorithm.

ALGORITHM 9.2

1. Solve decoupling Equations 9.96 and 9.97
2. Find coefficients $a_{i\text{r}}, b_{i\text{r}}, i = 1, 2, 3, 4$, by using Equations 9.105 and 9.106.
3. Solve the reduced-order algebraic Riccati equations (Equations 9.107 and 9.108), which leads to P_{ra} and P_{rb}.
4. Find the global solution of Riccati equation in terms of P_{ra} and P_{rb} by using Equation 9.104.
5. Find the optimal regulator gain from Equation 9.82 and the optimal performance criterion as $J_{\text{opt}} = 0.5x^{\text{T}}(t_0) P_{\text{r}}x(t_0)$.

Example 9.1

In order to demonstrate the efficiency of the proposed method, a discrete model (Katzberg 1977) is considered. The problem matrices are given by

$$
A = \begin{bmatrix} 0.964 & 0.184 & 0.017 & 0.019 \\ -0.342 & 0.802 & 0.162 & 0.179 \\ 0.016 & 0.019 & 0.983 & 0.181 \\ 0.144 & 0.179 & -0.163 & 0.820 \end{bmatrix}, \quad B = \begin{bmatrix} 0.019 & 0.001 \\ 0.180 & 0.019 \\ 0.005 & 0.019 \\ -0.054 & 0.181 \end{bmatrix}
$$

$$
Q = 0.1 I_4, \quad R = I_2, \quad \varepsilon = 0.329, \quad n_1 = 2, \quad n_2 = 2
$$

The optimal global solution of the discrete algebraic Riccati equation is obtained as

$$
P_{exact} = \begin{bmatrix} 1.3998 & -0.1261 & 0.1939 & -0.4848 \\ -0.1261 & 1.2748 & 0.5978 & 1.3318 \\ 0.1939 & 0.5978 & 1.1856 & 0.7781 \\ -0.4848 & 1.3318 & 0.7781 & 1.9994 \end{bmatrix}
$$

Solutions of the reduced-order algebraic Riccati equations obtained from Equations 9.107 and 9.108 are

$$
P_{ra} = \begin{bmatrix} 0.7664 & 0.0846 \\ 0.2287 & 0.3184 \end{bmatrix}, \quad P_{rb} = \begin{bmatrix} 1.2042 & 0.1840 \\ 1.1374 & 2.4557 \end{bmatrix}
$$

By using formula (Equation 9.104), the obtained solution for P_r is found to be identical to P_{exact} and between the solution of the proposed method and the exact one, which is obtained by using the classical global method for solving algebraic Riccati equation, given by

$$
P_{exact} - P_r = O(10^{-13})
$$

which is the standard MATLAB accuracy. Assuming the initial conditions as $x^T(t_0) = [1\ 1\ 1\ 1]$, the optimal performance value is $J_{opt} = 0.5 x^T(0) P_r x(0) = 5.2206$.

9.3.2 OPTIMAL KALMAN FILTERING

The continuous-time filtering problem of weakly coupled linear stochastic systems has been studied via the recursive approach by Shen and Gajić (1990). In this section, we solve the filtering problem of linear discrete-time weakly coupled systems using the problem formulation from Shen and Gajić (1990b). The method is based on the exact decomposition of the global weakly coupled discrete algebraic Riccati equation into reduced-order local algebraic Riccati equations. The optimal filter gain will be completely determined in terms of exact reduced-order continuous-time algebraic Riccati equations, based on the duality property between the optimal filter and regulator. Even more, we have obtained the exact expressions for the optimal reduced-order local filters both driven by the system measurements. This is an important advantage over the results of Shen and Gajić (1990a,b) where the local filters are driven by the innovation process so that the additional communication channels have to be used in order to construct the innovation process.

Consider the linear discrete-time invariant weakly coupled stochastic system

$$x_1(k+1) = A_1 x_1(k) + \varepsilon A_2 x_2(k) + G_1 w_1(k) + \varepsilon G_2 w_2(k)$$
$$x_2(k+1) = \varepsilon A_3 x_1(k) + A_4 x_2(k) + \varepsilon G_3 w_1(k) + G_4 w_2(k) \tag{9.114}$$
$$x_1(0) = x_{10}, \quad x_2(0) = x_{20}$$

with the corresponding measurements

$$y(k) = \begin{bmatrix} y_1(k) \\ y_2(k) \end{bmatrix} = \begin{bmatrix} C_1 & \varepsilon C_2 \\ \varepsilon C_3 & C_4 \end{bmatrix} \begin{bmatrix} x_1(k) \\ x_2(k) \end{bmatrix} + \begin{bmatrix} v_1(k) \\ v_2(k) \end{bmatrix} \tag{9.115}$$

where $x_i(k) \in R^{n_i}$ are state vectors, $w_i(k) \in R^{r_i}$ and $v_i(k) \in R^{l_i}$ are zero-mean stationary white Gaussian noise stochastic processes with intensities $W_i > 0$, $V_i > 0$, respectively, and $y_i(k) \in R^{l_i}$, $i = 1, 2$, are the system measurements. In the following A_i, G_i, C_i, $i = 1, 2, 3, 4$, are constant matrices.

The optimal Kalman filter, driven by the innovation process, is given by (Kwakernaak and Sivan 1972):

$$\hat{x}(k+1) = A\hat{x}(k) + K[y(k) - C\hat{x}(k)] \tag{9.116}$$

where

$$A = \begin{bmatrix} A_1 & \varepsilon A_2 \\ \varepsilon A_3 & A_4 \end{bmatrix}, \quad C = \begin{bmatrix} C_1 & \varepsilon C_2 \\ \varepsilon C_3 & C_4 \end{bmatrix}, \quad K = \begin{bmatrix} K_1 & \varepsilon K_2 \\ \varepsilon K_3 & K_4 \end{bmatrix} \tag{9.117}$$

The filter gain K is obtained from

$$K = AP_f C^T (V + CP_f C^T)^{-1}, \quad V = \begin{bmatrix} V_1 & 0 \\ 0 & V_2 \end{bmatrix} \tag{9.118}$$

where P_f is the positive semidefinite stabilizing solution of the discrete-time filter algebraic Riccati equation given by

$$P_f = AP_f A^T - AP_f C^T (V + CP_f C^T)^{-1} CP_f A^T + GWG^T \tag{9.119}$$

where

$$G = \begin{bmatrix} G_1 & \varepsilon G_2 \\ \varepsilon G_3 & G_4 \end{bmatrix}, \quad W = \begin{bmatrix} W_1 & 0 \\ 0 & W_2 \end{bmatrix} \tag{9.120}$$

In addition to the weak coupling assumption, Assumption 9.1, in order to be able to give estimates of the quantities to be involved in the derivations that follow, we need to assume that all matrices given in the problem formulation are $O(1)$, that is, we impose the following assumption.

Assumption 9.6 Matrices $A_i, C_i, G_i, i = 1, 2, 3, 4$, $W_j, V_j, j = 1, 2$, are $O(1)$.

Due to the weakly coupled structure of the problem matrices the required solution P_f has the form

$$P_f = \begin{bmatrix} P_{f1} & \varepsilon P_{f2} \\ \varepsilon P_{f2}^T & P_{f3} \end{bmatrix} \tag{9.121}$$

Partitioning the discrete-time filter Riccati equation (Equation 9.119), in the sense of weak coupling methodology, will produce a lot of terms and make the corresponding problem numerically inefficient, even though the problem order-reduction is achieved.

Using the decomposition procedure proposed from the previous section and the duality property between the optimal filter and regulator, we present a decomposition scheme such that the subsystem filters of weakly coupled discrete systems are completely decoupled and both of them are driven by system measurements. The new method is based on the exact decomposition technique, which is proposed in the previous section, for solving the regulator algebraic Riccati equation of weakly coupled discrete systems.

The results of interest which can be deduced from Section 9.3.1 are summarized in the form of the following lemma.

LEMMA 9.4

Consider the optimal closed-loop linear weakly coupled discrete system

$$\begin{aligned} x_1(k+1) &= (A_1 - B_1 F_1 - \varepsilon^2 B_2 F_3)x_1(k) \\ &\quad + \varepsilon(A_2 - B_1 F_2 - B_2 F_4)x_2(k) \\ x_2(k+1) &= \varepsilon(A_3 - B_3 F_1 - B_4 F_3)x_1(k) \\ &\quad + (A_4 - B_2 F_2 - \varepsilon^2 B_3 F_2)x_2(k) \end{aligned} \tag{9.122}$$

then there exists a nonsingular transformation $\mathbf{T_r}$

$$\begin{bmatrix} \zeta_1(k) \\ \zeta_2(k) \end{bmatrix} = \mathbf{T_r} \begin{bmatrix} x_1(k) \\ x_2(k) \end{bmatrix} \tag{9.123}$$

such that

$$\begin{aligned} \zeta_1(k+1) &= (a_{1r} + a_{2r} P_{ra})\zeta_1(k) \\ \zeta_2(k+1) &= (b_{1r} + b_{2r} P_{rb})\zeta_2(k) \end{aligned} \tag{9.124}$$

where P_{ra} and P_{rb} are the unique solutions of the exact reduced-order completely decoupled continuous-time algebraic Riccati equations

$$
\begin{aligned}
P_{ra}a_{1r} - a_{4r}P_{ra} - a_{3r} + P_{ra}a_{2r}P_{ra} &= 0 \\
P_{rb}b_{1r} - b_{4r}P_{rb} - b_{3r} + P_{rb}b_{2r}P_{rb} &= 0
\end{aligned}
\tag{9.125}
$$

Matrices $a_{ir}, b_{ir}, \; i = 1, 2, 3, 4$, can be found from Equations 9.105 and 9.106. The nonsingular transformation $\mathbf{T_r}$ is given by

$$
\mathbf{T_r} = (\Pi_{1r} + \Pi_{2r}P_r)
\tag{9.126}
$$

Moreover, the global solution P_r can be obtained from the reduced-order algebraic Riccati equations, that is

$$
P_r = \left(\Omega_{3r} + \Omega_{4r} \begin{bmatrix} P_{ra} & 0 \\ 0 & P_{rb} \end{bmatrix} \right) \left(\Omega_{1r} + \Omega_{2r} \begin{bmatrix} P_{ra} & 0 \\ 0 & P_{rb} \end{bmatrix} \right)^{-1}
\tag{9.127}
$$

Known matrices $\Omega_{ir}, \; i = 1, 2, 3, 4$, and Π_{1r}, Π_{2r}, are given in terms of the solutions of the decoupling equations defined in Section 9.3.1.

The desired decomposition of the Kalman filter (Equation 9.116) will be obtained by using duality between the optimal filter and regulator, and the decomposition method developed in Section 9.3.1. Consider the optimal closed-loop Kalman filter (Equation 9.116) driven by the system measurements, that is

$$
\begin{aligned}
\hat{x}_1(k+1) &= (A_1 - K_1C_1 - \varepsilon^2 K_2C_3)\hat{x}_1(k) \\
&\quad + \varepsilon(A_2 - K_1C_2 - K_2C_4)\hat{x}_2(k) + K_1y_1(k) + \varepsilon K_2y_2(k) \\
\hat{x}_2(k+1) &= \varepsilon(A_3 - K_3C_1 - K_4C_3)\hat{x}_1(k) \\
&\quad + (A_4 - K_4C_4 - \varepsilon^2 K_3C_2)\hat{x}_2(k) + \varepsilon K_3y_1(k) + K_4y_2(k)
\end{aligned}
\tag{9.128}
$$

By using Equations 9.114 and 9.115 and duality between the optimal filter and regulator, that is

$$
\begin{aligned}
A &\to A^T, \quad Q \to GWG^T, \quad B \to C^T \\
BR^{-1}B^T &\to C^T V^{-1} C
\end{aligned}
\tag{9.129}
$$

the filter "state-costate equation" can be defined as

$$
\begin{bmatrix} x(k+1) \\ \lambda(k+1) \end{bmatrix} = \mathbf{H_f} \begin{bmatrix} x(k) \\ \lambda(k) \end{bmatrix}
\tag{9.130}
$$

where

$$\mathbf{H_f} = \begin{bmatrix} A^{\mathrm{T}} + C^{\mathrm{T}}V^{-1}CA^{-1}GWG^{\mathrm{T}} & -C^{\mathrm{T}}V^{-1}CA^{-1} \\ -A^{-1}GWG^{\mathrm{T}} & A^{-1} \end{bmatrix} \tag{9.131}$$

Partitioning $\lambda(k)$ as $\lambda(k) = \begin{bmatrix} \lambda_1^{\mathrm{T}}(k) & \lambda_2^{\mathrm{T}}(k) \end{bmatrix}^{\mathrm{T}}$ with $\lambda_1(k) \in R^{n_1}$ and $\lambda_1(k) \in R^{n_2}$, Equation 9.130 can be rewritten as

$$\begin{bmatrix} x_1(k+1) \\ x_2(k+1) \\ \lambda_1(k+1) \\ \lambda_2(k+1) \end{bmatrix} = \begin{bmatrix} \overline{A}_{1f}^{\mathrm{T}} & \varepsilon\overline{A}_{3f}^{\mathrm{T}} & \overline{S}_{1f} & \varepsilon\overline{S}_{2f} \\ \varepsilon\overline{A}_{2f}^{\mathrm{T}} & \overline{A}_{4f}^{\mathrm{T}} & \varepsilon\overline{S}_{3f} & \overline{S}_{4f} \\ \overline{Q}_{1f} & \varepsilon\overline{Q}_{2f} & \overline{A}_{11f} & \varepsilon\overline{A}_{12f} \\ \varepsilon\overline{Q}_{3f} & \overline{Q}_{4f} & \varepsilon\overline{A}_{21f} & \overline{A}_{22f} \end{bmatrix} \begin{bmatrix} x_1(k) \\ x_2(k) \\ \lambda_1(k) \\ \lambda_2(k) \end{bmatrix} \tag{9.132}$$

Interchanging the second and third rows yields

$$\begin{bmatrix} x_1(k+1) \\ \lambda_1(k+1) \\ x_2(k+1) \\ \lambda_2(k+1) \end{bmatrix} = \begin{bmatrix} \overline{A}_{1f}^{\mathrm{T}} & \overline{S}_{1f} & \varepsilon\overline{A}_{3f}^{\mathrm{T}} & \varepsilon\overline{S}_{2f} \\ \overline{Q}_{1f} & \overline{A}_{11f} & \varepsilon\overline{Q}_{2f} & \varepsilon\overline{A}_{12f} \\ \varepsilon\overline{A}_{2f}^{\mathrm{T}} & \varepsilon\overline{S}_{3f} & \overline{A}_{4f}^{\mathrm{T}} & \overline{S}_{4f} \\ \varepsilon\overline{Q}_{3f} & \varepsilon\overline{A}_{21f} & \overline{Q}_{4f} & \overline{A}_{22f} \end{bmatrix} \begin{bmatrix} \varepsilon x_1(k) \\ \lambda_1(k) \\ x_2(k) \\ \lambda_2(k) \end{bmatrix}$$

$$= \begin{bmatrix} T_{1f} & \varepsilon T_{2f} \\ \varepsilon T_{3f} & T_{4f} \end{bmatrix} \begin{bmatrix} \varepsilon x_1(k) \\ \lambda_1(k) \\ x_2(k) \\ \lambda_2(k) \end{bmatrix} \tag{9.133}$$

where

$$T_{1f} = \begin{bmatrix} \overline{A}_{1f}^{\mathrm{T}} & \overline{S}_{1f} \\ \overline{Q}_{1f} & \overline{A}_{11f} \end{bmatrix}, \quad T_{2f} = \begin{bmatrix} \overline{A}_{3f}^{\mathrm{T}} & \overline{S}_{2f} \\ \overline{Q}_{2f} & \overline{A}_{12f} \end{bmatrix}$$

$$T_{3f} = \begin{bmatrix} \overline{A}_{2f}^{\mathrm{T}} & \overline{S}_{3f} \\ \overline{Q}_{3f} & A_{21f} \end{bmatrix}, \quad T_{4f} = \begin{bmatrix} \overline{A}_{4f}^{\mathrm{T}} & \overline{S}_{4f} \\ \overline{Q}_{4f} & \overline{A}_{22f} \end{bmatrix} \tag{9.134}$$

These matrices comprise the system matrix of a standard weakly coupled discrete system, so that the reduced-order decomposition can be achieved by applying the decoupling transformation to Equation 9.133, which yields two completely decoupled subsystems

$$\begin{bmatrix} \eta_1(k+1) \\ \eta_2(k+1) \end{bmatrix} = \begin{bmatrix} a_{1f} & a_{2f} \\ a_{3f} & a_{4f} \end{bmatrix} \begin{bmatrix} \eta_1(k) \\ \eta_2(k) \end{bmatrix} = [T_{1f} - \varepsilon^2 T_{2f}H_f] \begin{bmatrix} \eta_1(k) \\ \eta_2(k) \end{bmatrix} \tag{9.135}$$

$$\begin{bmatrix} \xi_1(k+1) \\ \xi_2(k+1) \end{bmatrix} = \begin{bmatrix} b_{1f} & b_{2f} \\ b_{3f} & b_{4f} \end{bmatrix} \begin{bmatrix} \xi_1(k) \\ \xi_2(k) \end{bmatrix} = [T_{4f} + \varepsilon^2 H_f T_{2f}] \begin{bmatrix} \xi_1(k) \\ \xi_2(k) \end{bmatrix} \tag{9.136}$$

Note that the decoupling transformation has the form of Equation 9.93 with H_f and L_f matrices obtained from Equations 9.96 and 9.97 with T_{if}'s taken from Equation 9.134. By duality and Lemma 9.4 the following reduced-order nonsymmetric algebraic Riccati equations exist:

$$P_{fa}a_{1f} - a_{4f}P_{fa} - a_{3f} + P_{fa}a_{2f}P_{fa} = 0 \qquad (9.137)$$

$$P_{fb}b_{1f} - b_{4f}P_{fb} - b_{3f} + P_{fb}b_{2f}P_{fb} = 0 \qquad (9.138)$$

By using the permutation matrix

$$\begin{bmatrix} x_1(k) \\ \lambda_1(k) \\ x_2(k) \\ \lambda_2(k) \end{bmatrix} = E \begin{bmatrix} x_1(k) \\ x_2(k) \\ \lambda_1(k) \\ \lambda_2(k) \end{bmatrix} = \begin{bmatrix} I_{n1} & 0 & 0 & 0 \\ 0 & 0 & I_{n1} & 0 \\ 0 & I_{n2} & 0 & 0 \\ 0 & 0 & 0 & I_{n2} \end{bmatrix} \begin{bmatrix} x_1(k) \\ x_2(k) \\ \lambda_1(k) \\ \lambda_2(k) \end{bmatrix} \qquad (9.139)$$

we can define

$$\Pi_f = \begin{bmatrix} \Pi_{1f} & \Pi_{2f} \\ \Pi_{3f} & \Pi_{4f} \end{bmatrix} = E^T \begin{bmatrix} I - \varepsilon^2 L_f H_f & -\varepsilon L_f \\ \varepsilon H_f & I \end{bmatrix} E \qquad (9.140)$$

Then, the desired transformation is given by

$$\mathbf{T_f} = (\Pi_{1f} + \Pi_{2f}P_f) \qquad (9.141)$$

The transformation $\mathbf{T_f}$ applied to the filter variables (Equation 9.128) as

$$\begin{bmatrix} \hat{\eta}_1 \\ \hat{\eta}_2 \end{bmatrix} = \mathbf{T_f^{-T}} \begin{bmatrix} \hat{x}_1 \\ \hat{x}_2 \end{bmatrix} \qquad (9.142)$$

produces

$$\begin{bmatrix} \hat{\eta}_1(k+1) \\ \hat{\eta}_2(k+1) \end{bmatrix} = \mathbf{T_f^{-T}} \begin{bmatrix} A_1 - K_1C_1 - \varepsilon^2 K_2C_3 & \varepsilon(A_2 - K_1C_2 - K_2C_4) \\ \varepsilon(A_3 - K_3C_1 - K_4C_3) & A_4 - K_4C_4 - \varepsilon^2 K_3C_2 \end{bmatrix} \mathbf{T_f^T} \begin{bmatrix} \hat{\eta}_1(k) \\ \hat{\eta}_2(k) \end{bmatrix}$$
$$+ \mathbf{T_f^{-T}} \begin{bmatrix} K_1 & \varepsilon K_2 \\ \varepsilon K_3 & K_4 \end{bmatrix} y(k) \qquad (9.143)$$

such that the complete closed-loop decomposition is achieved, that is

$$\begin{aligned} \hat{\eta}_1(k+1) &= (a_{1f} + a_{2f}P_{fa})^T \hat{\eta}_1(k) + \mathbf{K_1}y(k) \\ \hat{\eta}_2(k+1) &= (b_{1f} + b_{2f}P_{fb})^T \hat{\eta}_2(k) + \mathbf{K_2}y(k) \end{aligned} \qquad (9.144)$$

where

$$\begin{bmatrix} \mathbf{K_1} \\ \mathbf{K_2} \end{bmatrix} = \mathbf{T_f^{-T}} K \tag{9.145}$$

It is important to point out that the matrix P_f in Equation 9.141 can be obtained in terms of P_{fa} and P_{fb}, by using Equation 9.127, with $\Omega_{1f}, \Omega_{2f}, \Omega_{3f}, \Omega_{4f}$, obtained from

$$\Omega_f = \begin{bmatrix} \Omega_{1f} & \Omega_{2f} \\ \Omega_{3f} & \Omega_{4f} \end{bmatrix} = E^{\mathrm{T}} \begin{bmatrix} I & \varepsilon L_f \\ -\varepsilon H_f & I - \varepsilon^2 H_f L_f \end{bmatrix} E \tag{9.146}$$

A lemma dual to Lemma 1 can now be formulated as follows.

LEMMA 9.5

Given the closed-loop optimal Kalman filter (Equation 9.128) of a linear discrete weakly coupled system, there exists a nonsingular transformation matrix (Equation 9.141), which completely decouples Equation 9.128 into reduced-order local filters (Equation 9.144) both driven by the system measurements. Moreover, the decoupling transformation (Equation 9.141) and the filter coefficients given in Equations 9.135 and 9.136 can be obtained in terms of the exact reduced-order completely decoupled continuous-time Riccati equations (Equations 9.137 and 9.138).

It should be noted that the new filtering method allows complete decomposition and parallelism between local filters. The complete solution to our problem can be summarized in the form of the following algorithm.

ALGORITHM 9.3

1. Find T_{1f}, T_{2f}, T_{3f}, and T_{4f} from Equation 9.134.
2. Calculate L_f and H_f from Equations 9.96 and 9.97 with T_{if}'s obtained from Equation 9.134.
3. Find a_{if}, b_{if}, for $i = 1, 2, 3, 4$, from Equations 9.135 and 9.136.
4. Solve for P_{fa} and P_{fb} from Equations 9.137 and 9.138.
5. Find $\mathbf{T_f}$ from Equation 9.141 with P_f obtained from Equation 9.127.
6. Calculate $\mathbf{K_1}$ and $\mathbf{K_2}$ from Equation 9.145.
7. Find the local filter system matrices by using Equation 9.144.

9.3.3 LINEAR-QUADRATIC GAUSSIAN OPTIMAL CONTROL PROBLEM

This section presents the approach in the study of the LQG control problem of weakly coupled discrete systems when the performance index is defined in an infinite time period. The discrete-time LQG problem of weakly coupled systems has been studied via the recursive approach of Shen and Gajić (1990b). We will solve the LQG problem via the Hamiltonian approach by using the results obtained in previous sections. The discrete algebraic Riccati equation is first completely and exactly decomposed into two reduced-order continuous-time algebraic Riccati equations. The local filters will be driven by the system measurements, on the contrary to the work of Shen and Gajić (1990b) where the local filters are driven by the innovation process.

Consider the weakly coupled discrete-time linear stochastic control system represented by (Shen and Gajić 1990b)

$$
\begin{aligned}
x_1(k+1) &= A_1 x_1(k) + \varepsilon A_2 x_2(k) + B_1 u_1(k) + \varepsilon B_2 u_2(k) \\
&\quad + G_1 w_1(k) + \varepsilon G_2 w_2(k) \\
x_2(k+1) &= \varepsilon A_3 x_1(k) + A_4 x_2(k) + \varepsilon B_3 u_1(k) + B_4 u_2(k) \\
&\quad + \varepsilon G_3 w_1(k) + G_4 w_2(k)
\end{aligned}
\tag{9.147}
$$

$$
y(k) = \begin{bmatrix} y_1(k) \\ y_2(k) \end{bmatrix} = \begin{bmatrix} C_1 & \varepsilon C_2 \\ \varepsilon C_3 & C_4 \end{bmatrix} \begin{bmatrix} x_1(k) \\ x_2(k) \end{bmatrix} + \begin{bmatrix} v_1(k) \\ v_2(k) \end{bmatrix}
$$

with the performance criterion

$$
J = \frac{1}{2} E \left\{ \sum_{k=0}^{\infty} \left[z^{\mathrm{T}}(k) z(k) + u^{\mathrm{T}}(k) R u(k) \right] \right\}, \quad R > 0
\tag{9.148}
$$

where $x_i(k) \in R^{n_i}$, $i = 1, 2$, comprise state vectors, $u_i(k) \in R^{m_i}$, $i = 1, 2$, are control inputs, $y_i(k) \in R^{l_i}$, $i = 1, 2$, are observed outputs, $w_i(k) \in R^{r_i}$, $i = 1, 2$, and $v_i(k) \in R^{l_i}$, $i = 1, 2$, are independent zero-mean stationary Gaussian mutually uncorrelated white noise processes with intensities $W_i > 0$ and $V_i > 0$, $i = 1, 2$, respectively, and $z(k) \in R^s$ is the controlled output

$$
z(k) = D_1 x_1(k) + D_2 x_2(k)
\tag{9.149}
$$

All matrices are of appropriate dimensions and assumed to be constant and $O(1)$, which is consistent with Assumptions 9.5 and 9.6. In addition, we assume that the weak coupling Assumption 9.1 is satisfied.

The optimal control law of the system (Equation 9.147) with performance criterion (Equation 9.148) is given by (Kwakernaak and Sivan 1972)

$$
u(k) = -F\hat{x}(k)
\tag{9.150}
$$

with the time-invariant filter

$$
\hat{x}(k+1) = A\hat{x}(k) + Bu(k) + K[y(k) - C\hat{x}(k)]
\tag{9.151}
$$

where

$$A = \begin{bmatrix} A_1 & \varepsilon A_2 \\ \varepsilon A_3 & A_4 \end{bmatrix}, \quad B = \begin{bmatrix} B_1 & \varepsilon B_2 \\ \varepsilon B_3 & B_4 \end{bmatrix}$$

$$C = \begin{bmatrix} C_1 & \varepsilon C_2 \\ \varepsilon C_3 & C_4 \end{bmatrix}, \quad K = \begin{bmatrix} K_1 & \varepsilon K_2 \\ \varepsilon K_3 & K_4 \end{bmatrix} \tag{9.152}$$

The regulator gain F and filter gain K are obtained from

$$F = \left(R + B^T P_r B \right)^{-1} B^T P_r A \tag{9.153}$$

$$K = A P_f C^T \left(V + C P_f C^T \right)^{-1} \tag{9.154}$$

where P_r and P_f are positive semidefinite stabilizing solutions of the discrete-time algebraic regulator and filter Riccati equations, respectively, given by

$$P_r = D^T D + A^T P_r A - A^T P_r B \left(R + B^T P_r B \right)^{-1} B^T P_r A \tag{9.155}$$

$$P_f = A P_f A^T - A P_f C^T \left(V + C P_f C^T \right)^{-1} C P_f A^T + G W G^T \tag{9.156}$$

where

$$D = \begin{bmatrix} D_1 & \varepsilon D_2 \\ \varepsilon D_3 & D_4 \end{bmatrix}, \quad G = \begin{bmatrix} G_1 & \varepsilon G_2 \\ \varepsilon G_3 & G_4 \end{bmatrix} \tag{9.157}$$

The required solutions P_r and P_f have the forms

$$P_r = \begin{bmatrix} P_{r1} & \varepsilon P_{r2} \\ \varepsilon P_{r2}^T & P_{r3} \end{bmatrix}, \quad P_f = \begin{bmatrix} P_{f1} & \varepsilon P_{f2} \\ \varepsilon P_{f2}^T & P_{f3} \end{bmatrix} \tag{9.158}$$

In obtaining the required solutions of Equations 9.155 and 9.156 in terms of reduced-order problems Shen and Gajić (1990b) have used a bilinear transformation technique introduced to transform the discrete-time algebraic Riccati equation into the continuous-time algebraic Riccati equation. In this section, the exact decomposition method of the discrete algebraic regulator and filter Riccati equations produces two sets of two reduced-order nonsymmetric algebraic equations, that is, for the regulator

$$P_{ra} a_{1r} - a_{4r} P_{ra} - a_{3r} + P_{ra} a_{2r} P_{ra} = 0 \tag{9.159}$$

$$P_{rb} b_{1r} - b_{4r} P_{rb} - b_{3r} + P_{rb} b_{2r} P_{rb} = 0 \tag{9.160}$$

and for the filter

$$P_{fa}a_{1f} - a_{4f}P_{fa} - a_{3f} + P_{fa}a_{2f}P_{fa} = 0 \tag{9.161}$$

$$P_{fb}b_{1f} - b_{4f}P_{fb} - b_{3f} + P_{fb}b_{2f}P_{fb} = 0 \tag{9.162}$$

where the unknown coefficients are obtained using the procedures outlined in Sections 9.3.1 and 9.3.2. The Newton algorithm can be used efficiently in solving the reduced-order nonsymmetric Riccati equations (Equations 9.159 through 9.162).

It was shown in the previous section that the optimal global Kalman filter, based on the exact decomposition technique, is decomposed into reduced-order local optimal filters both driven by the system measurements. These local filters can be implemented independently and they are given by

$$\hat{\eta}_1(k+1) = (a_{1f} + a_{2f}P_{fa})^T \hat{\eta}_1(k) + \mathbf{K_1}y(k) + \mathbf{B_1}u(k)$$
$$\hat{\eta}_2(k+1) = (b_{1f} + b_{2f}P_{fb})^T \hat{\eta}_2(k) + \mathbf{K_2}y(k) + \mathbf{B_2}u(k) \tag{9.163}$$

where

$$\begin{bmatrix} \mathbf{B_1} \\ \mathbf{B_2} \end{bmatrix} = \mathbf{T}^{-T}B = (\Pi_{1f} + \Pi_{2f}P_f)^{-T}B \tag{9.164}$$

The optimal control in the new coordinates can be obtained as

$$u(k) = -F\hat{x}(k) = -F\mathbf{T_f^T}\begin{bmatrix} \hat{\eta}_1(k) \\ \hat{\eta}_2(k) \end{bmatrix} = -[\mathbf{F_1}\ \mathbf{F_2}]\begin{bmatrix} \hat{\eta}_1(k) \\ \hat{\eta}_2(k) \end{bmatrix} \tag{9.165}$$

where $\mathbf{F_1}$ and $\mathbf{F_2}$ are obtained from

$$[\mathbf{F_1}\ \mathbf{F_2}] = F\mathbf{T_f^T} = (R + B^T P_r B)^{-1} B^T P_r A(\Pi_{1f} + \Pi_{2f}P_f)^T \tag{9.166}$$

The optimal value of J is given by the very well-known form (Kwakernaak and Sivan 1972)

$$J_{opt} = \frac{1}{2}\text{tr}\left[D^T DP_f + P_r K(CP_f C^T + V)K^T\right] \tag{9.167}$$

where F, K, P_r, and P_r are obtained from Equations 9.153 through 9.156.

9.3.4 CASE STUDY: DISTILLATION COLUMN

In order to demonstrate the efficiency of the proposed method, we consider a real world control system—a fifth-order discrete model of a distillation column considered in Kautsky et al. (1985). This model is discretized with the sampling rate of $\Delta T = 0.1$ in Shen and Gajić (1990b) (see also, Gajić and Shen 1993, p. 153). The system matrices are given by

$$A = 10^{-3} \begin{bmatrix} 989.50 & 5.6382 & 0.2589 & 0.0125 & 0.0006 \\ 117.25 & 814.50 & 76.038 & 5.5526 & 0.3700 \\ 8.7680 & 123.87 & 750.20 & 107.96 & 11.245 \\ 0.9108 & 17.991 & 183.81 & 668.34 & 150.78 \\ 0.0179 & 0.3172 & 1.6974 & 13.298 & 985.19 \end{bmatrix}$$

$$B^{\mathrm{T}} = 10^{-3} \begin{bmatrix} 0.0192 & 6.0733 & 8.2911 & 9.1965 & 0.7025 \\ -0.0013 & -0.6192 & -13.339 & -18.442 & -1.4252 \end{bmatrix}$$

and the other matrices are chosen as

$$C = \begin{bmatrix} 1 & 1 & 0 & 0 & 0 \\ 0 & 0 & 1 & 1 & 1 \end{bmatrix}, \quad D^{\mathrm{T}}D = I_5, \quad R = I_2$$

It is assumed that $G = I_5$ and that the white noise processes are independent and have intensities

$$W = I_5, \quad V = 0.1 \times I_2$$

It is easy to see that this model possesses the weakly coupled structure with $n_1 = 2$, $n_2 = 3$, and $\varepsilon = 8.2911/13.339 = 0.62$.

The obtained solutions for the LQG control problem are summarized as the following. The completely decoupled filters driven by measurements $y(k)$ are given by

$$\hat{\eta}_1(k + 1) = \begin{bmatrix} 0.3495 & -0.5914 \\ -0.2126 & 0.4551 \end{bmatrix} \hat{\eta}_1(k) + \begin{bmatrix} 0.6426 & 0.0670 \\ 0.3277 & 0.0403 \end{bmatrix} y(k)$$
$$+ \begin{bmatrix} 0.0065 & -0.012 \\ 0.0027 & 0.0054 \end{bmatrix} u(k)$$

$$\hat{\eta}_2(k + 1) = \begin{bmatrix} 0.6607 & 0.0298 & -0.1464 \\ -0.0833 & 0.4020 & -0.1255 \\ -0.6043 & -0.6039 & 0.3915 \end{bmatrix} \hat{\eta}_2(k)$$
$$+ \begin{bmatrix} 0.0163 & 0.1037 \\ -0.0478 & 0.2649 \\ -0.1302 & 0.5799 \end{bmatrix} y(k) + \begin{bmatrix} 0.064 & -0.0080 \\ 0.0085 & -0.0165 \\ 0.0013 & -0.0103 \end{bmatrix} u(k)$$

The feedback control in the new coordinates is obtained as

$$u(k) = \begin{bmatrix} 0.3732 & 0.4318 \\ -0.4752 & -0.5981 \end{bmatrix} \hat{\eta}_1(k)$$
$$+ \begin{bmatrix} -0.0739 & -0.1755 & 0.4207 \\ 0.0961 & 0.2230 & -0.6128 \end{bmatrix} \hat{\eta}_2(k)$$

The difference of the performance criterion between the optimal value, $J_{\mathrm{opt}} = 216.572$, and the one of the proposed method, J, is given by $J - J_{\mathrm{opt}} = 2.6489 \times 10^{-11}$.

9.4 OPTIMAL CONTROL OF WEAKLY COUPLED SYSTEMS WITH N SUBSYSTEMS

In this section, we extend ideas from Sections 9.1 and 9.2 to weakly coupled linear continuous-time stochastic systems with N subsystems and solve the corresponding linear-quadratic optimal Gaussian control problem.

9.4.1 DECOUPLING OF THE ALGEBRAIC RICCATI EQUATION

Consider the linear-quadratic control problem defined by

$$\dot{x}(t) = Ax(t) + Bu(t)$$

$$J = \frac{1}{2} \int_0^\infty \left[x^{\mathrm{T}}(t)Qx(t) + u^{\mathrm{T}}(t)Ru(t) \right] dt, \quad Q \geq 0, \quad R > 0 \qquad (9.168)$$

where $x(t) \in \mathcal{R}^n$ is the state space vector partitioned consistently with N subsystems as $x^{\mathrm{T}}(t) = \left[x_1^{\mathrm{T}}(t), x_2^{\mathrm{T}}(t), \ldots, x_N^{\mathrm{T}}(t) \right]$, with dim $\{x_i(t)\} = n_i u(t)$ is the control input vector with components $u_i(t) \in \mathcal{R}^{m_i}$, $i = 1, 2, \ldots, N$. The quadratic performance criterion J has to be minimized by a suitable control action. The matrices A, B, Q, and R have the weakly coupled structure, that is

$$A = \begin{bmatrix} A_{11} & \varepsilon A_{12} & \cdots & \varepsilon A_{1N} \\ \varepsilon A_{21} & A_{22} & \cdots & \varepsilon A_{2N} \\ \cdots & \cdots & \cdots & \cdots \\ \varepsilon A_{N1} & \varepsilon A_{N2} & \cdots & A_{NN} \end{bmatrix}$$

$$B = \begin{bmatrix} B_{11} & \varepsilon B_{12} & \cdots & \varepsilon B_{1N} \\ \varepsilon B_{21} & B_{22} & \cdots & \varepsilon B_{2N} \\ \cdots & \cdots & \cdots & \cdots \\ \varepsilon B_{N1} & \varepsilon B_{N2} & \cdots & B_{NN} \end{bmatrix}$$

$$Q = \begin{bmatrix} Q_{11} & \varepsilon Q_{12} & \cdots & \varepsilon Q_{1N} \\ \varepsilon Q_{21} & Q_{22} & \cdots & \varepsilon Q_{2N} \\ \cdots & \cdots & \cdots & \cdots \\ \varepsilon Q_{N1} & \varepsilon Q_{N2} & \cdots & Q_{NN} \end{bmatrix}, \quad R = \begin{bmatrix} R_{11} & 0 & \cdots & 0 \\ 0 & R_{22} & \cdots & 0 \\ \cdots & \cdots & \cdots & \cdots \\ 0 & 0 & \cdots & R_{NN} \end{bmatrix} \qquad (9.169)$$

where ε is a small weak coupling parameter. Each diagonal block A_{ii}, B_{ii}, Q_{ii}, $i = 1, 2, \ldots, N$, is of dimension $n_i \times n_i$ and $\sum_{i=1}^N n_{ii} = n$. All nonzero elements in the above matrices are assumed to be $O(1)$ and the matrix A satisfies the weak coupling assumption of Chow and Kokotović (1983). Hence, the results presented in this chapter will be valid under the following assumption.

Assumption 9.7   The matrices A_{ii}, $i = 1, 2, \ldots, N$, are nonsingular with $\lambda_i(A_{ii}) = O(1)$ and $\det\{A_{ii}\} = O(1)$. In addition, all matrices A_{ij}, B_{ij}, Q_{ij} and R_{ii} and $O(1)$.

The optimal *closed-loop* control law of Equations 9.168 and 9.169 has the very well-known form

$$u(x(t)) = -R^{-1}B^T Px(t) = -Fx(t) \tag{9.170}$$

where P is the positive semidefinite stabilizing solution of the regulator algebraic Riccati equation

$$A^T P + PA + Q - PSP = 0, \quad S = B^T R^{-1} B \tag{9.171}$$

Note that the coefficient matrix S also has the weakly coupled structure

$$S = \begin{bmatrix} S_{11} & \varepsilon S_{12} & \cdots & \varepsilon S_{1N} \\ \varepsilon S_{21} & S_{22} & \cdots & \varepsilon S_{2N} \\ \cdots & \cdots & \cdots & \cdots \\ \varepsilon S_{N1} & \varepsilon S_{N2} & \cdots & S_{NN} \end{bmatrix} \tag{9.172}$$

It can be shown that the solution of Equation 9.171 also has weakly coupled structure compatibly partitioned as

$$P = \begin{bmatrix} P_{11} & \varepsilon P_{12} & \cdots & \varepsilon P_{1N} \\ \varepsilon P_{12}^T & P_{22} & \cdots & \varepsilon P_{2N} \\ \cdots & \cdots & \cdots & \cdots \\ \varepsilon P_{1N}^T & \varepsilon P_{2N}^T & \cdots & P_{NN} \end{bmatrix} \tag{9.173}$$

The optimal solution of the above defined linear-quadratic optimal control problem requires stabilizability–detectability of the triple $(A, B, \mathrm{Chol}(Q))$. However, for weakly coupled linear systems this condition can be replaced by stabilizability–detectability conditions imposed on the subsystems as stated in the following assumption.

Assumption 9.8   The triples $(A_{ii}, B_{ii}, \mathrm{Chol}(Q_{ii}))$, $i = 1, 2, \ldots, N$ are stabilizable–detectable.

The optimal *open-loop* control law of Equations 9.168 and 9.169 is given by

$$u(t) = -R^{-1}B^T p(t) \tag{9.174}$$

where $p(t) \in \mathcal{R}^n$ is a costate variable satisfying the Hamiltonian system

$$\begin{bmatrix} \dot{x}(t) \\ \dot{p}(t) \end{bmatrix} = \begin{bmatrix} A & -S \\ -Q & -A^T \end{bmatrix} \begin{bmatrix} x(t) \\ p(t) \end{bmatrix} \tag{9.175}$$

The methodology used in this section starts with partitioning the costate vector consistently to the state vector and rearranging rows in Equation 9.175 to form

$$
\begin{bmatrix} \dot{x}_1(t) \\ \dot{p}_1(t) \\ \vdots \\ \dot{x}_N(t) \\ \dot{p}_N(t) \end{bmatrix} = \begin{bmatrix} T_{11} & \varepsilon T_{12} & \cdots & \varepsilon T_{1N} \\ \varepsilon T_{21} & T_{22} & \cdots & \varepsilon T_{2N} \\ \cdots & \cdots & \cdots & \cdots \\ \varepsilon T_{N1} & \varepsilon T_{N2} & \cdots & T_{NN} \end{bmatrix} \begin{bmatrix} x_1(t) \\ p_1(t) \\ \vdots \\ x_N(t) \\ p_N(t) \end{bmatrix} \qquad (9.176)
$$

with

$$
T_{ij} = \begin{bmatrix} A_{ij} & -S_{ij} \\ -Q_{ij} & -A_{ij}^T \end{bmatrix}, \quad i, j = 1, 2, \ldots, N \qquad (9.177)
$$

The systems (Equation 9.176) will be a pure weakly coupled system without displaying multiple timescale phenomena (Chow and Kokotović 1983) under the following assumption.

Assumption 9.9 The matrices T_{ii}, $i = 1, 2, \ldots, N$, are nonsingular with $\lambda_i(T_{ii}) = O(1)$, that is, $\det\{T_{ii}\} = O(1)$.

It can be seen from Equation 9.177 that even for systems whose matrices A_{ii} are singular, Assumption 9.9 can be satisfied. Thus, Assumption 9.9 plays a central role in this section and it overrules Assumption 9.7.

The transformation of Gajić and Borno (2000) that relates the original weakly coupled linear system and a set of completely decoupled subsystems in the new coordinates, is given by

$$
\Gamma = \begin{bmatrix} I_{n_1} & \varepsilon L_{12} & \cdots & \varepsilon L_{1N} \\ \varepsilon L_{21} & I_{n_2} & \cdots & \varepsilon L_{2N} \\ \cdots & \cdots & \cdots & \cdots \\ \varepsilon L_{N1} & \varepsilon L_{N2} & \cdots & I_{n_N} \end{bmatrix} = I_n + \varepsilon \psi \qquad (9.178)
$$

with the obvious definition of ψ. Note that Γ is invertible matrix for small values of ε. This transformation offers the advantage that it exactly decomposes a higher-order linear weakly coupled system into N completely decoupled reduced order subsystems. In Equation 9.178, L_{ij}'s satisfy the following equations:

$$
L_{ij}T_{jj} - T_{ii}L_{ij} + T_{ij} + \varepsilon \left(\sum_{k=1, k \neq i, j}^{N} L_{ik}T_{kj} \right)
$$

$$
- \varepsilon^2 \left(\sum_{k=1, k \neq i, j}^{N} L_{ik}T_{ki} \right) L_{ij} = 0, \quad i, j = 1, 2, \ldots, N \qquad (9.179)
$$

The nonlinear algebraic equation (Equation 9.179) can be solved iteratively as reduced-order-independent linear Sylvester equations by using either the Newton method or fixed point iterations (see Section 5.2). The unique solution of Equation 9.179 exists under the following assumption, which in fact represents the condition for the existence of the unique solution of the Sylvester algebraic equation (Gajić and Qureshi 2008).

Assumption 9.10 The nonsingular matrices T_{ii} and T_{jj}, $i, j = 1, 2, \ldots, N$, $I \neq j$, have no eigenvalues in common.

The transformation (Equation 9.178) applied to Equation 9.176 produces N completely decoupled subsystems

$$\dot{n}_i(t) = D_i \eta_i(t), \quad i = 1, 2, \ldots, N \tag{9.180}$$

with

$$D_i = T_{ii} + \varepsilon^2 \sum_{j=1, j \neq i}^{N} L_{ij} T_{ji}, \quad i, j = 1, 2, \ldots, N \tag{9.181}$$

Let $\eta^T(t) = \begin{bmatrix} \eta_1^T(t) & \eta_2^T(t) & \cdots & \eta_N^T(t) \end{bmatrix}$, then

$$\dot{\eta}(t) = D\eta(t) \tag{9.182}$$

where $D = \text{diag}[D_1 \ D_2 \cdots D_N]$.

Introducing notation $s^T(t) = \begin{bmatrix} s_1^T(t) & s_1^T(t) & \cdots & s_N^T(t) \end{bmatrix}$ with

$$s_i(t) = \begin{bmatrix} x_i(t) \\ p_i(t) \end{bmatrix} \tag{9.183}$$

we have

$$\eta(t) = \Gamma s(t) \tag{9.184}$$

The rearrangement of variables in Equation 9.176 is done using a permutation matrix E of the form:

$$
\begin{bmatrix} x_1(t) \\ p_1(t) \\ x_2(t) \\ p_2(t) \\ \vdots \\ x_N(t) \\ p_N(t) \end{bmatrix}
=
\begin{bmatrix}
I_{n_1} & 0 & \cdots & 0 & 0 & 0 & \cdots & 0 \\
0 & 0 & \cdots & 0 & I_{n_1} & 0 & \cdots & 0 \\
0 & I_{n_2} & \cdots & 0 & 0 & 0 & \cdots & 0 \\
0 & 0 & \cdots & 0 & 0 & I_{n_2} & \cdots & 0 \\
\vdots & \vdots & \ddots & \vdots & \vdots & \vdots & \ddots & :0 \\
0 & 0 & \cdots & I_{n_N} & 0 & 0 & \cdots & 0 \\
0 & 0 & \cdots & 0 & 0 & 0 & \cdots & I_{n_N}
\end{bmatrix}
\begin{bmatrix} x_1(t) \\ x_2(t) \\ \vdots \\ x_N(t) \\ p_1(t) \\ p_2(t) \\ \vdots \\ p_N(t) \end{bmatrix}
$$

$$= E \begin{bmatrix} x(t) \\ p(t) \end{bmatrix} \tag{9.185}$$

Combining Equations 9.183 through 9.185, we obtain the relationship between the original and new coordinates

$$\begin{bmatrix} \eta_x(t) \\ \eta_p(t) \end{bmatrix} = E^{\mathrm{T}}\Gamma E \begin{bmatrix} x(t) \\ p(t) \end{bmatrix} = Z \begin{bmatrix} x(t) \\ p(t) \end{bmatrix} = \begin{bmatrix} Z_1 & Z_2 \\ Z_3 & Z_4 \end{bmatrix} \begin{bmatrix} x(t) \\ p(t) \end{bmatrix} \tag{9.186}$$

Since $p(t) = Px(t)$, where P satisfies the algebraic Riccati equation (Equation 9.171), it follows that

$$\eta_x(t) = (Z_1 + Z_2 P)x(t), \quad \eta_p(t) = (Z_3 + Z_4 P)x(t) \tag{9.187}$$

In the original coordinates, the required solution has the closed-loop nature. We have the same attribute for the new coordinates, that is

$$\eta_p(t) = \begin{bmatrix} P_1 & 0 & \cdots & \cdots & 0 \\ 0 & P_2 & 0 & \cdots & 0 \\ \vdots & \ddots & \ddots & \ddots & \vdots \\ 0 & \cdots & 0 & \ddots & 0 \\ 0 & \cdots & \cdots & 0 & P_N \end{bmatrix} \eta_x(t) = P_0 \eta_x(t) \tag{9.188}$$

with the obvious definition for P_0. Formulas (Equations 9.187 and 9.188) imply

$$P_0 = \begin{bmatrix} P_1 & 0 & \cdots & \cdots & 0 \\ 0 & P_2 & 0 & \cdots & 0 \\ \vdots & \ddots & \ddots & \ddots & \vdots \\ 0 & \cdots & 0 & \ddots & 0 \\ 0 & \cdots & \cdots & 0 & P_N \end{bmatrix} = (Z_3 + Z_4 P)(Z_1 + Z_2 P)^{-1} \tag{9.189}$$

Following the same logic, we can find P reversely as follows. From Equation 9.186 we have

$$\begin{bmatrix} x(t) \\ p(t) \end{bmatrix} = (E^{\mathrm{T}}\Gamma E)^{-1} \begin{bmatrix} \eta_x(t) \\ \eta_p(t) \end{bmatrix} = O \begin{bmatrix} \eta_x(t) \\ \eta_p(t) \end{bmatrix}$$

$$= \begin{bmatrix} O_1 & O_2 \\ O_3 & O_4 \end{bmatrix} \begin{bmatrix} \eta_x(t) \\ \eta_p(t) \end{bmatrix} \tag{9.190}$$

Using Equation 9.188 we obtain

$$x(t) = O_1 \eta_x(t) + O_2 \eta_p(t) = (O_1 + O_2 P_0)\eta_x(t)$$
$$p(t) = O_3 \eta_x(t) + O_4 \eta_p(t) = (O_3 + O_4 P_0)\eta_x(t) \tag{9.191}$$

Since $p(t) = Px(t)$ we have

$$P(O_1 + O_2 P_0)\eta_x(t) = (O_3 + O_4 P_0)\eta_x(t) \tag{9.192}$$

The last equality is valid for every $\eta_x(t)$, which implies

$$P = (O_3 + O_4 P_0)(O_1 + O_2 P_0)^{-1} \tag{9.193}$$

Note that

$$O = \begin{bmatrix} O_1 & O_2 \\ O_3 & O_4 \end{bmatrix} = E^{\mathrm{T}} \Gamma^{-1} E = Z^{-1} \tag{9.194}$$

Invertibility of the matrices defined in Equations 9.189 and 9.193 can be established for sufficiently small values of ε by closely examining these matrices. Namely, it can be shown that matrices in question are $O(\varepsilon)$ perturbations of an identity matrix, and hence invertible for sufficiently small values of ε.

Partitioning Equation 9.180 as $\eta_i^{\mathrm{T}}(t) = [\eta_{i1}(t) \ \eta_{i2}(t)]$, $i = 1, 2, \ldots, N$, we have

$$\begin{bmatrix} \dot{\eta}_{i1}(t) \\ \dot{\eta}_{i2}(t) \end{bmatrix} = D_i \begin{bmatrix} \eta_{i1}(t) \\ \eta_{i2}(t) \end{bmatrix} = \begin{bmatrix} D_{i1} & D_{i2} \\ D_{i3} & D_{i4} \end{bmatrix} \begin{bmatrix} \eta_{i1}(t) \\ \eta_{i2}(t) \end{bmatrix}, \quad i = 1, 2, \ldots, N \tag{9.195}$$

It can be shown that

$$\begin{aligned} D_{i1} &= A_{ii} + O(\varepsilon^2), \quad D_{i2} = -S_{ii} + O(\varepsilon^2) \\ D_{i3} &= -Q_{ii} + O(\varepsilon^2), \quad D_{i4} = -A_{ii}^{\mathrm{T}} + O(\varepsilon^2) \end{aligned} \tag{9.196}$$

Using Equation 9.188 in Equation 9.195 yields N reduced-order nonsymmetric subsystem algebraic Riccati equations

$$P_i D_{i1} - D_{i4} P_i - D_{i3} + P_i D_{i2} P_i = 0, \quad i = 1, 2, \ldots, N \tag{9.197}$$

Using Equation 9.196, it follows that these equations can be represented as

$$P_i A_{ii} + A_{ii}^{\mathrm{T}} P_i + Q_{ii} - P_i S_{ii} P_i + O(\varepsilon^2) = 0, \quad i = 1, 2, \ldots, N \tag{9.198}$$

The existence of the unique solutions of these equations for sufficiently small values of ε is guaranteed under Assumption 9.8. The $O(\varepsilon^2)$ perturbed equation (Equation 9.198), that is

$$P_i^{(0)} A_{ii} + A_{ii}^{\mathrm{T}} P_i^{(0)} + Q_{ii} - P_i^{(0)} S_{ii} P_i^{(0)} = 0, \quad i = 1, 2, \ldots, N \tag{9.199}$$

provides excellent initial conditions for the corresponding Newton algorithm for solving Equation 9.197 since P_i and $P_i^{(0)}$ are $O(\varepsilon^2)$ close.

The decoupled closed-loop subsystem in the new coordinates is

$$\dot{\eta}_{i1}(t) = (D_{i1} + D_{i2} P_i)\eta_{i1}(t), \quad i = 1, 2, \ldots, N \tag{9.200}$$

The problem of finding the corresponding optimal gains at the local levels compatibly to reduced-order algebraic Riccati equations will be discussed later in Section 9.4.3, where we will study the linear-quadratic optimal Gaussian stochastic control problem of weakly coupled systems.

9.4.2 KALMAN FILTERING FOR N WEAKLY COUPLED SUBSYSTEMS

In this section, we present a scheme for the Kalman filter of weakly coupled linear stochastic systems composed of N subsystems in terms of completely decoupled N local Kalman filters. The scheme is based on the main result from the previous section that can be summarized in the following lemma.

LEMMA 9.6

Consider the optimal closed loop linear systems

$$\dot{x}(t) = (A - BF)x(t) \tag{9.201}$$

Under Assumptions 9.7 through 9.10, there exists a nonsingular transformation M

$$\begin{bmatrix} z_1(t) \\ z_2(t) \\ \vdots \\ z_N(t) \end{bmatrix} = M \begin{bmatrix} x_1(t) \\ x_2(t) \\ \vdots \\ x_N(t) \end{bmatrix} \tag{9.202}$$

such that

$$\dot{z}_i(t) = (D_{i1} + D_{i2}P_i)z_i(t), \quad i - 1, 2, \ldots, N \tag{9.203}$$

where P_i's are the unique solutions of the N completely decoupled subsystem algebraic Riccati equations

$$P_i D_{i1} - D_{i4}P_i - D_{i3} + P_i D_{i2}P_i = 0, \quad i = 1, 2, \ldots, N \tag{9.204}$$

where D_{ij}, $i = 1, 2, \ldots, N, j = 1, 2, 3, 4$, are obtained from formulas (Equations 9.181 and 9.195). The nonsingular transformation M is given by

$$M = (Z_1 + Z_2 P) \tag{9.205}$$

where the global solution P is obtained from the solutions of the reduced-order subsystem algebraic Riccati equations using formula (Equation 9.193). The matrices Z_1 and Z_2 are obtained from Equation 9.194 in terms of solutions of the Chang transformation decoupling algebraic equation (Equation 9.179).

We consider now the corresponding filtering problem of continuous-time linear stochastic systems. Dynamics of such systems are represented by the weakly coupled differential equation and measurements

$$\dot{x}(t) = Ax(t) + Gw(t)$$
$$y(t) = Cx(t) + v(t)$$

(9.206)

where $x(t)$ is n-dimensional state vector partitioned consistently with N subsystems as $x^T(t) = \begin{bmatrix} x_1^T(t) \ x_2^T(t) \ \cdots \ x_N^T(t) \end{bmatrix}$, with dim $\{x_i(t)\} = n_i$, and $y(t)$ vector of dimension q are the system measurements partitioned as $y^T(t) = \begin{bmatrix} y_1^T(t) \ y_2^T(t) \ \cdots \ y_N^T(t) \end{bmatrix}$ with dim $\{y_i(t)\} = q_i$. The vectors $w(t)$ and $v(t)$, respectively of dimensions r and q, represent zero-mean Gaussian white noise stationary uncorrelated stochastic processes partitioned consistently as $w^T(t) = \begin{bmatrix} w_1^T(t) \ w_2^T(t) \ \cdots \ w_N^T(t) \end{bmatrix}$ and $v^T(t) = \begin{bmatrix} v_1^T(t) \ v_2^T(t) \ \cdots \ v_N^T(t) \end{bmatrix}$ with dim$\{w_i(t)\} = r_i$ and dim$\{v_i(t)\} = q_i$. The intensities matrices (spectral densities) are $W = \text{diag } \{W_1, W_2, \ldots, W_N\} \geq 0$ and $V = \text{diag } \{V_1, V_2, \ldots, V_N\} > 0$, implying $W_i \geq 0$ and $V_i > 0$. The matrices G and C have the weakly coupled structure, that is

$$G = \begin{bmatrix} G_{11} & \varepsilon G_{12} & \cdots & \varepsilon G_{1N} \\ \varepsilon G_{21} & G_{22} & \cdots & \varepsilon G_{2N} \\ \cdots & \cdots & \cdots & \cdots \\ \varepsilon G_{N1} & \varepsilon G_{N2} & \cdots & G_{NN} \end{bmatrix}$$

$$C = \begin{bmatrix} C_{11} & \varepsilon C_{12} & \cdots & \varepsilon C_{1N} \\ \varepsilon C_{21} & C_{22} & \cdots & \varepsilon C_{2N} \\ \cdots & \cdots & \cdots & \cdots \\ \varepsilon C_{N1} & \varepsilon C_{N2} & \cdots & C_{NN} \end{bmatrix}$$

(9.207)

which is stated in the following assumption

Assumption 9.11   All matrices $G_{ij}, C_{ij}, i, j = 1, 2, \ldots, N$, and $W_i, V_i, i = 1, 2, \ldots, N$, are $O(1)$.

The optimal closed-loop Kalman filter driven by the system measurements is given by

$$\dot{\hat{x}}(t) = (A - KC)\hat{x}(t) + Ky(t)$$

(9.208)

where K is the optimal Kalman filter gain obtained from

$$K = P_f C^T V^{-1}$$

(9.209)

with P_f representing the positive semidefinite stabilizing solution of the algebraic filter Riccati equation

$$AP_f + P_f A^T + GWG^T - P_f C^T V^{-1} C \, P_f = 0$$

(9.210)

Such a solution exists under the following assumption.

Assumption 9.12 The triples (A_{ii}, G_{ii}, C_{ii}), $I = 1, 2, \ldots, N$, are stabilizable–detectable.

The optimal Kalman filter again is consistently partitioned according to the system weakly coupled structure as

$$
K = \begin{bmatrix}
K_{11} & \varepsilon K_{12} & \cdots & \varepsilon K_{1N} \\
\varepsilon K_{21} & K_{22} & \cdots & \varepsilon K_{2N} \\
\cdots & \cdots & \cdots & \cdots \\
\varepsilon K_{N1} & \varepsilon K_{N2} & \cdots & K_{NN}
\end{bmatrix} \tag{9.211}
$$

The desired closed-loop decomposition of the Kalman filter (Equation 9.208) will be obtained using Lemma 9.6. By the duality property between the optimal filter and regulator, the filter Riccati equation (Equation 9.210) can be solved via the same decomposition technique presented in Section 9.4.1 for solving the regulator Riccati equation (Equation 9.171) with the following change of variables:

$$
\begin{aligned}
A &\rightarrow A^{\mathrm{T}}, \quad Q \rightarrow GWG^{\mathrm{T}}, \quad F \rightarrow K^{\mathrm{T}} \\
S &= BR^{-1}B^{\mathrm{T}} \rightarrow C^{\mathrm{T}}V^{-1}C
\end{aligned} \tag{9.212}
$$

Using the results from the previous section and established dualities (Equation 9.212), we first form the matrices

$$
T^{\mathrm{f}}_{ij} = \begin{bmatrix}
A^{\mathrm{T}}_{ji} & -C^{\mathrm{T}}V^{-1}C \\
-GWG^{\mathrm{T}} & -A_{ij}
\end{bmatrix}, \quad i, j = 1, 2, \ldots, N \tag{9.213}
$$

These matrices satisfy the following assumptions (consistent with Assumption 9.9 and 9.10).

Assumption 9.13 The matrices T^{f}_{ii}, $i = 1, 2, \ldots, N$, are nonsingular with $\lambda_i(T^{\mathrm{f}}_{ii}) = O(1)$, that is, $\det\{T^{\mathrm{f}}_{ii}\} = O(1)$.

Assumption 9.14 The matrices T^{f}_{ii} and T^{f}_{jj}, $i, j = 1, 2, \ldots, N$, $i \neq j$, have no eigenvalues in common.

Let us define

$$
\begin{aligned}
Z_{\mathrm{f}} = E^{\mathrm{T}}\Gamma_{\mathrm{f}}E &= \begin{bmatrix}
Z^{\mathrm{f}}_1 & Z^{\mathrm{f}}_2 \\
Z^{\mathrm{f}}_3 & Z^{\mathrm{f}}_4
\end{bmatrix} \Rightarrow Z^{-1}_{\mathrm{f}} = (E^{\mathrm{T}}\Gamma_{\mathrm{f}}E)^{-1} \\
&= O_{\mathrm{f}} = \begin{bmatrix}
O^{\mathrm{f}}_1 & O^{\mathrm{f}}_2 \\
O^{\mathrm{f}}_3 & O^{\mathrm{f}}_4
\end{bmatrix}
\end{aligned} \tag{9.214}
$$

Then using Lemma 9.6, the desired transformation is given by

$$
M_{\mathrm{f}} = (Z^{\mathrm{f}}_1 + Z^{\mathrm{f}}_2 P_{\mathrm{f}}) \tag{9.215}
$$

where P_f is the solution of global filter algebraic Riccati equation (Equation 9.210) which can be obtained via formula dual to Equation 9.193 in terms of reduced-order subsystem filter algebraic Riccati equations dual to Equation 9.197. The transformation M_f applied to the filter variables defined by Equation 9.208 as

$$\hat{z}(t) = M_f^{-T}\hat{x}(t) \tag{9.216}$$

produces

$$\dot{\hat{z}}(t) = M_f^{-T}(A - KC)M_f^T \hat{z}(t) + M_f^{-T}Ky(t) \tag{9.217}$$

such that the complete closed-loop decomposition is achieved, that is

$$\dot{\hat{z}}_i(t) = \left(D_{i1}^f + D_{i2}^f P_i^f\right)\hat{z}_i(t) + K_i y(t) \tag{9.218}$$

with

$$\begin{bmatrix} D_{i1}^f & D_{i2}^f \\ D_{i3}^f & D_{i4}^f \end{bmatrix} = \left(T_{ii}^f + \varepsilon^2 \sum_{j=1,\, j\neq i}^{N} L_{ij}^f T_{ji}^f \right) \tag{9.219}$$

$$\begin{bmatrix} K_1 \\ K_2 \\ \vdots \\ K_N \end{bmatrix} = M_f^{-T}K, \quad i, j = 1, 2, \dots, N, \quad K_i \in R^{n_i \times q} \tag{9.220}$$

$$P_i^f D_{i1}^f - P_i^f D_{i4}^f - D_{i3}^f + P_i^f D_{i2}^f P_i^f = 0, \quad i = 1, 2, \dots, N \tag{9.221}$$

Matrices L_{ij}^f represent solutions of Equation 9.179 with the coefficient matrices equal to those defined in Equation 9.213.

9.4.3 LINEAR-QUADRATIC GAUSSIAN OPTIMAL CONTROL

Consider the linear-quadratic Gaussian optimal control problem defined by the system state space equation

$$\dot{x}(t) = Ax(t) + Bu(t) + Gw(t) \tag{9.222}$$

with a performance criterion to be minimized by appropriately choosing $u(t)$, which in fact represents the stochastic version of the quadratic performance criterion defined in Equation 9.168

$$J = \lim_{t_f \to \infty} \left\{ \frac{1}{2t_f} \int_0^{t_f} \left[x^T(t)Qx(t) + u^T(t)Ru(t) \right] dt \right\}, \quad Q \geq 0, \quad R > 0 \tag{9.223}$$

The optimal solution to Equation 9.222 and 9.223 is given by the well-known separation principle as

$$u_{\text{opt}} = -F\hat{x}(t) \tag{9.224}$$

where F is defined in Equation 9.170 and $\hat{x}(t)$ is the optimal estimate of the system state variables as determined by the Kalman filter (Equation 9.208). By using Equation 9.216 that relates the optimal estimates in the new and the original coordinates we have

$$
\begin{aligned}
u_{\text{opt}} &= -F\hat{x}(t) = -FM_{\text{f}}\hat{z}(t) = -F_{\text{f}}\hat{z}(t) \\
&= -F_1^{\text{f}}\hat{z}_1(t) - F_2^{\text{f}}\hat{z}_2(t) - \cdots - F_N^{\text{f}}\hat{z}_N(t) \\
&= u_{\text{opt}}^1 + u_{\text{opt}}^2 + \cdots + u_{\text{opt}}^N
\end{aligned}
\tag{9.225}
$$

The optimal filter gain in the new coordinates can be expressed in terms of quantities obtained in Sections 9.4.1 and 9.4.2 as shown in Lim and Gajić (1999)

$$F_f = R^{-1}B^{\text{T}}P\left(Z_1^f + Z_2^f P_f\right) \tag{9.226}$$

The presented scheme has a nice fully decoupled form for the optimal local filters and a simple additive form for the optimal controller in terms of local optimal controllers.

The optimal performance value is given by

$$J_{\text{opt}} = \text{tr}\left\{PW + P_{\text{f}}F^{\text{T}}RF\right\} = \text{tr}\left\{PKVK^{\text{T}} + P_fQ\right\} \tag{9.227}$$

The presented methodology is applied in Lim and Gajić (1999) to a real world example, the 17th-order three-stand cold-rolling mill, a system that is composed of three weakly coupled subsystems.

9.5 CONCLUSION

In this chapter, optimal steady-state, closed-loop control problems of continuous- and discrete-time weakly coupled systems are solved by way of the subsystem level reduced-order nonsymmetric algebraic Riccati equations. The methodology is extended to the Kalman filtering problem of weakly coupled subsystems and the corresponding linear-quadratic Gaussian optimal control problem. The generalization of the main techniques to weakly coupled linear systems composed of N subsystems is also presented. Since the decomposed Riccati equations are completely independent, and the local filters work in parallel, the processing time for the optimal control and filtering problems is reduced. The results presented in this chapter are mostly based on the work by Su and Gajić (1992), Aganović and Gajić (1995), Lim and Gajić (1990), and Gajić and Borno (2000).

APPENDIX 9.1

According to Equations 9.20 and 9.25, it can be seen that

$$\Pi = \begin{bmatrix} I_{n_1} + O(\varepsilon^2) & O(\varepsilon) & O(\varepsilon^2) & O(\varepsilon) \\ O(\varepsilon) & I_{n_2} & O(\varepsilon) & 0 \\ O(\varepsilon^2) & O(\varepsilon) & I_{n_1} + O(\varepsilon^2) & O(\varepsilon) \\ O(\varepsilon) & 0 & O(\varepsilon) & I_{n_2} \end{bmatrix}$$

and

$$\Omega = \begin{bmatrix} I_{n_1} & O(\varepsilon) & 0 & O(\varepsilon) \\ O(\varepsilon) & I_{n_2} + O(\varepsilon^2) & O(\varepsilon) & O(\varepsilon^2) \\ 0 & O(\varepsilon) & I_{n_1} & O(\varepsilon) \\ O(\varepsilon) & O(\varepsilon^2) & O(\varepsilon) & I_{n_2} + O(\varepsilon^2) \end{bmatrix}$$

Therefore

$$(\Pi_1 + \Pi_2 P) = I_{n_1 + n_2} + O(\varepsilon)$$

$$\left(\Omega_1 + \Omega_2 \begin{bmatrix} P_1 & 0 \\ 0 & P_2 \end{bmatrix} \right) = I_{n_1 + n_2} + O(\varepsilon)$$

There exists $\varepsilon_1 > 0$ such that; $\forall \varepsilon \leq \varepsilon_1$ the required matrices are invertible.

REFERENCES

Ackerson, G. and K. Fu, On the state estimation in switching environments, *IEEE Transactions on Automatic Control*, AC-15, 10–17, 1970.

Aganović, Z. and Z. Gajić, New filtering method for linear weakly coupled stochastic systems, *Journal of Guidance, Control, and Dynamics*, 18, 630–633, 1995.

Avramović, B., Subspace iteration approach to the time scale separation, *Proceedings of the Conference on Decision and Control*, Fort Lauderdale, FL, 684–687, 1979.

Bunse-Gernster, A., R. Byrns, and V. Mehrmann, A chart for numerical methods for structured eigenvalue problem, *SIAM Journal of Matrix Analysis and Applications*, 13, 419–453, 1992.

Chow, J. and P. Kokotović, Sparsity and time scales, *Proceedings of the American Control Conference*, San Francisco, CA, 656–661, 1983.

Dorato, P. and A. Levis, Optimal linear regulators: The discrete time case, *IEEE Transactions on Automatic Control*, AC-16, 613–620, 1970.

Gajić, Z. and I. Borno, General transformation for block diagonalization of weakly coupled linear systems composed of N subsystems, *IEEE Transactions on Circuits and Systems—I: Fundamental Theory and Applications*, 47, 909–912, 2000.

Gajić, Z. and M. Qureshi, *Lyapunov Matrix Equation in System Stability and Control*, Dover Publications, New York, 2008.

Gajić, Z. and X. Shen, Decoupling transformation for weakly coupled linear systems, *International Journal of Control*, 50, 1517–1523, 1989.

Gajić, Z., D. Petkovski, and X. Shen, *Singularly Perturbed and Weakly Coupled Linear Control Systems—A Recursive Approach*, Springer-Verlag, Lecture Notes in Control and Information Sciences, p. 140, New York, 1990.

Gajić, Z. and X. Shen, *Parallel Algorithm for Optimal Control of Large Scale Linear Systems*, Springer Verlag, London, 1993.

Grimbble, M., Linear quadratic Gaussian/loop transfer recovery design for a helicopter in low-speed flight, *Journal of Guidance, Control, and Dynamics*, 16, 754–761, 1993.

Katzberg, J., Structured feedback control of discrete linear stochastic systems with quadratic cost, *IEEE Transactions on Automatic Control*, AC–22, 232–236, 1977.

Kautsky, J., N. Nichols, and P. Van Douren, Robust pole assignment in linear state feedback, *International Journal of Control*, 41, 1129–1155, 1985.

Kwakernaak, H. and R. Sivan, *Linear Optimal Control Systems*, Wiley-Interscience, New York, 1972.

Lewis, F., *Optimal Control*, Wiley, New York, 1986.

10 Eigenvector Method for the Hamiltonian Approach

10.1 INTRODUCTION

In this chapter, we show how to decompose the weakly coupled algebraic Riccati equation and the corresponding linear-quadratic optimal control problem at steady state in terms of reduced-order subproblems by using the eigenvector approach. The eigenvector approach should be used for decomposition of weakly coupled control systems in the cases when the weak coupling parameter ε is not sufficiently small. In such cases, the decomposition methods based on series expansions, fixed point iterations, and Newton iterations either fail to produce solutions of the corresponding algebraic equations or display very slow convergence. In addition, the eigenvector approach provides new tools and a novel insight into the nature of the decomposition problem and finds all required solutions without solving the corresponding subsystem Riccati equations.

The research of this chapter is motivated by the existence of a transformation for decomposition of the weakly coupled algebraic Riccati equation and the corresponding linear-quadratic optimal control and filtering problems (Su and Gajić 1992; Gajić and Shen 1993). This transformation is valid for any value of a weak coupling parameter ε. The algebraic equations comprising the transformation in Su and Gajić (1992) have the structure of general nonsquare Riccati equations, which for sufficiently small values of a weak coupling parameter ε can be solved by performing iterations on systems of linear algebraic equations (e.g., fixed point iterations and Newton iterations). However, these iterative algorithms depend highly on the initial guesses and, when the weak coupling parameter ε is not sufficiently small, there is no guarantee that the above methods will find the solutions, hence they will not provide the desired decomposition. Moreover, the upper bound of the small weak coupling parameter ε for which the corresponding algebraic equations can be solved by any of the above iterative methods is highly problem dependent. In this chapter, we present the eigenvector method for solving the corresponding algebraic equations of weakly coupled linear systems that produces the required solutions for any value of the weak coupling parameter ε. The eigenvector method will produce the desired solutions and allow the reduced-order decomposition of the optimal control (and filtering) tasks. Hence, they can be performed independently at subsystem levels, producing reduction in off-line and online computational requirements. In addition, this method provides a new insight of the decomposition procedure, and produces novel methods and algorithms for the optimal controller design on the subsystem level.

10.2 DECOMPOSITION OF WEAKLY COUPLED ALGEBRAIC RICCATI EQUATION

In Su and Gajić (1992), a powerful transformation for the exact decomposition of the algebraic Riccati equation of weakly coupled systems is obtained so that the optimal control and filtering tasks can be solved exactly and performed independently for subsystems (Gajić and Shen 1993). Before the results in Su and Gajić (1992) became available, control engineers were not able to decompose exactly weakly coupled optimal control systems.

The weakly coupled linear control system under consideration is given by

$$\dot{x}_1(t) = A_1 x_1(t) + \varepsilon A_2 x_2(t) + B_1 u_1(t) + \varepsilon B_2 u_2(t)$$
$$\dot{x}_2(t) = \varepsilon A_3 x_1(t) + A_4 x_2(t) + \varepsilon B_3 u_1(t) + B_4 u_2(t)$$
(10.1)

where ε is the small weak coupling parameter and $x_1(t) \in \mathfrak{R}^{n_1}$ and $x_2(t) \in \mathfrak{R}^{n_2}$ are system state space variables ($n_1 + n_2 = n$ is the system order). Matrices A_i, $i = 1, 2, 3, 4$, are constant with elements bounded by $O(1)$. It is assumed that magnitudes of all the system eigenvalues are $O(1)$, that is $|\lambda_j| = O(1)$, $j = 1, 2$, $3, \ldots, n$, implying that matrices A_1 and A_4 are nonsingular with $\det(A_1) = O(1)$ and $\det(A_4) = O(1)$. This is the standard assumption for weakly coupled linear systems, which also corresponds to the block diagonal dominance of the system matrix A (Chow and Kokotović 1983). Hence, the main results presented in this chapter are valid under the following assumption.

Assumption 10.1 The magnitudes of the system eigenvalues are $O(1)$, that is, $|\lambda_j| = O(1)$, $j = 1, 2, 3, \ldots, n$, implying that matrices A_1 and A_4 are nonsingular with $\det(A_1) = O(1)$ and $\det(A_4) = O(1)$.

This assumption in fact indicates block diagonal dominance of the system matrix. It states the condition that guarantees that weak connections among the subsystems will indeed imply weak dynamic coupling. Note that when this assumption is not satisfied, the system defined in Equation 10.1, in addition of weak coupling can also display multiple timescale phenomena (singular perturbations), as considered in Phillips and Kokotović (1981), Delebeque and Quadrant (1981), and Chow (1982). The quadratic performance criterion to be minimized, associated with Equation 10.1, is given as

$$J = \frac{1}{2} \int_0^\infty \left(x^{\mathrm{T}}(t) Q x(t) + u^{\mathrm{T}}(t) R_C u(t) \right) dt, \quad Q \geq 0, \quad R_C > 0$$
(10.2)

Let P denote the solution of the algebraic Riccati equation corresponding to the weakly coupled control system. The corresponding Riccati equation is given in Kokotović et al. (1969)

$$A^{\mathrm{T}}P + PA + Q - PSP = 0, \quad P = \begin{bmatrix} P_1 & \varepsilon P_2 \\ \varepsilon P_2^{\mathrm{T}} & P_3 \end{bmatrix} \tag{10.3}$$

where

$$A = \begin{bmatrix} A_1 & \varepsilon A_2 \\ \varepsilon A_3 & A_4 \end{bmatrix}, \quad Q = \begin{bmatrix} Q_1 & \varepsilon Q_2 \\ \varepsilon Q_2^{\mathrm{T}} & Q_3 \end{bmatrix}, \quad R_C = \begin{bmatrix} R_{C_1} & 0 \\ 0 & R_{C_2} \end{bmatrix}$$

$$S = \begin{bmatrix} S_1 & \varepsilon S_2 \\ \varepsilon S_2^{\mathrm{T}} & S_3 \end{bmatrix} = BR_C^{-1}B^{\mathrm{T}}, \quad B = \begin{bmatrix} B_1 & \varepsilon B_2 \\ \varepsilon B_3 & B_4 \end{bmatrix} \tag{10.4}$$

The optimal control is given in terms of P as

$$u(t) = -R_C^{-1}B^{\mathrm{T}}Px(t) = -\begin{bmatrix} F_1 & \varepsilon F_2 \\ \varepsilon F_3 & F_4 \end{bmatrix}x(t), \quad x^{\mathrm{T}}(t) = \begin{bmatrix} x_1^{\mathrm{T}}(t) & x_2^{\mathrm{T}}(t) \end{bmatrix}^{\mathrm{T}} \tag{10.5}$$

For the optimal control problem defined by Equations 10.1 through 10.5, the exact subsystem decomposition result of the algebraic Riccati equation, as obtained in Su and Gajić (1992), is presented in the next lemma.

LEMMA 10.1 *Consider the closed-loop system*

$$\begin{bmatrix} \dot{x}_1(t) \\ \dot{x}_2(t) \end{bmatrix} = \begin{bmatrix} A_1 - B_1F_1 - \varepsilon^2 B_2F_3 & \varepsilon(A_2 - B_1F_2 - B_2F_4) \\ \varepsilon(A_3 - B_3F_1 - B_4F_3) & A_4 - B_4F_4 - \varepsilon^2 B_3F_2 \end{bmatrix} \begin{bmatrix} x_1(t) \\ x_2(t) \end{bmatrix} \tag{10.6}$$

There exists a nonsingular transformation T such that

$$\begin{bmatrix} \xi_1(t) \\ \xi_2(t) \end{bmatrix} = T \begin{bmatrix} x_1(t) \\ x_2(t) \end{bmatrix} \Rightarrow \begin{array}{l} \dot{\xi}_1(t) = (a_1 + a_2P_1)\xi_1(t) \\ \dot{\xi}_2(t) = (b_1 + b_2P_2)\xi_2(t) \end{array} \tag{10.7}$$

where P_1 and P_2 are the unique solutions of the reduced order subsystem algebraic Riccati equations given by

$$\begin{array}{l} P_1a_1 - a_4P_1 - a_3 + P_1a_2P_1 = 0 \\ P_2b_1 - b_4P_2 - b_3 + P_2b_2P_2 = 0 \end{array} \tag{10.8}$$

where matrices a_i and b_i, $i = 1, \ldots, 4$ are obtained from

$$R_1 = \begin{bmatrix} a_1 & a_2 \\ a_3 & a_4 \end{bmatrix} = T_1 - \varepsilon^2 LT_3, \quad R_2 = \begin{bmatrix} b_1 & b_2 \\ b_3 & b_4 \end{bmatrix} = T_4 + \varepsilon^2 T_3L \tag{10.9}$$

with

$$T_1 = \begin{bmatrix} A_1 & -S_1 \\ -Q_1 & -A_1^T \end{bmatrix}, \quad T_2 = \begin{bmatrix} A_2 & -S_2 \\ -Q_2 & -A_3^T \end{bmatrix}$$

$$T_3 = \begin{bmatrix} A_3 & -S_2^T \\ -Q_2^T & -A_2^T \end{bmatrix}, \quad T_4 = \begin{bmatrix} A_4 & -S_3 \\ -Q_3 & -A_4^T \end{bmatrix} \tag{10.10}$$

The solution of the original global algebraic Riccati Equation 10.3 can be obtained from

$$P = \left\{ \Omega_3 + \Omega_4 \begin{bmatrix} P_1 & 0 \\ 0 & P_2 \end{bmatrix} \right\} \left\{ \Omega_1 + \Omega_2 \begin{bmatrix} P_1 & 0 \\ 0 & P_2 \end{bmatrix} \right\}^{-1} \tag{10.11}$$

where

$$\begin{bmatrix} \Omega_1 & \Omega_2 \\ \Omega_3 & \Omega_4 \end{bmatrix} = \Omega = E^T \begin{bmatrix} I & -\varepsilon L \\ \varepsilon H & I - \varepsilon^2 HL \end{bmatrix}^{-1} \tag{10.12}$$

with

$$E = \begin{bmatrix} I_{n_1} & 0 & 0 & 0 \\ 0 & 0 & I_{n_1} & 0 \\ 0 & I_{n_2} & 0 & 0 \\ 0 & 0 & 0 & I_{n_2} \end{bmatrix}. \tag{10.13}$$

The matrices L and H satisfy the transformation equations from Gajić and Shen (1989)

$$T_1 L + T_2 - L T_4 - \varepsilon^2 L T_3 L = 0$$
$$H(T_1 - \varepsilon^2 L T_3) + T_3 - (T_4 + \varepsilon^2 T_3 L)H = 0 \tag{10.14}$$

The decomposition transformation T is given by

$$T = (\Pi_1 + \Pi_2 P) \tag{10.15}$$

with

$$\Pi = \begin{bmatrix} \Pi_1 & \Pi_2 \\ \Pi_3 & \Pi_4 \end{bmatrix} = \Omega^{-1} \tag{10.16}$$

All steps in Lemma 10.1 can easily be computed by using any appropriate mathematical package. The reduced-order subsystem algebraic Riccati Equations 10.8 can be solved in terms of Lyapunov iterations, which is in fact the Newton method for solving Equation 10.8 as demonstrated in Su and Gajić (1992). The initial conditions for the Newton method are obtained from the $O(\varepsilon^2)$-approximate subsystems algebraic Riccati equations derived in Su and Gajić (1992)

$$A_1^T P_{1(0)} + P_{1(0)} A_1 + Q_1 - P_{1(0)} S_1 P_{1(0)} = 0$$
$$A_4^T P_{2(0)} + P_{2(0)} A_4 + Q_3 - P_{2(0)} S_3 P_{2(0)} = 0 \tag{10.17}$$

The unique positive semidefinite stabilizing solutions of the above algebraic Riccati equations exist under stabilizability–detectability conditions imposed on subsystems (Kwakernaak and Sivan 1972), which are the standard assumptions in the theory of weakly coupled systems. The following assumption is required.

Assumption 10.2 The triples $(A_1, B_1, \text{Chol}(Q_1))$ and $(A_4, B_2, \text{Chol}(Q_3))$ are stabilizable–detectable.

It has been shown in Su and Gajić (1992) that under the same assumption the unique stabilizing solutions of the reduced-order subsystem algebraic Riccati Equations 10.8 exist for sufficiently small values of the weak coupling parameter ε.

The transformation Equations 10.14 can be efficiently solved as linear algebraic equations using either fixed point iterations or Newton method as demonstrated in Gajić and Shen (1993). Solvability of Equation 10.14 requires that matrices T_1 and T_2 have no eigenvalues in common (Gajić and Shen 1993).

10.3 EIGENVECTOR METHOD FOR NONSYMMETRIC (NONSQUARE) ALGEBRAIC RICCATI EQUATION

The eigenvector method for solving the algebraic square and symmetric Riccati equation dates back to MacFarlane (1963), Medanić (1982), and Fath (1969). The main results for numerical solution of the symmetric square algebraic Riccati equation by the eigenvector method were obtained in Van Dooren (1981). A detailed survey of the eigenvector numerical methods for solving the algebraic Riccati equations can be found in Bunse-Gerstner et al. (1992), and Bunse-Gerstner and Fassbender (1997). Analytical studies of the general nonsquare algebraic Riccati equations were reported in Clements and Anderson (1976) and Medanić (1982).

Without loss of generality, we will present results for the square nonsymmetric algebraic Riccati equation (ARE). The same approach can be used for the general nonsquare algebraic Riccati equation. The algebraic square nonsymmetric matrix Riccati equation is given by

$$AX + XB + C + XDX = 0 \tag{10.18}$$

where all matrices are constant and square of dimensions (n, n). Consider now the following $(2n, 2n)$ matrix:

$$R = \begin{bmatrix} B & D \\ -C & -A \end{bmatrix} \tag{10.19}$$

Note that R is a Hamiltonian matrix since it satisfies $(JR)^* = JR$, where the symbol asterisk "$*$" stands for complex–conjugate transpose and $J = \begin{bmatrix} O & I \\ -I & O \end{bmatrix}$. Such matrices have the eigenvalues occurring in pairs $\lambda(R)$ and $-\lambda^*(R)$.

Now, calculate $2n$ eigenvalues of R, $\lambda_i = a_i + jb_i$ and all the corresponding $2n$ eigenvectors, $v_i = x_i + jy_i$, where $i = 1, 2, \ldots, 2n$. Arrange in the $(2n, 2n)$ matrix M all real eigenvectors (i.e., real vectors $v_i = x_i$) and for each complex–conjugate pair use consecutively the real and imaginary parts of one eigenvector only, i.e., the real vectors x_i and y_i. (Note that there are many ways to form the matrix M.) The matrix M is a "real number" eigenvector matrix (Bingulac and VanLandingham 1993) and it satisfies,

$$RM = MK = [M_1 \ M_2] \begin{bmatrix} K_1 & 0 \\ 0 & K_2 \end{bmatrix} \tag{10.20}$$

where the $(2n, n)$ matrix M_1 contains the first n columns of M and the matrix M_2 contains the remaining n columns of M. When the eigenvalues of the matrix R are real then, $K = \text{diag}(\lambda_1, \lambda_2, \lambda_3, \ldots, \lambda_i, \ldots, \lambda_{2n}) = \text{diag}(a_1, a_2, a_3, \ldots, a_i, \ldots, a_{2n})$. In general, matrix R may have the complex–conjugate pairs of eigenvalues in which case the matrix K becomes a "block diagonal" matrix and Equation 10.20 may be rewritten as

$$RM_1 = M_1 K_1, \quad RM_2 = M_2 K_2 \tag{10.21}$$

By partitioning the matrix M_1 as

$$M_1 = \begin{bmatrix} M_{11} \\ M_{21} \end{bmatrix} \tag{10.22}$$

where M_{11} and M_{21} are (n, n) matrices, we get from Equation 10.21

$$\begin{aligned} BM_{11} + DM_{21} &= M_{11} K_1 \\ -CM_{11} - AM_{21} &= M_{21} K_1 \end{aligned} \tag{10.23}$$

Rearranging the last two equations and by using the substitution

$$X = M_{21} M_{11}^{-1} \tag{10.24}$$

results in

$$AX + XB + C + XDX = 0 \tag{10.25}$$

Thus, solutions of Equation 10.18 are given by $X = M_{21}M_{11}^{-1} = X_1$. Note that following the same procedure leads to the solutions $X_2 = M_{22}M_{12}^{-1}$, too. For each selection of n vectors from M into M_1, a different matrix X satisfying Equation 10.18 will be obtained. In other words, there are many solutions to algebraic equation (Equation 10.18). The same statement holds for different choices of the matrix M_2 and the corresponding solutions of Equation 10.18 obtained from $X_2 = M_{22}M_{12}^{-1}$. The solutions X_1 and X_2 are valid under the assumption that the matrices M_{11} and M_{12} are nonsingular.

It has been seen that by using the similarity transformation composed of the eigenvectors of the matrix R, this matrix is put into diagonal form (or Jordan form in the case of multiple eigenvalues) defined by Equation 10.20. Another similarity transformation from Smith (1987) that puts the matrix R into block-diagonal form is also known in the literature. Let

$$T_{1t} = \begin{bmatrix} I & 0 \\ X & I \end{bmatrix} \tag{10.26}$$

where X is a solution of Equation 10.18; then

$$T_{1t}^{-1}RT_{1t} = \begin{bmatrix} B + DX & D \\ 0 & -(A + XD) \end{bmatrix} \tag{10.27}$$

Moreover, this upper block diagonal matrix can be put into block diagonal form by using another similarity transformation defined by

$$T_{2t} = \begin{bmatrix} I & Y \\ 0 & I \end{bmatrix} \tag{10.28}$$

where Y satisfies the algebraic Lyapunov (Sylvester) equation

$$(B + DX)Y + Y(A + DX) + D = 0 \tag{10.29}$$

The unique solution of Equation 10.29 is guaranteed under the assumption that matrices $B + DX$ and $-(A + DX)$ have no eigenvalues in common (Gajić and Qureshi 1995). The second transformation produces

$$T_{2t}^{-1}T_{1t}^{-1}RT_{1t}T_{2t} = T_D^{-1}RT_D = \begin{bmatrix} B + DX & 0 \\ 0 & -(A + XD) \end{bmatrix} \tag{10.30}$$

where

$$T_D = \begin{bmatrix} I & Y \\ X & I + XY \end{bmatrix}, \quad T_D^{-1} = \begin{bmatrix} I + YX & -Y \\ -X & I \end{bmatrix} \tag{10.31}$$

The similarity transformation defined by Equations 10.26, 10.28, and 10.31 is valid for both nonsquare and square algebraic Riccati equations.

For the purpose of our chapter, Lemma 10.2 (together with Lemma 10.3 given below) plays a fundamental role.

LEMMA 10.2

Let the matrix X be a solution of Equation 10.18 obtained by using Equation 10.24 with $\left[M_{11}^{T} M_{21}^{T}\right]^{T}$ consisting of l eigenvectors spanning the stable subspace of R and $n - l$ eigenvectors from the corresponding unstable subspace. Then, the matrix $B + DX$ as defined in Equations 10.27 and 10.30 will have l stable and $n - l$ unstable eigenvalues corresponding to the eigenvectors used in $\left[M_{11}^{T} M_{21}^{T}\right]^{T}$.

This lemma is just a special case of a more general theorem proved in Clements and Anderson (1976). See also Theorem 10.1 in Medanić (1982).

10.4   EXACT DECOMPOSITION ALGORITHM FOR WEAKLY COUPLED SYSTEMS

It can be seen from Lemma 10.1 that in order to solve the linear-quadratic optimal control problem of weakly coupled systems given by Equations 10.1 through 10.5 in terms of reduced-order problems, one has in addition to solving algebraic Equation 10.14 (whose solutions comprise the desired transformation) to solve the reduced-order algebraic Riccati equation (Equation 10.8). Note that the equations for L, P_1, P_2 also have forms of nonsymmetric algebraic Riccati equations. The equation for H in Equation 10.14 is a linear equation, hence its solution is straightforward.

In the following, we will show that by using the eigenvector method presented in the previous section all three equations for L, P_1, P_2 can be solved at the same time avoiding solving the reduced-order algebraic Riccati equation (Equation 10.8).

Before we present the decomposition algorithm, we first establish some features of the matrices related with the L-equation in Equation 10.14. This equation has the form of the general nonsymmetric algebraic Riccati equation (Equation 10.18) with the corresponding matrix R_L given by

$$R_L = \begin{bmatrix} -T_4 & -\varepsilon^2 T_3 \\ -T_2 & -T_1 \end{bmatrix} \tag{10.32}$$

It can be seen that the matrix R_L has the eigenvalues that are $O(\varepsilon^2)$ perturbations of the eigenvalues of the matrices $-T_1$ and $-T_4$. Since $-T_1$ and $-T_4$ are Hamiltonian matrices that have the eigenvalues symmetrically distributed with respect to the imaginary axis with no imaginary axis eigenvalues (under Assumption 10.2), it follows that R_L is a nonsingular matrix.

The following lemma gives more information about the eigenvalues of the matrix R_L and it guarantees the existence of the stable subspace solutions.

LEMMA 10.3

The eigenvalues of the matrix R_L are symmetric with respect to the imaginary axis.

Proof It is well known that the matrix R_P formed of the coefficients of the algebraic Riccati equation (Equation 10.3) as

$$R_P = \begin{bmatrix} A & -S \\ -Q & -A^T \end{bmatrix} \tag{10.33}$$

under stabilizability–detectability conditions of a triple $(A, \text{Chol}(S), \text{Chol}(Q))$ has the eigenvalues symmetrically distributed with respect to the imaginary axis (Kwakernaak and Sivan 1972). By introducing a permutation matrix of the form

$$E = \begin{bmatrix} 0 & 0 & 0 & -\frac{1}{\varepsilon}I_{n_1} \\ 0 & -I_{n_2} & 0 & 0 \\ 0 & 0 & \frac{1}{\varepsilon}I_{n_1} & 0 \\ I_{n_2} & 0 & 0 & 0 \end{bmatrix} \tag{10.34}$$

it can be shown by matrix multiplications that the following holds:

$$R_L = (E^{-1} R_P E)^T \tag{10.35}$$

This implies that $\lambda(R_L) = \lambda(R_P)$, which establishes the result stated in Lemma 10.3. Note that the symmetry of the eigenvalues $\lambda(R_L)$ ensures that there are exactly $(n_1 + n_2)$ stable eigenvalues of the matrix R_L. ∎

Now we are ready to formulate the order-reduction algorithms for weakly coupled systems based on the eigenvector approach.

ALGORITHM 10.1

Step 1: Find the solution matrix L from Equation 10.14 via the eigenvector approach applied to the matrix R_L defined in Equation 10.32. Let a solution be obtained from a collection of n_2 eigenvectors spanning the stable subspace and n_2 eigenvectors spanning the corresponding unstable subspace. This solution for L will be called the admissible solution. (Note that there are many solutions for L.)

Step 2: Use the solution obtained in Step 1 in order to solve the algebraic Lyapunov equation for Y_L, that has the form of Equation 10.29

$$(B + DX)Y_L + Y_L(A + DX) + D$$
$$= -(T_4 + \varepsilon^2 T_3 L)Y_L + Y_L(T_1 - \varepsilon^2 LT_3) - \varepsilon^2 T_3 = 0 \tag{10.36}$$

and apply the transformation defined in Equations 10.30 and 10.31 to matrix R_L given in Equation 10.32. This leads to

$$T_D^{-1} R_L T_D = \begin{bmatrix} -R_2 & 0 \\ 0 & -R_1 \end{bmatrix}, \quad T_D = \begin{bmatrix} I & Y_L \\ L & I + LY_L \end{bmatrix} \tag{10.37}$$

Note that Y_L can be calculated directly. See comments in Step 2 of Algorithm 10.2. Note also that ordering of the matrices after the transformation is applied that produces R_2 in the upper-left corner and R_1 in the bottom-right one. See also the transformations given in Equations 10.42 and 10.45.

Step 3: Partition matrices R_1 and R_2 as

$$R_1 = \begin{bmatrix} a_1 & a_2 \\ a_3 & a_4 \end{bmatrix}, \quad R_2 = \begin{bmatrix} b_1 & b_2 \\ b_3 & b_4 \end{bmatrix} \tag{10.38}$$

where a_i's, $i = 1, 2, 3, 4$, are of dimensions (n_1, n_1) and b_j's, $j = 1, 2, 3, 4$, are of dimensions (n_2, n_2). Define the reduced-order subsystem algebraic Riccati equations as given in Equation 10.8, that is

$$P_1 a_1 - a_4 P_1 - a_3 + P_1 a_2 P_1 = 0, \quad \text{and} \quad P_2 b_1 - b_4 P_2 - b_3 + P_2 b_2 P_2 = 0$$

Form the algebraic Lyapunov equations corresponding to Equation 10.29 as

$$(a_1 + a_2 P_1)Y_1 - Y_1(a_4 - P_1 a_2) + a_2 = 0$$
$$(b_1 + b_2 P_2)Y_2 - Y_2(b_4 - P_2 b_2) + b_2 = 0 \tag{10.39}$$

Then, the similarity transformations, obtained from the solutions of Equation 10.8, defined by

$$T_{1d} = \begin{bmatrix} I & Y_1 \\ P_1 & I + P_1 Y_1 \end{bmatrix}, \quad T_{2d} = \begin{bmatrix} I & Y_2 \\ P_2 & I + P_2 Y_2 \end{bmatrix} \tag{10.40}$$

will block diagonalize, respectively matrices R_1 and R_2, in the spirit of Equations 10.29 through 10.31; hence

$$T_{1d}^{-1} R_1 T_{1d} = \begin{bmatrix} a_1 + a_2 P_1 & 0 \\ 0 & a_4 - P_1 a_2 \end{bmatrix}$$
$$T_{2d}^{-1} R_2 T_{2d} = \begin{bmatrix} b_1 + b_2 P_2 & 0 \\ 0 & b_4 - P_2 b_2 \end{bmatrix} \tag{10.41}$$

Thus, a successive application of transformations in Equations 10.37 and 10.41 produces in the new coordinates a four-block block-diagonal form in which closed-loop subsystems' state and costate variables are completely decoupled. Note that the transformations defined in Equations 10.37 and 10.41 can be put in a compact form leading to

$$R_{LD} = T_{LD}^{-1}R_L T_{LD} = - \begin{bmatrix} b_1 + b_2 P_2 & 0 & 0 & 0 \\ 0 & b_4 - P_2 b_2 & 0 & 0 \\ 0 & 0 & a_1 + a_2 P_1 & 0 \\ 0 & 0 & 0 & a_4 - P_1 a_2 \end{bmatrix} \quad (10.42)$$

where

$$T_{LD} = T_D \begin{bmatrix} T_{2d} & 0 \\ 0 & T_{1d} \end{bmatrix} = \begin{bmatrix} T_{2d} & Y_L T_{1d} \\ LT_{2d} & T_{1d} + LY_L T_{1d} \end{bmatrix} \quad (10.43)$$

Step 4: From the obtained values for P_1 and P_2 calculate the solution of the global algebraic Riccati Equation 10.3 by using Equation 10.11.

It can be observed that in Step 3 of Algorithm 10.1, we have to solve the reduced-order subsystem nonsymmetric square algebraic Riccati equation (Equation 10.8). That can be done either by using the Newton method with appropriately chosen initial conditions (which are $O(\varepsilon^2)$ apart from the exact solutions as demonstrated in Su and Gajić (1992)) or by using the eigenvector approach presented in this chapter.

In the following, we present another more powerful algorithm whose important feature is that the main results of Lemma 10.1 are obtained without the need to solve independently the reduced-order subsystem algebraic Riccati equation (Equation 10.8). These solutions are obtained as a by-product of the decomposition algorithm.

ALGORITHM 10.2

Step 1: Equal to Step 1 of Algorithm 10.1.

Step 2: Equal to Step 2 of Algorithm 10.1. Note that the matrix Y_L, here and in the Step 2 of Algorithm 10.1, can be calculated directly by using a simple equality

$$Y_L = M_{12}(M_{22} - LM_{12})^{-1} \quad (10.44)$$

where the matrices M_{ij} result from the partition of the "real number" eigenvector matrix M from Equation 10.20. This way we avoid solving the Lyapunov equation (Equation 10.36).

Step 3: Calculate the $(2n_1, 2n_1)$ dimensional eigenvector matrix for the first subsystem M_{1D}, and the $(2n_2, 2n_2)$ dimensional eigenvector matrix for the second subsystem M_{2D}, using the formula

$$T_D^{-1}[M_1 M_2] = \begin{bmatrix} M_{2D} & 0 \\ 0 & M_{1D} \end{bmatrix} \tag{10.45}$$

where $[M_1 M_2]$ is the eigenvector matrix of R_L obtained in Step 1 according to Equation 10.20. Step 3 can be justified as given in Appendix 10.1.

Step 4: Having obtained the subsystems' eigenvector matrices M_{1D} and M_{2D} we calculate the (n_1, n_1) dimensional solution P_1 and the (n_2, n_2) dimensional solution P_2 from

$$P_1 = M_{211}M_{111}^{-1}, \quad P_2 = M_{212}M_{112}^{-1} \tag{10.46}$$

where the required matrices are obtained by appropriately partitioning matrices M_{1D} and M_{2D}, that is

$$M_{1D} = \begin{bmatrix} M_{111} & M_{121} \\ M_{211} & M_{221} \end{bmatrix}, \quad M_{2D} = \begin{bmatrix} M_{112} & M_{122} \\ M_{212} & M_{222} \end{bmatrix} \tag{10.47}$$

Step 5: Partition appropriately matrices R_1 and R_2 obtained in Step 2 according to Equation 10.38 and form the subsystem optimal feedback matrices, respectively, given by

$$a_1 + a_2 P_1, \quad b_1 + b_2 P_2 \tag{10.48}$$

Step 6: Use the values for P_1 and P_2 obtained in Step 4 to calculate the solution of the global algebraic Riccati Equation 10.3 by using Equation 10.11.

Once the subsystem eigenvectors are known, Step 4 of Algorithm 10.2 finds the solutions of the subsystem algebraic Riccati equations by the eigenvector method as described at the beginning of this section. Step 6 finds the solution of the global algebraic Riccati equations that is used to find the optimal value of the performance criterion, and Step 5 produces the reduced-order independent subsystems as given by Equation 10.7.

In the next section, we solve several different examples in order to demonstrate the presented procedure.

10.5 EXAMPLES

Example 10.1

Consider the fourth-order real world example, a satellite control problem considered in Su and Gajić (1992). Problem matrices are given as

$$A_1 = \begin{bmatrix} 0 & 0.667 \\ -0.667 & 0 \end{bmatrix}, \quad A_2 = 0, \quad A_3 = 0, \quad A_4 = \begin{bmatrix} 0 & 1.53 \\ -1.53 & 0 \end{bmatrix},$$

$$B_1 = \begin{bmatrix} 0 \\ 1 \end{bmatrix}, \quad B_2 = \frac{1}{\varepsilon}\begin{bmatrix} 0.2 \\ 0 \end{bmatrix}, \quad B_3 = \frac{1}{\varepsilon}\begin{bmatrix} 0.4 \\ 0 \end{bmatrix}, \quad B_4 = \begin{bmatrix} 0 \\ 1 \end{bmatrix}$$

The penalty matrices are chosen as identities, $Q = I_4$, $R_C = I_2$, and $\varepsilon = 0.2$.

The particular feature of this example is that for the coupling parameter $\varepsilon = 0.2$, all the eigenvalues of the Hamiltonian matrix R_L associated with nonsymmetric ARE (Equation 10.14) for L, are the complex–conjugate pairs symmetric with respect to the origin. Eigenvalues of the matrix R_L, presented here in two columns due to the space constraints, are

$$\Lambda_L = \begin{bmatrix} -0.74 + 1.52i & -0.74 - 1.52i & 0.74 + 1.52i & 0.74 - 1.52i \\ -0.67 + 0.62i & -0.67 - 0.62i & 0.67 + 0.62i & 0.67 - 0.62i \end{bmatrix}^T$$

As mentioned before, starting from the initial guess ($\varepsilon = 0$), and using Newtonian iterations we will be able to find one single solution for L only. For $n = 4$ and all eigenvalues of R_L being complex there are six solutions of nonsymmetric ARE (Equation 10.14) for L. In fact when there are less distinct real eigenvalues of the matrix R_L than the complex–conjugate ones (i.e., when $r < n$, where r is the number of the real eigenvalues, here $n = 4$ and $r = 0$) one should use two different formulas for calculation of the total number of solutions S, depending on whether n is an even or an odd number. For even n and when ($n \geq r \geq 0$) there is

$$S = \sum_{i=0}^{\frac{r}{2}} \binom{r}{2i}\binom{\frac{2n-r}{2}}{\frac{n-2i}{2}} = \binom{4}{2} = 6$$

solution matrices L. (More on the number of solutions to the ARE [Equation 10.14], i.e. [Equation 10.18] can be found in Appendix 10.2.) As stated above, not all of these matrices L will solve the problem of exact decomposition for this weakly coupled system. According to Lemma 10.2, only the matrices L obtained by using Equation 10.24 with $[M_{11}^T M_{21}^T]^T$ consisting of two eigenvectors spanning the stable subspace of R_L and two eigenvectors from the corresponding unstable subspace are the admissible ones. There are four such admissible matrices L. The decoupling matrix L with the minimal norm is given below

$$L = \begin{bmatrix} 0.00 & 0.01 & -0.22 & 0.00 \\ 0.32 & 0.00 & 0.00 & 0.98 \\ 0.20 & 0.00 & 0.00 & -0.03 \\ 0.00 & 0.50 & -0.31 & 0.00 \end{bmatrix}$$

The unique stable solutions to the reduced-order subsystem algebraic Riccati equation (Equation 10.8) (matrices P_1 and P_2) are associated with each admissible matrix L, and by using Equation 10.11 they produce the unique solution matrix P to the global Riccati equation (Equation 10.3). The three matrices above, associated with the matrix L, having the minimal norm are

$$P_1 = \begin{bmatrix} 2.2201 & 0.4589 \\ 0.4410 & 1.2748 \end{bmatrix} \quad P_2 = \begin{bmatrix} 1.5056 & 0.1947 \\ 0.2282 & 1.2782 \end{bmatrix}$$

$$P = \begin{bmatrix} 2.2437 & 0.4622 & 0.1361 & -0.1074 \\ 0.4622 & 1.3456 & -0.2091 & -0.2475 \\ 0.1361 & -0.2091 & 1.5375 & 0.2482 \\ -0.1074 & -0.2475 & 0.2482 & 1.3396 \end{bmatrix}$$

Note that the global solution P is the unique one, meaning that it is obtained using any out of 4 admissible matrices L. At the same time, the solutions for P_1 and P_2 are dependent on the particular admissible solution matrix L used.

Example 10.2

The system matrices of the fifth-order chemical plant considered in Gomathi et al. (1980) are given below. The analysis of the discretized version of this plant can be found in Gajić and Shen (1993).

$$A = \begin{bmatrix} -0.11 & 0.06 & 0 & 0 & 0 \\ 1.31 & -2.13 & 0.98 & 0 & 0 \\ 0 & 1.60 & -3.15 & 1.55 & 0 \\ 0 & 0.04 & 2.63 & -4.26 & 1.86 \\ 0 & 0.00 & 0 & 0.16 & -0.16 \end{bmatrix}, \quad B = \begin{bmatrix} 0 & 0 \\ 0.06 & 0 \\ 0.08 & -0.14 \\ 0.10 & -0.21 \\ 0.01 & -0.01 \end{bmatrix}$$

The penalty matrices are chosen as identities, $Q = I_5$, $R_C = I_2$. The coupling parameter ε is built into the problem and it was estimated from the strongest coupled matrix B as $\varepsilon = 0.68$.

This specific real world example is composed of the second-order subsystem $(n_1 = 2)$ and of the third-order one $(n_2 = 3)$. The Hamiltonian matrix R_L associated with the decoupling Equation 10.14 for L now has 10 real eigenvalues. Hence, there are 210 choices for the eigenvector matrix M_1 which produce solutions to Equation 10.14. However, in accordance with Lemma 10.2, only 100 of them are admissible solutions for the purpose of the reduced order design. Below we present the decoupling matrix L which is equal to the one obtained by Newton's method. This Newton's method iterative solution is obtained after seven iterations as

$$L = \begin{bmatrix} -0.08 & -0.06 & -0.31 & -0.01 & -0.01 & -0.01 \\ 0.58 & 0.59 & -0.58 & 0.01 & 0.01 & -0.00 \\ 5.48 & 4.52 & 27.68 & -2.67 & -3.58 & -1.78 \\ -0.07 & -0.12 & 0.38 & 0.97 & 1.65 & -0.15 \end{bmatrix}$$

The corresponding (L dependent) subsystem Riccati equations produce the solutions

$$P_1 = \begin{bmatrix} 47.65 & 4.80 \\ 1.15 & 0.25 \end{bmatrix}, \quad P_2 = \begin{bmatrix} 0.50 & 0.30 & 0.79 \\ 0.25 & 0.27 & 0.71 \\ 1.18 & 1.05 & 9.98 \end{bmatrix}$$

The unique global solution P obtained by using P_1 and P_2 in Equation 10.11 is given as

$$P = \begin{bmatrix} 30.17 & 2.39 & 1.73 & 12.45 \\ 2.39 & 0.58 & 0.36 & 1.45 \\ 1.73 & 0.36 & 0.47 & 1.08 \\ 12.45 & 1.45 & 1.08 & 13.82 \end{bmatrix}$$

The matrix Y_L needed for the block-diagonalization of both the Hamiltonian matrix R_L and its eigenvector matrix M is given below

$$Y_l = \begin{bmatrix} 0.618 & -0.230 & -0.002 & 0.001 \\ 0.739 & -0.404 & -0.002 & 0.002 \\ 0.570 & 0.062 & -0.002 & -0.001 \\ 1.237 & -0.017 & 0.018 & -0.139 \\ 0.929 & -0.029 & 0.012 & -0.146 \\ 8.885 & 0.170 & 0.099 & 0.241 \end{bmatrix}$$

Note that this matrix can be calculated by using either Equation 10.36 or Equation 10.44 given in Algorithms 10.1 and 10.2, respectively.

The block-diagonalized Hamiltonian matrix R_{LD} obtained by Equation 10.42 in terms of the completely decoupled state and costate variables of the first subsystem and the second subsystem is given below

$$R_{LD} = \begin{bmatrix} 2.54 & -2.17 & 0.71 & 0 & 0 & 0 & 0 & 0 & 0 & 0 \\ -2.61 & 4.27 & -1.74 & 0 & 0 & 0 & 0 & 0 & 0 & 0 \\ 0.00 & -0.16 & 0.17 & 0 & 0 & 0 & 0 & 0 & 0 & 0 \\ 0 & 0 & 0 & -2.53 & 3.69 & -0.10 & 0 & 0 & 0 & 0 \\ 0 & 0 & 0 & 1.53 & -4.28 & 0.16 & 0 & 0 & 0 & 0 \\ 0 & 0 & 0 & -0.10 & 1.71 & -0.17 & 0 & 0 & 0 & 0 \\ 0 & 0 & 0 & 0 & 0 & 0 & 0.12 & -0.15 & 0 & 0 \\ 0 & 0 & 0 & 0 & 0 & 0 & -1.31 & 2.78 & 0 & 0 \\ 0 & 0 & 0 & 0 & 0 & 0 & 0 & 0 & -0.11 & 2.75 \\ 0 & 0 & 0 & 0 & 0 & 0 & 0 & 0 & 0.06 & -2.78 \end{bmatrix}$$

Example 10.3

Here, it is shown how the proposed eigenvector approach, developed for weakly coupled systems, can be applied to the solution of the optimal controller design of any standard linear control system (as well as to nonweakly coupled system, $\varepsilon = 1$).

The application of the proposed method will be of particular interest for the system of higher order, that is, when the Hamiltonian matrix associated with the global Riccati equation (Equation 10.3) $R_P(2n, 2n)$ is of a very high order. In such a situation (and after proper partitioning of the problem matrices), in the first step of Algorithm 10.1, calculation of the decoupling matrix L would be done by Newton's iterative method. The rest of the algorithm remains the same. In this way, instead of working with a $(2n, 2n)$ Hamiltonian matrices in solving nonsymmetric Riccati equation (Equation 10.3), in the rest of the algorithm we work with the subsystem Riccati equation (Equation 10.8). Note, however, that the Newton iterative method may fail or that it can find a non-admissible solution L. In such a case, the first step should be repeated with a new partitioning of the system matrices. In the case of the system presented below, Newton method failed for the following partitionings $(n_1 = 2, n_2 = 5)$, $(n_1 = 4, n_2 = 3)$, and $(n_1 = 6, n_2 = 1)$.

Here we consider a seventh-order system of a Saturn V booster from McBrinn and Roy (1972) represented by the following system matrices:

$$A = \begin{bmatrix} 0 & 1 & 0 & 0 & 0 & 0 & 0 \\ 0 & 0 & 0.20 & -0.05 & -0.002 & 2.6 & 0 \\ -0.014 & 1 & -0.041 & 0.0002 & -0.015 & -0.033 & 0 \\ 0 & 0 & 0 & 0 & 1 & 0 & 0 \\ 0 & 0 & 0 & -45 & -0.13 & 255 & 0 \\ 0 & 0 & 0 & 0 & 0 & 0 & 1 \\ 0 & 0 & 0 & 0 & 0 & -50 & -10 \end{bmatrix}$$

$$B = [0\ 0\ 0\ 0\ 0\ 0\ 1]^T$$

In order to apply the decoupling algorithm presented here, the input matrix B was augmented with a zero column. The penalty matrices are chosen as identities, $Q = I_7$, $R_C = I_2$. In this case of the standard linear system, we take the coupling parameter $\varepsilon = 1$. It is interesting to note that by applying the eigenvector method proposed here, the global unique solution P of the Riccati equation (Equation 10.3) was obtained by any partitioning of the given system ($1 \leq n_1 \leq 6$, $6 \geq n_2 \geq 1$). Here, we present some of the resulting matrices obtained from the partitioning ($n_1 = 3$, $n_2 = 4$). Note that such a partitioning results in the nonsquare algebraic Riccati equation (Equation 10.14) for the rectangular $(2n_1, 2n_2)$ decoupling matrix L.

In this particular example and for a given partitioning (3, 4), the Newton iteration scheme has calculated the matrix L after two iteration steps only. There are two real eigenvalues $(r = 2)$ and six complex-conjugate pairs of the eigenvalues of the matrix R_L associated with ARE (Equation 10.14). While applying the eigenvector method, 34 arrangements of matrix M_1 provide the solutions matrices L of this ARE. Again, there are less admissible matrices L (the ones that at the same time solve the problem of the exact decomposition of this standard control system). Here, the matrix M_1 comprised of $n_2 =$ four stable and four unstable eigenvectors can be composed in nine different manners. Hence, there will be nine admissible solutions for the matrix L. Below, the matrix L obtained by the Newton method, together with a part of the other solutions matrices associated with it, are presented

$$
L = \begin{bmatrix}
0.006 & 0.000 & 0.048 & 0.009 & 0.000 & 0.005 & 0.000 & 0.001 \\
-0.002 & 0.001 & -0.456 & -0.046 & -0.005 & 0.000 & -0.001 & 0.000 \\
-0.009 & 0.000 & 0.054 & 0.010 & 0.000 & 0.005 & 0.000 & 0.001 \\
0.000 & 0.000 & 0.008 & 0.002 & 0.000 & 0.000 & 0.000 & 0.000 \\
-0.007 & 0.000 & -0.050 & -0.009 & 0.000 & -0.005 & 0.000 & -0.001 \\
0.000 & -0.000 & -0.010 & -0.000 & -0.000 & 0.000 & 0.000 & 0.000
\end{bmatrix}
$$

$$
Y_L = \begin{bmatrix}
0.000 & 0.000 & 0.000 & 0.000 & -0.005 & 0.000 \\
0.000 & 0.005 & -0.000 & 0.005 & 0.000 & 0.005 \\
0.000 & 0.000 & 0.000 & 0.000 & -0.001 & 0.000 \\
0.000 & 0.001 & 0.000 & 0.001 & 0.000 & 0.001 \\
0.000 & -0.007 & 0.000 & -0.006 & 0.002 & 0.009 \\
0.000 & 0.000 & -0.000 & 0.000 & -0.001 & 0.000 \\
0.008 & -0.050 & -0.010 & -0.048 & 0.456 & -0.054 \\
0.002 & -0.009 & -0.000 & -0.009 & 0.046 & -0.010
\end{bmatrix}
$$

$$
P_1 = \begin{bmatrix}
240.95 & 51.651 & -190.72 \\
58.772 & 419.93 & 126.67 \\
-187.69 & 132.99 & 224.26
\end{bmatrix}
$$

$$
P_2 = \begin{bmatrix}
14.87 & -0.422 & -70.96 & -6.246 \\
-0.422 & 0.366 & 9.503 & 0.246 \\
-70.96 & 9.503 & 563.0 & 35.72 \\
-6.246 & 0.246 & 35.72 & 3.132
\end{bmatrix}
$$

$$
P = \begin{bmatrix}
242.01 & 67.636 & -184.86 & -1.752 & 0.2438 & 34.926 & 2.4852 \\
67.636 & 550.11 & 174.18 & 10.468 & 2.3336 & 298.87 & 22.014 \\
-184.86 & 174.18 & 239.27 & 6.0269 & 0.7733 & 96.017 & 7.1450 \\
-1.752 & 10.468 & 6.0269 & 15.106 & -0.377 & -65.25 & -5.824 \\
0.2438 & 2.3336 & 0.7733 & -0.377 & 0.3756 & 10.773 & 0.3392 \\
34.926 & 298.87 & 96.017 & -65.25 & 10.773 & 725.45 & 47.689 \\
2.4852 & 22.014 & 7.1450 & -5.824 & 0.3392 & 47.689 & 4.0135
\end{bmatrix}
$$

Example 10.4

The F-4 fighter aircraft analyzed in Section 3.2.1 is investigated in this example ($\varepsilon = 0.291349$). Note that the regularization has been used in Section 3.2.1, meaning the matrix A has been added small amount to diagonal elements. This regularization step is neither needed nor used while applying the eigenvector approach here. Below are shown the matrix L obtained by the Newton method and the unique global solution P associated with and calculated by the presented eigenvector method

$$
L = \begin{bmatrix}
-0.0212 & -0.2842 & -0.7320 & -4.9267 \\
0.0357 & 0.0183 & 1.3971 & 0.3321 \\
0.0033 & 0.0037 & -0.0067 & 0.0012 \\
0.0019 & 0.0508 & 0.0024 & 0.0309 \\
-0.0019 & -0.0489 & -0.0012 & -0.0008 \\
0.0034 & -0.0024 & -0.0005 & -0.0132 \\
0.0009 & 0.0248 & 0.0338 & 0.5111 \\
0.0000 & 0.0025 & 0.0015 & 0.0402
\end{bmatrix}
$$

$$
P = \begin{bmatrix}
0.8241 & 1.4557 & -2.6959 & 0.7042 & -0.0214 & 0.1016 \\
1.4557 & 7.6274 & -5.9707 & 1.0176 & -0.1507 & 0.1521 \\
-2.6959 & -5.9707 & 24.9889 & -2.2546 & 0.0818 & -0.3677 \\
0.7042 & 1.0176 & -2.2546 & 1.6154 & -0.0132 & 0.0874 \\
-0.0214 & -0.1507 & 0.0818 & -0.0132 & 0.0340 & -0.0019 \\
0.1016 & 0.1521 & -0.3677 & 0.0874 & -0.0019 & 0.0753
\end{bmatrix}
$$

The next example shows both the versatility and power of the proposed eigenvector approach for the decomposition of (not only weakly coupled) control systems. In Example 10.3, it has already been shown that the proposed algorithm can be used to obtain the solution of the optimal controller design of any standard linear control system, including the ones having $\varepsilon = 1$, and below we consider the application of the proposed eigenvector method to the linear quadratic optimal control problem for multiparameter systems (Mukaidani et al. 2003).

Example 10.5

In Mukaidani et al. (2003), a new method is developed to design a *near-optimal* controller for multiparameter singularly perturbed systems in which N lower-level fast subsystems are interconnected through a higher-level slow subsystem which does not depend on the unknown small parameters. One such system is shown below. Consider the optimal control problem given in Mukaidani et al. (2003) as

$$
\dot{x}_0(t) = \begin{bmatrix} 0 & 1 \\ -1 & -2 \end{bmatrix} x_0(t) + \begin{bmatrix} 0 \\ 2 \end{bmatrix} x_1(t) + \begin{bmatrix} 0 \\ 3 \end{bmatrix} x_2(t) + \sum_{j=1}^{2} \begin{bmatrix} 0 \\ 1 \end{bmatrix} u_j(t),
$$

$$
\varepsilon_1 \dot{x}_1(t) = [1 \; 0.2] x_0(t) + u_1(t),
$$
$$
\varepsilon_2 \dot{x}_2(t) = [1 \; 0.3] x_0(t) + u_2(t),
$$

with a performance index

$$
J = \frac{1}{2} \int_0^\infty \left(x_0^T(t) x_0(t) + 2 \sum_{j=1}^{2} \left\{ x_j^T(t) x_j(t) + u_j^T(t) u_j(t) \right\} \right) dt
$$

The proposed eigenvector algorithm will be applied to the (just slightly rearranged) system below, and by using $\varepsilon = 1$ and any (desired) values for ε_1 and ε_2,

$$\dot{v}_1(t) = \begin{bmatrix} 0 & 1 \\ -1 & -2 \end{bmatrix} v_1(t) + \varepsilon \begin{bmatrix} 0 & 0 \\ 2 & 3 \end{bmatrix} v_2(t) + \begin{bmatrix} 0 \\ 1 \end{bmatrix} u_1(t) + \varepsilon \begin{bmatrix} 0 \\ 1 \end{bmatrix} u_2(t),$$

$$\dot{v}_2(t) = \varepsilon \begin{bmatrix} \frac{1}{\varepsilon_1} & \frac{0.2}{\varepsilon_1} \\ \frac{1}{\varepsilon_2} & \frac{0.3}{\varepsilon_2} \end{bmatrix} v_1(t) + \begin{bmatrix} 0 & 0 \\ 0 & 0 \end{bmatrix} v_2(t) + \varepsilon \begin{bmatrix} \frac{1}{\varepsilon_1} \\ 0 \end{bmatrix} u_1(t) + \begin{bmatrix} 0 \\ \frac{1}{\varepsilon_2} \end{bmatrix} u_2(t),$$

$$\text{with } Q = \begin{bmatrix} 1 & 0 & 0 & 0 \\ 0 & 1 & 0 & 0 \\ 0 & 0 & 2 & 0 \\ 0 & 0 & 0 & 2 \end{bmatrix}, \quad R = \begin{bmatrix} 2 & 0 \\ 0 & 2 \end{bmatrix}.$$

where the vectors $v(t)$ have obvious meaning as given here,

$$v_1(t) = x_0(t), \quad v_2(t) = \begin{bmatrix} x_1(t) & x_2(t) \end{bmatrix}^\mathsf{T}.$$

The matrices L, P, and controller matrix K for $\varepsilon_1 = 0.1$ and $\varepsilon_2 = 0.1$, are shown below. Note that, unlike in Mukaidani et al. (2003), where the approximate, i.e., a *near optimal*, solution is given, the solution presented here is an *exact* solution up to the machine precision.

$$L = \begin{bmatrix} 0.0149 & 0.0238 & -0.0306 & -0.0180 \\ 0.0732 & 0.0479 & -0.8517 & -1.3380 \\ 0.1761 & 0.1705 & 0.0419 & 0.0351 \\ 0.0157 & 0.0265 & 0.1108 & 0.0815 \end{bmatrix}$$

$$P = \begin{bmatrix} 2.5907 & 0.6331 & 0.2253 & 0.2639 \\ 0.6331 & 0.3773 & 0.0703 & 0.1126 \\ 0.2253 & 0.0703 & 0.2058 & 0.0106 \\ 0.2639 & 0.1126 & 0.0106 & 0.2190 \end{bmatrix}$$

$$K = \begin{bmatrix} 1.4430 & 0.5403 & 1.0644 & 0.1094 \\ 1.6360 & 0.7516 & 0.0883 & 1.1514 \end{bmatrix}$$

When the values for ε_1 and ε_2 are $\varepsilon_1 = 0.01$ and $\varepsilon_2 = 0.5$, exact matrices L, P, and controller matrix K are shown below

$$L = \begin{bmatrix} 0.0001 & 0.4179 & -0.0053 & 0.1722 \\ 0.0095 & 0.5084 & -0.9752 & -1.5620 \\ 0.0200 & -0.5570 & -0.0120 & 1.7010 \\ 0.0038 & -0.3959 & 0.0075 & 0.5117 \end{bmatrix}$$

$$P = \begin{bmatrix} 2.6015 & 0.6803 & 0.0264 & 0.8061 \\ 0.6803 & 0.4210 & 0.0082 & 0.3772 \\ 0.0264 & 0.0082 & 0.0201 & 0.0037 \\ 0.8061 & 0.3772 & 0.0037 & 1.2223 \end{bmatrix}$$

$$K = \begin{bmatrix} 1.6616 & 0.6184 & 1.0081 & 0.3754 \\ 1.1462 & 0.5877 & 0.0078 & 1.4109 \end{bmatrix}$$

Example 10.6

Now, we study the real example (already shown as Case Study 3.1.3) representing the portion of the Serbian power grid in isolated operations composed of two hydropower plants. All the system matrices of the linear model, as well as all the other relevant parameters are shown previously, and for the sake of brevity they will not be repeated here. This example is chosen to show impeccable numerical properties of the eigenvector method introduced here. Namely, the multimachine power system has the specially structured matrix B with a consequence that the feedback control law affects only slightly some of the very small eigenvalues, and the system will remain almost marginally stable under feedback control. This problem has been facilitated by choosing a specially designed performance criterion in Chapter 3. In addition, and in order to apply the reduced-order algorithm proposed in Chapter 3, some elements of the matrix A must have been balanced. (See all the details as well as full description of the problem in Chapter 3.) The eigenvector decomposition method does not have any particular difficulties here, and it solves the controller design problem for any value of the weak coupling parameter ε.

Below, the feedback control matrices K for four values of ε (0.01, 0.1, 0.5, and 1) are shown.

$\varepsilon = 0.01$:

$$K = \begin{bmatrix} -0.0033 & -0.2060 & 0.0545 & -0.0437 & -0.0323 & -0.0042 & -0.0061 \\ 0.0002 & 2.1205 & 0.0284 & -0.8205 & -0.0141 & -0.0783 & -0.0013 \end{bmatrix}$$

$\varepsilon = 0.1$:

$$K = \begin{bmatrix} -0.0334 & -3.0154 & 0.5294 & 0.0042 & -0.3048 & -0.0045 & -0.2900 \\ 0.0002 & 7.7065 & 0.0926 & -2.5443 & -0.0043 & -2.3935 & -0.0014 \end{bmatrix}$$

$\varepsilon = 0.5$:

$$K = \begin{bmatrix} -0.151 & -11.522 & 2.757 & 0.198 & -1.113 & 0.800 & -5.165 \\ -0.002 & 6.999 & 0.203 & -1.839 & 0.046 & -8.598 & 0.247 \end{bmatrix}$$

$\varepsilon = 1$:

$$K = \begin{bmatrix} -0.240 & -16.884 & 6.046 & 0.104 & -1.522 & 0.688 & -13.771 \\ -0.004 & 5.749 & 0.396 & -1.482 & 0.016 & -13.792 & 0.212 \end{bmatrix}$$

10.6 CONCLUSION

In this chapter, we have used the eigenvector method to solve the order-reduction problem of linear weakly coupled systems. We have also shown how the decomposition task of the standard linear optimal control problem can be solved by using the same approach. In this way, calculation of the solution for higher order Riccati equation (Equation 10.3) can be replaced by solving the lower order ones. So far, all existing methods based on recursive methods for solving nonsquare Riccati equations have been able to solve exactly this problem up to the problem dependent "sufficiently small" parameter ε. The eigenvector approach shown should be used for decomposition of weakly coupled control systems in the cases when the decomposition methods based on series expansions, fixed point iterations, and Newton iterations fail to produce solutions of the corresponding algebraic equations. In addition, the eigenvector approach provides both new tools and novel insights into the nature of the decomposition problem and finds the solutions without solving the corresponding subsystem Riccati equations. It should be emphasized that the results obtained present the exact system decomposition and that the solution for the global algebraic Riccati equation is also exactly obtained (up to the computer's precision).

APPENDIX 10.1   JUSTIFICATION OF STEP 3 OF ALGORITHM 10.2

Step 3 of Algorithm 10.2 can be justified from the following facts. By applying the transformation (Equation 10.30) to the L-equation in Equation 1014, we can write

$$T_D^{-1} R_L T_D = \begin{bmatrix} B + DL & 0 \\ 0 & -(A + LD) \end{bmatrix} \qquad (10.49)$$

Since the similarity transformation preserves the eigenvalues, the subsystem eigenvectors satisfy

$$\begin{bmatrix} B + DL & 0 \\ 0 & -(A + LD) \end{bmatrix} \begin{bmatrix} M_{2D} & 0 \\ 0 & M_{1D} \end{bmatrix} = \begin{bmatrix} M_{2D} & 0 \\ 0 & M_{1D} \end{bmatrix} \begin{bmatrix} K_1 & 0 \\ 0 & K_2 \end{bmatrix}$$

or

$$\begin{bmatrix} M_{2D} & 0 \\ 0 & M_{1D} \end{bmatrix}^{-1} \begin{bmatrix} B + DL & 0 \\ 0 & -(A + LD) \end{bmatrix} \begin{bmatrix} M_{2D} & 0 \\ 0 & M_{1D} \end{bmatrix} = \begin{bmatrix} K_1 & 0 \\ 0 & K_2 \end{bmatrix}$$

Finally, from the last two equations it follows that

$$\begin{bmatrix} M_{2D} & 0 \\ 0 & M_{1D} \end{bmatrix}^{-1} T_D^{-1} R_L T_D \begin{bmatrix} M_{2D} & 0 \\ 0 & M_{1D} \end{bmatrix} = \begin{bmatrix} K_1 & 0 \\ 0 & K_2 \end{bmatrix} \qquad (10.50)$$

Also, it is known from Equation 10.20 that

$$[M_1 \ M_2]^{-1} R [M_1 \ M_2] = \begin{bmatrix} K_1 & 0 \\ 0 & K_2 \end{bmatrix} \qquad (10.51)$$

It follows from the last two formulas that

$$T_D \begin{bmatrix} M_{2D} & 0 \\ 0 & M_{1D} \end{bmatrix} = [M_1 \ M_2] \qquad (10.52)$$

which indicates that the subsystems eigenvectors can be obtained by using the same formula, that is

$$\begin{bmatrix} M_{2D} & 0 \\ 0 & M_{1D} \end{bmatrix} = T_D^{-1}[M_1 M_2] \tag{10.53}$$

APPENDIX 10.2 ON THE NUMBER OF SOLUTIONS TO NONSYMMETRIC ARE

In Example 10.1, we have given an expression for the number of solutions for a special case when there are less distinct real eigenvalues of the matrix R_L than the complex-conjugate ones, for an even n and when $(n \geq r \geq 0)$. This equation states that there will not be a unique solution of the general Riccati equation. Depending upon the configuration of the matrix R_L eigenvalues, there are few more possibilities. For an odd n and for $(n \geq r \geq 0)$ in the case that there are less distinct real eigenvalues of the matrix R_L than the complex-conjugate ones, the number of solutions matrices S equals

$$S = \sum_{i=1}^{\frac{r}{2}} \binom{r}{2i-1} \binom{\frac{2n-r}{2}}{\frac{n-(2i-1)}{2}} \tag{10.54}$$

If there are r $(2n \geq r \geq n)$, distinct real eigenvalues of the matrix R_L, the total number S of different (n, n) solution matrices to (10.14) equals

$$S = \sum_{i=0}^{\frac{2n-r}{2}} \binom{\frac{2n-r}{2}}{i} \binom{r}{n-2i} \tag{10.55}$$

In a more general setting, the matrix X in ARE (Equation 10.18) is a rectangular or nonsquare (m, n) matrix. This fact will not alter the eigenvector method (EM) presented above and the solutions X will be found by following the very same procedure. For $m \geq r \geq n$, and when there are r, $(m+n \geq r \geq m)$, distinct real eigenvalues of the matrix R associated with the ARE (10.18), the number of solutions S for an (m, n) matrix X equals

$$S = \sum_{i=0}^{\frac{m+n-r}{2}} \binom{\frac{m+n-r}{2}}{i} \binom{r}{n-2i} \tag{10.56}$$

REFERENCES

Bingulac, S. and H. VanLandingham, *Algorithms for Computer-Aided Design of Multivariable Control Systems*, Marcel Dekker, New York, 1993.

Bunse-Gerstner, A., R. Byers, and V. Mehrmann, A chart of numerical methods for structured eigenvalue problem, *SIAM Journal on Matrix Analysis and Applications*, 13, 1992: 419–453.

Bunse-Gerstner, A. and H. Fassbender, A Jacobi-like method for solving algebraic Riccati equations on parallel computers, *IEEE Transactions on Automatic Control*, AC-42, 1997: 1071–1094.

Clements, D. and B. Anderson, Polynomial factorization via the Riccati equation, *SIAM Journal on Applied Mathematics*, 31, 1976: 179–205.

Chow, J., *Time-Scale Modeling of Dynamic Networks with Applications to Power Systems*, Springer Verlag, New York, 1982.

Chow, J. and P. Kokotović, Sparsity and time scales, *Proceedings of the American Control Conference*, San Francisco, CA, 1983: 656–661.

Delebecque, F. and J. Quadrat, Optimal control of Markov-chains admitting strong and weak interactions, *Automatica*, 17, 1981: 281–296.

Gajić, Z. and X. Shen, Decoupling transformation of weakly coupled linear systems, *International Journal of Control*, 50, 1989: 1517–1523.

Gajić, Z. and X. Shen, *Parallel Algorithms for Optimal Control of Large Scale Linear Systems*, Springer Verlag, London, 1993.

Gajić, Z. and M. Qureshi, *Lyapunov Matrix Equation in Systems Stability and Control*, Academic Press, San Diego, 1995.

Fath, A., Computational aspects of the linear optimal regulator problem, *IEEE Transactions on Automatic Control*, AC-14, 1969: 547–550.

Gomathi, K., S.S. Prabhu, and M.A. Pai, A suboptimal controller for minimum sensitivity of closed loop eigenvalues to parameter variations, *IEEE Transactions on Automatic Control*, AC-25, 1980: 587–588.

Kokotović, P., W. Perkins, J. Cruz, and G. D'Ans, ε-coupling approach for near optimum design of large scale linear systems, *IEE Proceedings Part D: Control Theory and Applications*, 116, 1969: 889–892.

Kwakernaak, H. and R. Sivan, *Linear Optimal Control Systems*, Wiley, New York, 1972.

MacFarlane, A., An eigenvector solution to the optimal linear regulator problem, *Journal of Electronics and Control*, 14, 1963: 643–654.

McBrinn, D. and R. Roy, Stabilization of linear multivariable systems by output feedback, *IEEE Transactions on Automatic Control*, AC-17, 1972: 243–245.

Medanić, J., Geometric properties and invariant manifolds of the Riccati equation, *IEEE Transactions on Automatic Control*, AC-27, 1982: 670–677.

Mukaidani, H., H. Xu, and K. Mizukami, New results for near-optimal control of linear multiparameter singularly perturbed systems, *Automatica*, 39, 2003: 2157–2167.

Phillips, R. and P. Kokotović, A singular perturbation approach to modeling and control of Markov chains, *IEEE Transactions on Automatic Control*, AC-26, 1981: 1087–1094.

Porter, J., Matrix quadratic equations, *SIAM Journal on Applied Mathematics*, 14, 1966: 496–501.

Smith, D., Decoupling and order reduction via the Riccati transformation, *SIAM Review*, 29, 1987: 91–113.

Su, W. and Z. Gajić, Decomposition method for solving weakly coupled algebraic Riccati equation, *AIAA Journal of Guidance, Control, and Dynamics*, 15, 1992: 536–538.

Van Dooren, P., A generalized eigenvalue approach for solving Riccati equations, *SIAM Journal on Scientific Computing*, 2, 1981: 121–135.

Part III

---

*Bilinear Weakly Coupled
Control Systems*

11 Optimal Control of Bilinear Weakly Coupled Systems

11.1 INTRODUCTION

The major importance of bilinear systems indeed lies in their applications to the real world systems as demonstrated in some economic processes, ecology processes, socioeconomic processes, and many biological processes, such as the population dynamics of biological species, water balance, and temperature regulation in human body, control of carbon dioxide in lungs, blood pressure, immune system, cardiac regulator, etc. (Figalli et al. 1984; Mohler 1991). These bilinear systems are linear in control and linear in state but not jointly linear in state and control. It is important to understand their real properties or to guarantee the global stability or to improve the performance by applying the various control techniques to bilinear systems rather than their linearized systems since the linearization of bilinear systems loses its natural properties (Mohler 1991; Cebuhar and Costanza 1984; Hoffer and Tibken 1988; Aganović and Gajić 1993, 1995).

Many real physical systems are naturally weakly coupled such as power systems, communication satellites, helicopters, chemical reactors, electrical networks, flexible space structures, and mechanical systems in modal coordinates. The weakly coupled linear systems were introduced to the control audience by Kokotović et al. (1969). Since then many theoretical aspects for weakly coupled linear systems have been studied. These results lead to the reduction in the size of the required computations and allow parallel processing. Practically, the optimal control is obtained in the form of a feedback law, with the feedback gains calculated from two independent reduced-order optimal control problems (Gajić and Shen 1989, 1992). By using these results, the optimal control problems for weakly coupled bilinear systems have been studied (Aganović and Gajić 1993, 1995). The results of Aganović and Gajić (1995) are based on the idea of the recursive reduced-order scheme for solving the algebraic Riccati equation of weakly coupled systems (Gajić and Shen 1989, 1992) and the recursive scheme for the optimal control of general bilinear system (Cebuhar and Costanza 1984). However, the weak coupling theory has been studied so far mostly for the linear control scheme. By using the idea of the reduced-order scheme for solving the algebraic Riccati equation of weakly coupled systems, we extend the obtained results to a nonlinear optimal control scheme which solves the algorithm of two independent reduced-order Hamilton–Jacobi–Bellman (HJB) equations for a weakly coupled bilinear system (Kim and Lim 2006, 2007).

The solutions of HJB equations of bilinear and nonlinear systems can be hardly found (Kirk 1970), and thus we find approximate solutions using the successive Galerkin approximation (SGA) reported in Beard (1995), Beard et al. (1996), and Beard and McLain (1998). However, the SGA method has the difficulty that the complexity of computations increases according to the order of a system or a state variable. Specifically, for using the SGA method, we need N basis functions and must compute n-dimensional integrals, where n is the order of the system. Moreover, the number of those computations increases in proportion to $O(N^3)$. Therefore, we deal with two reduced-order HJB equations so that the optimal control is designed from the solutions of two independent reduced-order HJB equations using the SGA method. Then, n_1- and n_2-dimensional integrals are computed in parallel, and the number of computations is greatly decreased, where $n = n_1 + n_2$.

Recently, robust control is issued and developed by many researchers for linear systems (Doyle et al. 1989; Xie and Carlos 1991; Xie and Carlos 1992; Zhou et al. 1996). But in the class of bilinear and nonlinear systems, because conditions for the solvability of the robust H_∞ control design problem are hard, still there are a lot of problems to be developed. For bilinear and nonlinear systems with parameter uncertainties, the H_∞ optimal control problem can be reduced to the solution of the HJI equation, which is a nonlinear partial differential equation (PDE) van der Schaft (1992). The solution of a nonlinear PDE is extremely difficult to find so that researchers have looked for methods of obtaining its approximate solution. Specially, the practical method named SGA was proposed in Beard (1995) and Beard et al. (1996) to improve a stabilizing feedback control. The problem of a stabilizing H_∞ control can be reduced to solving a first-order linear PDE called the Generalized-Hamilton–Jacobi–Isaacs (GHJI) equation (Beard and McLain 1998). An interesting fact is that when the process is iterated, the solution to the GHJI equation converges uniformly to the solution of the HJI equation which solves the H_∞ optimal control problem (Beard and McLain 1998). Also, Beard (1995) shows how to find a uniform approximation such that the approximate controls are still stable on a specified set using SGA. However, the SGA method has the difficulty that the complexity of computations increases according to the order of the system. That is, to use the SGA method, we need N basis functions and must compute n-dimensional integrals, where n is the order of the system. Moreover, the number of those computations increases according to $O(N^3)$. Therefore, we deal with two reduced-order HJI equations in this chapter. Hence the robust H_∞ control law is designed from the solutions of two independent reduced-order HJI equations using the SGA method. Then, n_1- and n_2-dimensional integrals are computed in parallel, and the number of computations is greatly decreased, where $n = n_1 + n_2$. In addition, a duel successive algorithm (Algorithm 11.2) is proposed as a heuristic formulation, and it is a modification addressed in the successive approximation reported in Cebuhar and Costanza (1984), Hoffer and Tibken (1988), and Beard and McLain (1998). Since the GHJI equations are partial differential equations, we can hardly solve them. Therefore, we propose the alternative method (Algorithm 11.3) using Galerkin's approximation. In Algorithm 11.3, only linear equations remain to be solved.

In this chapter we will study an algorithm for the closed-loop parallel optimal control of weakly coupled bilinear systems with respect to certain performance criteria using the SGA. By using weak coupling theory, the optimal control can be obtained from two reduced-order optimal control problems in parallel. Two optimal control laws are constructed in terms of the approximate solution of two independent HJB equations using SGA. The characteristic of the algorithm is the design of the closed-loop optimal control law for weakly coupled bilinear systems using the SGA method and reduces the computational complexity when the SGA method is applied to the higher order systems. Along the same line, we present a new algorithm for the closed-loop H_∞ composite control of weakly coupled bilinear systems with time-varying parameter uncertainties and exogenous disturbance using the SGA. It is shown that by using weak coupling theory, the robust H_∞ control can be obtained from two reduced-order robust H_∞ control problems in parallel. The H_∞ control theory guarantees the robust closed-loop performance but the resulting problem is difficult to solve for uncertain bilinear systems. In order to overcome the difficulties inherented in the H_∞ control problem, two H_∞ control laws are constructed in terms of the approximate solution of two independent HJI equations using the SGA method. A real physical bilinear model of a paper making machine is solved to demonstrate the efficiency of the proposed parallel algorithms.

11.2 OPTIMAL CONTROL FOR WEAKLY COUPLED BILINEAR SYSTEMS USING SGA

In this section, we will study an algorithm for the closed-loop parallel optimal control of weakly coupled bilinear systems with respect to certain performance criteria using the SGA. By using weak coupling theory, the optimal control can be obtained from two reduced-order optimal control problems in parallel.

11.2.1 PROBLEM FORMULATION

Consider the weakly coupled bilinear system represented by

$$
\begin{bmatrix} \dot{x}_1(t) \\ \dot{x}_2(t) \end{bmatrix} = \begin{bmatrix} A_1 & \varepsilon A_2 \\ \varepsilon A_3 & A_4 \end{bmatrix} \begin{bmatrix} x_1(t) \\ x_2(t) \end{bmatrix} + \begin{bmatrix} B_1 & \varepsilon B_2 \\ \varepsilon B_3 & B_4 \end{bmatrix} \begin{bmatrix} u_1(t) \\ u_2(t) \end{bmatrix}
$$
$$
+ \left\{ \begin{bmatrix} x_1(t) \\ x_2(t) \end{bmatrix} \begin{bmatrix} M_a & \varepsilon M_b \\ \varepsilon M_c & M_d \end{bmatrix} \right\} \begin{bmatrix} u_1(t) \\ u_2(t) \end{bmatrix} \tag{11.1}
$$

with an initial condition:

$$
\begin{bmatrix} x_1(t_0) \\ x_2(t_0) \end{bmatrix} = \begin{bmatrix} x_1^0 \\ x_2^0 \end{bmatrix}
$$

where $x_1(t) \in R^{n_1}$, $x_2(t) \in R^{n_2}$, $u_1(t) \in R^{m_1}$, $u_2(t) \in R^{m_2}$, and ε is a small coupling parameter, with

$$\left\{ \begin{bmatrix} x_1(t) \\ x_2(t) \end{bmatrix} \begin{bmatrix} M_a & \varepsilon M_b \\ \varepsilon M_c & M_d \end{bmatrix} \right\} = \sum_{i=1}^{n_1} x_{1i}(t) \begin{bmatrix} M_{ai} & \varepsilon M_{bi} \\ \varepsilon M_{ci} & 0 \end{bmatrix}$$

$$+ \sum_{j=n_1+1}^{n_1+n_2} x_{2(j-n_1)}(t) \begin{bmatrix} 0 & \varepsilon M_{bj} \\ \varepsilon M_{cj} & M_{dj} \end{bmatrix} \tag{11.2}$$

where $M_{ai} \in R^{n_1 \times m_1}$, $M_{bi} \in R^{n_1 \times m_2}$, $M_{ci} \in R^{n_2 \times m_1}$, $M_{di} \in R^{n_2 \times m_2}$. Moreover, $x(t) = \begin{bmatrix} x_1^T(t) & x_2^T(t) \end{bmatrix}^T$ is a state variable and $u(t) = \begin{bmatrix} u_1^T(t) & u_2^T(t) \end{bmatrix}^T$ is a control input.

A quadratic cost functional associated with Equation 11.1 to be minimized has the following form

$$J = \frac{1}{2} \int_0^\infty \left(x^T(t)Qx(t) + u^T(t)Ru(t) \right) dt \tag{11.3}$$

with $Q \geq 0$, $R > 0$ possessing the weakly coupling structures, that is

$$Q = \begin{bmatrix} Q_1 & \varepsilon Q_2 \\ \varepsilon Q_2^T & Q_3 \end{bmatrix}, \quad R = \begin{bmatrix} R_1 & 0 \\ 0 & R_2 \end{bmatrix} \tag{11.4}$$

For computational simplification, we introduce the following notation:

$$\tilde{B}(x(t)) = \begin{bmatrix} B_1 & \varepsilon B_2 \\ \varepsilon B_3 & B_4 \end{bmatrix} + \left\{ \begin{bmatrix} x_1(t) \\ x_2(t) \end{bmatrix} \begin{bmatrix} M_a & \varepsilon M_b \\ \varepsilon M_c & M_d \end{bmatrix} \right\}$$

$$= \begin{bmatrix} \tilde{B}_1(x_1(t)) & \varepsilon \tilde{B}_2(x(t)) \\ \varepsilon \tilde{B}_3(x(t)) & \tilde{B}_4(x_2(t)) \end{bmatrix} \tag{11.5}$$

Setting $\varepsilon^2 = 0$, we can get the following $O(\varepsilon^2)$ approximation:

$$S(x(t)) = \tilde{B}(x(t))R^{-1}\tilde{B}(x(t))^T = \begin{bmatrix} S_1(x_1(t)) & \varepsilon S_2(x(t)) \\ \varepsilon S_2^T(x(t)) & S_3(x_2(t)) \end{bmatrix} \tag{11.6}$$

From the result of Kirk (1970), we can derive the following state-dependent Riccati equation for the weakly coupled bilinear system Equation 11.1 with respect to the performance criterion Equation 11.3.

$$PA + A^TP - PS(x)P + Q = 0 \tag{11.7}$$

Moreover, optimal control law is given by

$$u^*(t) = -R^{-1}\tilde{B}(x(t))^TPx(t) \tag{11.8}$$

where P is partitioned as

$$P = \begin{bmatrix} P_1 & \varepsilon P_2 \\ \varepsilon P_2^T & P_3 \end{bmatrix} \tag{11.9}$$

Partitioning the state-dependent Riccati equation (Equation 11.7) according to Equations 11.6 and 11.9, and setting $\varepsilon^2 = 0$, we get an $O(\varepsilon^2)$ approximation of Equation 11.7 in terms of two reduced-order, decoupled Riccati equations:

$$P_1 A_1 + A_1^T P_1 + Q_1 - P_1 S_1(x_1(t)) P_1 = 0 \tag{11.10}$$

$$P_3 A_4 + A_4^T P_3 + Q_3 - P_3 S_3(x_2(t)) P_3 = 0 \tag{11.11}$$

and nonsymmetric Riccati equation with no input:

$$\{A_1 - S_1(x_1(t)) P_1\}^T P_2 + P_2 \{A_4 - S_3(x_2(t)) P_3\}$$
$$+ P_1 A_2 + A_3^T P_3 - P_1 S_2(x(t)) P_3 + Q_2 = 0 \tag{11.12}$$

A detailed description of a reduced-order scheme can be found in Aganović and Gajić (1995). Since Equations 11.10 through 11.12 are state-dependent Riccati equations, they have no analytical solutions.

Focusing on the nonlinear optimal control, we deal with HJB equations rather than Riccati equations. HJB equations corresponding to Equations 11.10 and 11.11 are given by

$$\frac{\partial J_1^T}{\partial x_1} A_1 x_1(t) + \frac{1}{2} x_1^T(t) Q_1 x_1(t) - \frac{1}{2} \frac{\partial J_1^T}{\partial x_1} S_1(x_1(t)) \frac{\partial J_1}{\partial x_1} = 0 \tag{11.13}$$

$$\frac{\partial J_2^T}{\partial x_2} A_4 x_2(t) + \frac{1}{2} x_2^T(t) Q_3 x_2(t) - \frac{1}{2} \frac{\partial J_2^T}{\partial x_2} S_3(x_2(t)) \frac{\partial J_2}{\partial x_2} = 0 \tag{11.14}$$

where
$$\partial J_1 / \partial x_1 = P_1 x_1(t)$$
$$\partial J_2 / \partial x_2 = P_3 x_2(t)$$

Moreover, denoting $\partial J_3 / \partial x_1 = P_2 x_2(t)$ and $\{\partial J_3 / \partial x_2\}^T = x_1^T(t) P_2$, we obtain the following equation equivalent to Equation 11.12 after substitutions:

$$x_1^T(t) \{A_1 - S_1(x_1(t)) P_1\}^T \frac{\partial J_3}{\partial x_1} + \frac{\partial J_3^T}{\partial x_2} \{A_4 - S_3(x_2(t)) P_3\} x_2(t)$$
$$x_1^T(t) \{P_1 A_2 + A_3^T P_3 - P_1 S_2(x(t)) P_3 + Q_2\} x_2(t) = 0 \tag{11.15}$$

Unfortunately, they still have no analytical solutions. However, we can obtain approximate solutions of Equations 11.13 and 11.14 using the SGA. If the solutions of Equations 11.13 and 11.14 are found, then the solution of Equation 11.15 can easily be found using the Galerkin approximation.

11.2.2 Design of Optimal Control Law for Weakly Coupled Bilinear Systems Using SGA

In order to design the optimal control law $u^*(t)$, we present the scheme to find solutions of Equations 11.13 through 11.15 using the SGA method.

Assumption 11.1 Ω_1 and Ω_2 are compact sets of R^{n_1} and R^{n_2}, respectively. The state $x_1(t)$ and $x_2(t)$ are bounded on Ω_1 and Ω_2, respectively.

Under the above Assumption 11.1, we define the GHJB equations for weakly coupled bilinear systems.

DEFINITION 11.1

If initial control laws, $\tilde{u}_1^{(0)}(t): R^{m_1} \times \Omega_1 \to R$, $\tilde{u}_2^{(0)}(t): R^{m_2} \times \Omega_2 \to R$, are admissible and functions, $J_1^{(i)}: R \times \Omega_1 \to R$, $J_2^{(i)}: R \times \Omega_2 \to R$, satisfy the following GHJB equations, written by $\mathrm{GHJB}\left(J_1^{(i)}, \tilde{u}_1^{(i)}\right) = 0$, namely

$$
\frac{\partial J_1^{(i)\mathrm{T}}}{\partial x_1} A_1 x_1(t) + \frac{1}{2} x_1^\mathrm{T}(t) Q_1 x_1(t) + \frac{1}{2} \frac{\partial J_1^{(i-1)\mathrm{T}}}{\partial x_1} S_1(x_1(t)) \frac{\partial J_1^{(i-1)}}{\partial x_1}
$$
$$
- \frac{\partial J_1^{(i)\mathrm{T}}}{\partial x_1} S_1(x_1(t)) \frac{\partial J_1^{(i-1)}}{\partial x_1} = 0 \tag{11.16}
$$

then ith subcontrol law is given by

$$
\tilde{u}_1^{(i)}(t) = -R_1^{-1} \tilde{B}_1^\mathrm{T}(x_1(t)) \frac{\partial J_1^{(i)}}{\partial x_1} \tag{11.17}
$$

and $\mathrm{GHJB}\left(J_2^{(i)}, \tilde{u}_2^{(i)}\right) = 0$, namely

$$
\frac{\partial J_2^{(i)\mathrm{T}}}{\partial x_2} A_4 x_2(t) + \frac{1}{2} x_2^\mathrm{T}(t) Q_3 x_2(t) + \frac{1}{2} \frac{\partial J_2^{(i-1)\mathrm{T}}}{\partial x_2} S_3(x_2(t)) \frac{\partial J_2^{(i-1)}}{\partial x_2}
$$
$$
- \frac{\partial J_2^{(i)}}{\partial x_2} S_3(x_2(t)) \frac{\partial J_2^{(i-1)}}{\partial x_2} = 0 \tag{11.18}
$$

then ith subcontrol law is given by

$$
\tilde{u}_2^{(i)}(t) = -R_2^{-1} \tilde{B}_4^\mathrm{T}(x_2(t)) \frac{\partial J_2^{(i)}}{\partial x_2} \tag{11.19}
$$

where i is the iteration number.

Using Galerkin's projection method, we seek an approximate solution, $J_N^{(i)}$ to the equation GHJB $(J^{(i)}, u^{(i)}(t)) = 0$ in the compact set Ω, by letting

$$J_N^{(i)}(x(t)) = \sum_{j=1}^{N} c_j^{(i)} \phi_j(x(t)) \tag{11.20}$$

Substituting this expression into the GHJB equation results in an approximation error:

$$\text{Error} = \text{GHJB}\left(\sum_{j=1}^{N} c_j^{(i)} \phi_j, \, u^{(i)}(t)\right) \tag{11.21}$$

The coefficients c_j are determined by setting the projection of the error, Equation 11.21, on the finite basis, $\{\phi_j\}_1^N$, to zero $\forall x(t) \in \Omega$:

$$\left\langle \text{GHJB}\left(\sum_{j=1}^{N} c_j^{(i)} \phi_j, u^{(i)}(t)\right), \phi_n \right\rangle_\Omega = 0, \quad n = 1, \ldots, N \tag{11.22}$$

which are N equations in N unknowns. We define

$$\Phi_N(x(t)) \equiv (\phi_1(x(t)), \ldots, \phi_N(x(t)))^{\mathrm{T}} \tag{11.23}$$

and let $\nabla \Phi_N$ be the Jacobian of Φ_N. If $\eta \colon R^N \to R^N$ is a vector-valued function, then we introduce the notation:

$$\langle \eta, \Phi_N \rangle_\Omega \equiv \begin{bmatrix} \langle \eta_1, \phi_1 \rangle_\Omega & \cdots & \langle \eta_N, \phi_1 \rangle_\Omega \\ \vdots & \ddots & \vdots \\ \langle \eta_1, \phi_N \rangle_\Omega & \cdots & \langle \eta_N, \phi_N \rangle_\Omega \end{bmatrix} \tag{11.24}$$

where the inner product is defined as

$$\langle f, g \rangle_\Omega \equiv \int_\Omega f(x)g(x)\mathrm{d}x \tag{11.25}$$

and

$$J_N \equiv \mathbf{c}_N^{\mathrm{T}} \Phi_N \tag{11.26}$$

where

$$\mathbf{c}_N(x(t)) \equiv (c_1, \ldots, c_N)^{\mathrm{T}} \tag{11.27}$$

Given an initial control $\tilde{u}_1^{(0)}(t)$, we compute an approximation to its cost $J_{1N_1}^{(0)} = \mathbf{c}_{1N_1}^{T(0)}\Phi_{1N_1}$ where $\mathbf{c}_{1N_1}^{(0)}$ is the solution of Galerkin approximation of GHJB equation (Equation 11.22), i.e.

$$a_1^{(0)}\mathbf{c}_{1N_1}^{(0)} + b_1^{(0)} = 0 \tag{11.28}$$

where

$$a_1^{(0)} = \left\langle \nabla_1\Phi_{1N_1}A_1x_1(t),\, \Phi_{1N_1}\right\rangle_{\Omega_1} + \left\langle \nabla_1\Phi_{1N_1}\tilde{B}_1(x_1(t))\tilde{u}_1^{(0)}(t),\, \Phi_{1N_1}\right\rangle_{\Omega_1}$$

$$b_1^{(0)} = \frac{1}{2}\left\langle x_1^T(t)Q_1x_1(t),\, \Phi_{1N_1}\right\rangle_{\Omega_1} + \frac{1}{2}\left\langle \tilde{u}_1^{(0)^T}(t)R_1\tilde{u}_1^{(0)}(t),\, \Phi_{1N_1}\right\rangle_{\Omega_1}$$

We can compute the updated control law that is based on the approximated solution, $J_{1N_1}^{(i-1)}$:

$$\begin{aligned} \tilde{u}_{1N_1}^{(i)}(t) &= -R_1^{-1}\tilde{B}_1^T(x_1(t))\frac{\partial J_{1N_1}^{(i-1)}}{\partial x_1} \\ &= -R_1^{-1}\tilde{B}_1^T(x_1(t))\nabla_1\Phi_{1N_1}^T\mathbf{c}_{1N_1}^{(i-1)} \end{aligned} \tag{11.29}$$

Then we can obtain the approximation:

$$J_{1N_1}^{(i)} = \mathbf{c}_{1N_1}^{T(i)}\Phi_{1N_1} \tag{11.30}$$

where $\mathbf{c}_{1N_1}^{(i)}$ is the solution to

$$a_1^{(i)}\mathbf{c}_{1N_1}^{(i)} + b_1^{(i)} = 0 \tag{11.31}$$

where

$$\begin{aligned} a_1^{(i)} &= \left\langle \nabla_1\Phi_{1N_1}A_1x_1(t),\, \Phi_{1N_1}\right\rangle_{\Omega_1} \\ &\quad - \left\langle \nabla_1\Phi_{1N_1}S_1(x_1(t))\nabla_1\Phi_{1N_1}^T\mathbf{c}_{1N_1}^{(i-1)},\, \Phi_{1N_1}\right\rangle_{\Omega_1} \end{aligned}$$

$$\begin{aligned} b_1^{(i)} &= \frac{1}{2}\left\langle x_1^T(t)Q_1x_1(t),\, \Phi_{1N_1}\right\rangle_{\Omega_1} \\ &\quad + \frac{1}{2}\left\langle \mathbf{c}_{1N_1}^{(i-1)^T}\nabla_1\Phi_{1N_1}S_1(x_1(t))\nabla_1\Phi_{1N_1}^T\mathbf{c}_{1N_1}^{(i-1)},\, \Phi_{1N_1}\right\rangle_{\Omega_1} \end{aligned}$$

and i is the iteration number.

Similarly, given an initial control $u_2^{(0)}(t)$, we can compute an approximation to its cost $J_{2N_2}^{(0)} = \mathbf{c}_{2N_2}^{T(0)}\Phi_{2N_2}$ where $\mathbf{c}_{2N_2}^{(0)}$ is the solution of Galerkin approximation of GHJB equation (Equation 11.18).

The following theorem shows the existence of an unique solution of SGA.

THEOREM 11.1

Suppose that $\{\phi_j\}_1^N$ is linearly independent and $\partial\phi_j/\partial x \neq 0$, then there exists a unique solution, c_N.

Proof Suppose that $\{\phi_j\}_1^N$ are linearly independent, then Φ_N are linearly independent. Suppose $\partial\phi_j/\partial x \neq 0$, such that $\nabla\Phi_N \neq 0$, then linearly independent Φ_N implies that $\nabla\Phi_N$ is linearly independent. This implies that $\langle\nabla\Phi_N Ax(t), \Phi_N\rangle_\Omega - \langle\nabla\Phi_N S(x(t))\nabla\Phi_N^T c_N, \Phi_N\rangle_\Omega$ is invertible. This implies that $a_1^{(i)}$ is invertible in Equation 11.31 for all i. Therefore, there exists a unique solution to a linear equation (Equation 11.31). ∎

From the solutions of Galerkin approximations of GHJB equations (Equations 11.16 and 11.18), P_1 and P_3 can be determined. Then, we can obtain the approximate solution of Equation 11.15.

Defining $J_{3N_3} = c_{3N_3}^T \Phi_{N_3}$, we denote that $\partial J_3/\partial x_1 = \nabla_1 \Phi_{3N_3}^T c_{3N_3}$ and $\partial J_3/\partial x_2 = \nabla_2\Phi_{3N_3}^T c_{3N_3}$. Using these notations, we can derive the Galerkin approximation of Equation 11.15 as follows:

$$a_3 c_{3N_3} + b_3 = 0 \tag{11.32}$$

where

$$a_3 = \langle\nabla_1\Phi_{3N}\{A_1 - S_1(x_1(t))P_1\}x_1(t), \Phi_{3N}\rangle_{\Omega_3}$$
$$+ \langle\nabla_2\Phi_{3N}\{A_4 - S_3(x_2(t))P_3\}x_2(t), \Phi_{3N}\rangle_{\Omega_3}$$
$$b_3 = \langle x_1^T(t)\{P_1A_2 + A_3^T P_3 - P_1 S_2(x(t))P_3 + Q_2\}x_2(t), \Phi_{3N}\rangle_{\Omega_3}$$

In this case, $\Omega_3 = \Omega_1 \cup \Omega_2$, and P_3 can be determined without an iterative step.

Now, we present a new algorithm to design an optimal control law with two independent reduced-order HJB equations (Equations 11.13 and 11.14) and an Equation 11.15 using the SGA method for weakly coupled bilinear systems.

ALGORITHM 11.1

Initial Step
Compute

$$a_1^{(0)} = \langle\nabla_1\Phi_{1N_1}A_1x_1(t), \Phi_{1N_1}\rangle_{\Omega_1} + \langle\nabla_1\Phi_{1N_1}\tilde{B}_1(x_1(t))\tilde{u}_1^{(0)}(t), \Phi_{1N_1}\rangle_{\Omega_1}$$
$$b_1^{(0)} = \frac{1}{2}\langle x_1^T(t)Q_1x_1(t), \Phi_{1N_1}\rangle_{\Omega_1} + \frac{1}{2}\langle\tilde{u}_1^{T(0)}(t)R_1\tilde{u}_1^{(0)}(t), \Phi_{1N_1}\rangle_{\Omega_1}$$

and

$$a_2^{(0)} = \left\langle \nabla_2 \Phi_{2N_2} A_4 x_2(t),\ \Phi_{2N_2} \right\rangle_{\Omega_2} + \left\langle \nabla_2 \Phi_{2N_2} \tilde{B}_3(x_2(t)) \tilde{u}_2^{(0)}(t),\ \Phi_{2N_2} \right\rangle_{\Omega_2}$$

$$b_2^{(0)} = \frac{1}{2} \left\langle x_2^T(t) Q_3 x_2(t),\ \Phi_{2N_2} \right\rangle_{\Omega_2} + \frac{1}{2} \left\langle \tilde{u}_2^{T(0)}(t) R_2 \tilde{u}_2^{(0)}(t),\ \Phi_{2N_2} \right\rangle_{\Omega_2}$$

Find $\mathbf{c}_{1N_1}^{(0)}$ and $\mathbf{c}_{2N_2}^{(0)}$ satisfying the following linear equations:

$$a_1^{(0)} \mathbf{c}_{1N_1}^{(0)} + b_1^{(0)} = 0$$
$$a_2^{(0)} \mathbf{c}_{2N_2}^{(0)} + b_2^{(0)} = 0$$

Set $i = 1$.

Iterative Step

Improved controllers are given by

$$\tilde{u}_{1N_1}^{(i)}(t) = -R_1^{-1} \tilde{B}_1^T(x_1(t)) \nabla_1 \Phi_{1N_1}^T \mathbf{c}_{1N_1}^{(i-1)}$$
$$\tilde{u}_{2N_2}^{(i)}(t) = -R_2^{-1} \tilde{B}_3^T(x_2(t)) \nabla_2 \Phi_{2N_2}^T \mathbf{c}_{2N_2}^{(i-1)}$$

Compute

$$a_1^{(i)} = \left\langle \nabla_1 \Phi_{1N_1} A_1 x_1(t),\ \Phi_{1N_1} \right\rangle_{\Omega_1}$$
$$\quad - \left\langle \nabla_1 \Phi_{1N_1} S_1(x_1(t)) \nabla_1 \Phi_{1N_1}^T \mathbf{c}_{1N_1}^{(i-1)},\ \Phi_{1N_1} \right\rangle_{\Omega_1}$$

$$b_1^{(i)} = \frac{1}{2} \left\langle x_1^T(t) Q_1 x_1(t),\ \Phi_{1N_1} \right\rangle_{\Omega_1}$$
$$\quad + \frac{1}{2} \left\langle \mathbf{c}_{1N_1}^{T(i-1)} \nabla_1 \Phi_{1N_1} S_1(x_1(t)) \nabla_1 \Phi_{1N_1}^T \mathbf{c}_{1N_1}^{(i-1)},\ \Phi_{1N_1} \right\rangle_{\Omega_1}$$

and

$$a_2^{(i)} = \left\langle \nabla_2 \Phi_{2N_2} A_4 x_2(t),\ \Phi_{2N_2} \right\rangle_{\Omega_2}$$
$$\quad - \left\langle \nabla_2 \Phi_{2N_2} S_3(x_2(t)) \nabla_2 \Phi_{2N_2}^T \mathbf{c}_{2N_2}^{(i-1)},\ \Phi_{2N_2} \right\rangle_{\Omega_2}$$

$$b_2^{(i)} = \frac{1}{2} \left\langle x_2^T(t) Q_3 x_2(t),\ \Phi_{2N_2} \right\rangle_{\Omega_2}$$
$$\quad + \frac{1}{2} \left\langle \mathbf{c}_{2N_2}^{T(i-1)} \nabla_2 \Phi_{2N_2} S_3(x_2(t)) \nabla_2 \Phi_{2N_2}^T \mathbf{c}_{2N_2}^{(i-1)},\ \Phi_{2N_2} \right\rangle_{\Omega_2}$$

Find $\mathbf{c}_{1N_1}^{(i)}$ and $\mathbf{c}_{2N_2}^{(i)}$ satisfying the following linear equations:

$$a_1^{(i)} \mathbf{c}_{1N_1}^{(i)} + b_1^{(i)} = 0$$
$$a_2^{(i)} \mathbf{c}_{2N_2}^{(i)} + b_2^{(i)} = 0$$

Set $i = i + 1$.

Final Step
Determine P_1 and P_3, and compute

$$a_3 = \left\langle \nabla_1 \Phi_{3N} \{A_1 - S_1(x_1(t))P_1\}x_1(t),\ \Phi_{3N} \right\rangle_{\Omega_3}$$
$$+ \left\langle \nabla_2 \Phi_{3N} \{A_4 - S_3(x_2(t))P_3\}x_2(t),\ \Phi_{3N} \right\rangle_{\Omega_3}$$
$$b_3 = \left\langle x_1^T(t)\{P_1 A_2 + A_3^T P_3 - P_1 S_2(x(t))P_3 + Q_2\}x_2(t),\ \Phi_{3N} \right\rangle_{\Omega_3}$$

Find $\mathbf{c}_{3N_3}$ satisfying the following linear equations:

$$a_3 \mathbf{c}_{3N_3} + b_3 = 0$$

Determine P_2, then the approximate parallel optimal control law is given by

$$u_{pN}(t) = -R^{-1}\tilde{B}(x(t))^T \begin{bmatrix} P_1 & \varepsilon P_2 \\ \varepsilon P_2^T & P_3 \end{bmatrix} x(t)$$

The following theorem shows that the approximate parallel optimal control law, $u_{pN}(t)$, designed by the proposed algorithm converges to the optimal control law, $u^*(t)$.

THEOREM 11.2

For any small positive constant α, we can choose N for a sufficiently large i to satisfy

$$\left\| u^*(t) - u_{pN}^{(i)}(t) \right\| < \alpha \tag{11.33}$$

Proof It was proved that $u^*(t)$ converges to $u_N(t)$ pointwise on Ω for finite N in Beard (1995), where $u_N(t)$ is a control law designed using the SGA. It implies that for a sufficiently large i, we can choose N satisfying $\left\| u_p(t) - u_{pN}^{(i)}(t) \right\| < \tilde{\alpha}$, where $u_p(t)$ is the parallel control law obtained by the reduced-order scheme for weakly coupled bilinear systems and $\tilde{\alpha}$ is a small positive constant. By the help of weakly coupling theory, $u_p(t) = u^*(t) + O(\varepsilon^2)$. This implies that for any small positive constant α, we can choose N for a sufficiently large i satisfying Equation 11.33. ∎

11.2.3 CASE STUDY: A PAPER MAKING MACHINE

In order to demonstrate the efficiency of the proposed method for the parallel optimal control of weakly coupled bilinear systems using the SGA, we have run a fourth-order real example, a paper making machine control problem, done in Ying et al. (1992).

The problem matrices have the following values:

$$A = \begin{bmatrix} -1.93 & 0 & 0 & 0 \\ 0.394 & -0.426 & 0 & 0 \\ 0 & 0 & -0.63 & 0 \\ 0.095 & -0.103 & 0.413 & -0.426 \end{bmatrix},$$

$$B = \begin{bmatrix} 1.274 & 1.274 \\ 0 & 0 \\ 1.34 & -0.65 \\ 0 & 0 \end{bmatrix}, \quad M_1 = \begin{bmatrix} 0 & 0 \\ 0 & 0 \\ 0.755 & 0.366 \\ 0 & 0 \end{bmatrix},$$

$$M_2 = M_4 = \begin{bmatrix} 0 & 0 \\ 0 & 0 \\ 0 & 0 \\ 0 & 0 \end{bmatrix}, \quad M_3 = \begin{bmatrix} 0 & 0 \\ 0 & 0 \\ -0.718 & -0.718 \\ 0 & 0 \end{bmatrix},$$

$$Q = \begin{bmatrix} 1 & 0 & 0.13 & 0 \\ 0 & 1 & 0 & 0.09 \\ 0.13 & 0 & 0.1 & 0 \\ 0 & 0.09 & 0 & 0.2 \end{bmatrix}, \quad R = \begin{bmatrix} 1 & 0 \\ 0 & 1 \end{bmatrix}.$$

Initial states are chosen as $x(t_0) = [3.7 \ 3.2 \ 4 \ 2.8]^T$. The simulation results are presented in the Figures 11.1 through 11.5, where the dashed lines are the trajectories that are obtained from full-order SGA method and the solid lines are the

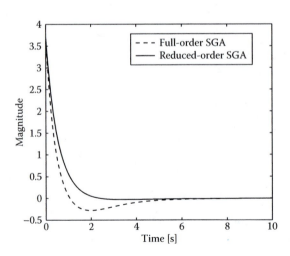

FIGURE 11.1 Trajectories of $x_1(t)$.

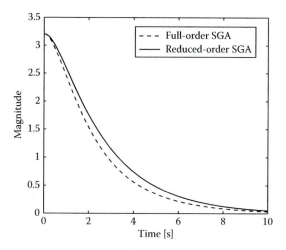

FIGURE 11.2 Trajectories of $x_2(t)$.

trajectories that are obtained from the proposed algorithm. Figure 11.5 shows that the performance criterion trajectory of the proposed algorithm is better than that of the full-order SGA method, because errors of the full-order SGA method are bigger than those of the proposed algorithm. In the full-order SGA method, eight-dimensional basis is used and four-dimensional integrals of $8 \times (1 + 8 + 64) = 584$ times are performed. But, in the proposed algorithm, we can use only three-dimensional basis and compute two-dimensional integrals of $3 \times (1 + 3 + 9) = 39$ times for each reduced-order problem in parallel, and compute four-dimensional integrals of $8 \times (1 + 8) = 72$ times based on eight-dimensional basis for the problem according to Equation 11.15. Therefore, the computational complexity is greatly reduced.

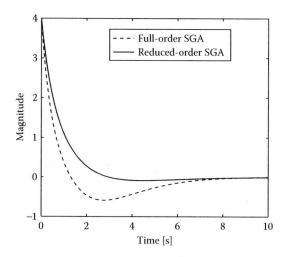

FIGURE 11.3 Trajectories of $x_3(t)$.

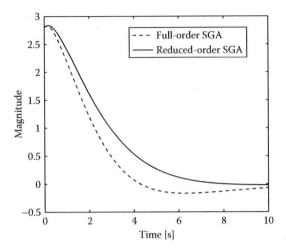

FIGURE 11.4 Trajectories of $x_4(t)$.

11.3 ROBUST H_∞ CONTROL FOR WEAKLY COUPLED BILINEAR SYSTEMS WITH PARAMETER UNCERTAINTIES USING SGA

In this section, we present a new algorithm for the closed-loop H_∞ composite control of weakly coupled bilinear systems with time-varying parameter uncertainties and exogenous disturbance using the SGA. It is shown that by using weak coupling theory, the robust H_∞ control can be obtained from two reduced-order robust H_∞ control problems in parallel.

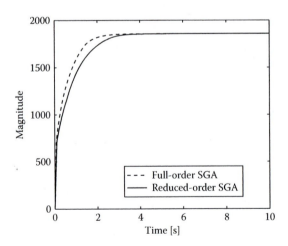

FIGURE 11.5 Values of the performance criterion.

11.3.1 PROBLEM FORMULATION

The weakly coupled bilinear system with time-varying parameter uncertainties and exogenous disturbance under consideration is represented by

$$
\begin{bmatrix} \dot{x}_1(t) \\ \dot{x}_2(t) \end{bmatrix} = \left\{ \begin{bmatrix} A_1 & \varepsilon A_2 \\ \varepsilon A_3 & A_4 \end{bmatrix} + \begin{bmatrix} \Delta A_1 & \varepsilon \Delta A_2 \\ \varepsilon \Delta A_3 & \Delta A_4 \end{bmatrix} \right\} \begin{bmatrix} x_1(t) \\ x_2(t) \end{bmatrix}
$$
$$
+ \begin{bmatrix} H_1 & \varepsilon H_2 \\ \varepsilon H_3 & H_4 \end{bmatrix} \begin{bmatrix} \omega_1(t) \\ \omega_2(t) \end{bmatrix} + \begin{bmatrix} B_1 & \varepsilon B_2 \\ \varepsilon B_3 & B_4 \end{bmatrix} \begin{bmatrix} u_1(t) \\ u_2(t) \end{bmatrix}
$$
$$
+ \left\{ \begin{bmatrix} x_1(t) \\ x_2(t) \end{bmatrix} \begin{bmatrix} M_a & \varepsilon M_b \\ \varepsilon M_c & M_d \end{bmatrix} \right\} \begin{bmatrix} u_1(t) \\ u_2(t) \end{bmatrix} \tag{11.34}
$$

$$
z(t) = \begin{bmatrix} Cx(t) \\ Du(t) \end{bmatrix} \tag{11.35}
$$

with an initial condition

$$
\begin{bmatrix} x_1(t_0) \\ x_2(t_0) \end{bmatrix} = \begin{bmatrix} x_1^0 \\ x_2^0 \end{bmatrix}
$$

and

$$
\left\{ \begin{bmatrix} x_1(t) \\ x_2(t) \end{bmatrix} \begin{bmatrix} M_a & \varepsilon M_b \\ \varepsilon M_c & M_d \end{bmatrix} \right\} = \sum_{i=1}^{n_1} x_{1i}(t) \begin{bmatrix} M_{ai} & \varepsilon M_{bi} \\ \varepsilon M_{ci} & 0 \end{bmatrix}
$$
$$
+ \sum_{j=n_1+1}^{n_1+n_2} x_{2(j-n_1)}(t) \begin{bmatrix} 0 & \varepsilon M_{bj} \\ \varepsilon M_{cj} & M_{dj} \end{bmatrix} \tag{11.36}
$$

where

$$
x_1(t) \in R^{n_1}
$$
$$
x_2(t) \in R^{n_2}
$$
$$
u_1(t) \in R^{m_1}
$$
$$
u_2(t) \in R^{m_2}
$$
$$
M_{ai} \in R^{n_1 \times m_1}
$$
$$
M_{bi} \in R^{n_1 \times m_2}
$$
$$
M_{ci} \in R^{n_2 \times m_1}
$$
$$
M_{di} \in R^{n_2 \times m_2},
$$

$x(t) = \begin{bmatrix} x_1^T(t) & x_2^T(t) \end{bmatrix}^T$ is a state variable, $u(t) = \begin{bmatrix} u_1^T(t) & u_2^T(t) \end{bmatrix}^T$ is a control input, $z(t) \in R^q$ is a controlled output, and $\omega(t) = \begin{bmatrix} \omega_1^T(t) & \omega_2^T(t) \end{bmatrix}^T \in R^p$ is an exogenous disturbance. A_i, B_i, H_i, C, D are constant matrices of appropriate dimensions, and ε is a small coupling parameter. In addition, ΔA_i represents the uncertainty in the system and satisfies the following assumption.

Assumption 11.2

$$\begin{bmatrix} \Delta A_1 & \varepsilon \Delta A_2 \\ \varepsilon \Delta A_3 & \Delta A_4 \end{bmatrix} = \begin{bmatrix} E_1 & \varepsilon E_2 \\ \varepsilon E_3 & E_4 \end{bmatrix} Q(t) \begin{bmatrix} F_1 & \varepsilon F_2 \\ \varepsilon F_3 & F_4 \end{bmatrix} \tag{11.37}$$

where
   E_i and F_i are known real constant matrices with appropriate dimensions
   $Q(t)$ is an unknown matrix function with Lebesgue measurable elements such
      that $Q(t)^T Q(t) \leq I$

A quadratic cost functional associated with Equations 11.34 and 11.35 to be minimized has the following form

$$J = \frac{1}{2} \int_0^\infty \left(z^T(t)z(t) - \gamma^2 \omega^T(t)\omega(t) \right) dt \tag{11.38}$$

where γ is a positive design parameter.
   For computational simplification, denote the following notations

$$\tilde{B}(x(t)) = \begin{bmatrix} B_1 & \varepsilon B_2 \\ \varepsilon B_3 & B_4 \end{bmatrix} + \left\{ \begin{bmatrix} x_1(t) \\ x_2(t) \end{bmatrix} \begin{bmatrix} M_a & \varepsilon M_b \\ \varepsilon M_c & M_d \end{bmatrix} \right\}$$

$$= \begin{bmatrix} \tilde{B}_1(x_1(t)) & \varepsilon \tilde{B}_2(x(t)) \\ \varepsilon \tilde{B}_3(x(t)) & \tilde{B}_4(x_2(t)) \end{bmatrix} \tag{11.39}$$

and without loss of generality, we assume that $C^T C = \begin{bmatrix} C_1 & \varepsilon C_2 \\ \varepsilon C_2^T & C_3 \end{bmatrix}$ and $D^T D = I$.
   By the help of van der Schaft (1992) and Xie and Desouza (1991), we can derive the following state-dependent Riccati equation for the weakly coupled bilinear system Equations 11.34 and 11.35 with respect to the performance criterion equation (Equation 11.38).

$$PA + A^T P - P \left\{ \tilde{B}(x(t))\tilde{B}(x(t))^T - \gamma^{-2} HH^T - \sigma EE^T \right\} P$$

$$+ C^T C + \frac{1}{\sigma} F^T F + \delta I = 0 \tag{11.40}$$

where
   $\sigma > 0$ is a design parameter
   δ is a sufficiently small positive constant

Moreover, H_∞ control law is given by

$$u^*(t) = -\tilde{B}(x(t))^{\mathrm{T}} P x(t) \tag{11.41}$$

and the disturbance is given by

$$\omega^*(t) = \gamma^{-2} H^{\mathrm{T}} P x(t) \tag{11.42}$$

where P is partitioned as

$$P = \begin{bmatrix} P_1 & \varepsilon P_2 \\ \varepsilon P_2^{\mathrm{T}} & P_3 \end{bmatrix} \tag{11.43}$$

Setting $\varepsilon^2 = 0$, we can get the following $O(\varepsilon^2)$ approximations:

$$\begin{aligned} S(x(t)) &= \tilde{B}(x(t))\tilde{B}(x(t))^{\mathrm{T}} - \gamma^{-2} H H^{\mathrm{T}} - \sigma E E^{\mathrm{T}} \\ &= \begin{bmatrix} S_1(x_1(t)) & \varepsilon S_2(x(t)) \\ \varepsilon S_2^{\mathrm{T}}(x(t)) & S_3(x_2(t)) \end{bmatrix} \end{aligned} \tag{11.44}$$

$$T = C^{\mathrm{T}} C + \frac{1}{\sigma} F^{\mathrm{T}} F = \begin{bmatrix} T_1 & \varepsilon T_2 \\ \varepsilon T_2^{\mathrm{T}} & T_3 \end{bmatrix} \tag{11.45}$$

Partitioning the state-dependent Riccati equation (Equation 11.40) according to Equations 11.44 and 11.45, and setting $\varepsilon^2 = 0$, we get an $O(\varepsilon^2)$ approximation of Equation 11.40 in terms of two reduced-order, decoupled Riccati equations:

$$P_1 A_1 + A_1^{\mathrm{T}} P_1 - P_1 S_1(x_1(t)) P_1 + T_1 + \delta I = 0 \tag{11.46}$$

$$P_3 A_4 + A_4^{\mathrm{T}} P_3 - P_3 S_3(x_2(t)) P_3 + T_3 + \delta I = 0 \tag{11.47}$$

and nonsymmetric Riccati equation with no input and no disturbance:

$$\begin{aligned} \{A_1 - S_1(x_1(t)) P_1\}^{\mathrm{T}} P_2 &+ P_2 \{A_4 - S_3(x_2(t)) P_3\} \\ &+ P_1 A_2 + A_3^{\mathrm{T}} P_3 - P_1 S_2(x(t)) P_3 + T_2 = 0 \end{aligned} \tag{11.48}$$

A detailed description of reduced-order scheme can be found in Aganović and Gajić (1995). Since Equations 11.46 and 11.47 are state-dependent Riccati equations, they have no analytical solution.

Focusing on the nonlinear H_∞ control in this section, we deal with HJI equations rather than Riccati equations. The HJI equations corresponding to Equations 11.46 and 11.47 are given by

$$\frac{\partial J_1^{\mathrm{T}}}{\partial x_1} A_1 x_1(t) + \frac{1}{2} x_1^{\mathrm{T}}(t)(T_1 + \delta I) x_1(t) - \frac{1}{2} \frac{\partial J_1^{\mathrm{T}}}{\partial x_1} S_1(x_1(t)) \frac{\partial J_1}{\partial x_1} = 0 \tag{11.49}$$

$$\frac{\partial J_2^{\mathrm{T}}}{\partial x_2} A_4 x_2(t) + \frac{1}{2} x_2^{\mathrm{T}}(t)(T_3 + \delta I)x_2(t) - \frac{1}{2}\frac{\partial J_2^{\mathrm{T}}}{\partial x_2} S_3(x_2(t))\frac{\partial J_2}{\partial x_2} = 0 \qquad (11.50)$$

where
$$\partial J_1 / \partial x_1 = P_1 x_1(t)$$
$$\partial J_2 / \partial x_2 = P_3 x_2(t)$$

Moreover, denoting $\partial J_3/\partial x_1 = P_2 x_2(t)$ and $\{\partial J_3/\partial x_2\}^{\mathrm{T}} = x_1^{\mathrm{T}}(t)P_2$, we obtain the following equation equivalent to Equation 11.48 after substitutions:

$$x_1^{\mathrm{T}}(t)\{A_1 - S_1(x_1(t))P_1\}^{\mathrm{T}}\frac{\partial J_3}{\partial x_1} + \frac{\partial J_3^{\mathrm{T}}}{\partial x_2}\{A_4 - S_3(x_2(t))P_3\}x_2(t)$$
$$x_1^{\mathrm{T}}(t)\{P_1 A_2 + A_3^{\mathrm{T}}P_3 - P_1 S_2(x(t))P_3 + T_2\}x_2(t) = 0 \qquad (11.51)$$

Unfortunately, they still have no analytical solution. However, we can obtain approximate solutions of Equations 11.49 and 11.50 using successive Galerkin approximation. If the solutions of Equations 11.49 and 11.50 are found, then the solution of Equation 11.51 can easily be found using the Galerkin approximation.

11.3.2 Design of H_∞ Control Law for Weakly Coupled Bilinear Systems with Parameter Uncertainties Using SGA

In order to design the H_∞ control law $u^*(t)$, we present the scheme to find solutions of Equations 11.49 and 11.50 using the SGA method.

Under Assumption 11.1, the successive approximation, which is a duel iteration in policy space to solve HJI equations, is proposed as follows.

ALGORITHM 11.2:   Duel Successive Approximation

Let an initial control law, $\tilde{u}_1^{(0)}(t): R^{m_1} \times \Omega_1 \to R$, be stabilizing for the system $\dot{x}_1(t) = A_1 x_1(t) + \tilde{B}(x_1(t))\tilde{u}_1(x_1(t))$ with no uncertainty and no disturbance (i.e., $\Delta A_1 = 0$, $\omega_1^{(0,0)} = 0$).
   Obtain $J_1^{(1,0)}$ from

$$\frac{\partial J_1^{(1,0)^{\mathrm{T}}}}{\partial x_1}\{A_1 x_1(t) + \tilde{B}_1(x_1(t))\tilde{u}_1^{(0)}(t)\} + \frac{1}{2}x_1^{\mathrm{T}}(t)C_1 x_1(t) + \frac{1}{2}\tilde{u}_1^{(0)^{\mathrm{T}}}(t)\tilde{u}_1^{(0)}(t) = 0 \quad (11.52)$$

While $\left\| J_1^{(i,j)} - J_1^{(i-1,j)} \right\| > \alpha$
   Set $j = 0$ and $\omega_1^{(i,0)}(t) = 0$.
   While $\left\| J_1^{(i,j)} - J_1^{(i,j-1)} \right\| > \alpha$

Obtain $J_1^{(i,j)}$ from the GHJI equation defined as

$$\frac{\partial J_1^{(i,j)^T}}{\partial x_1} A_1 x_1(t) + \frac{1}{2} \frac{\partial J_1^{(i,j-1)^T}}{\partial x_1} S_1(x_1(t)) \frac{\partial J_1^{(i,j-1)}}{\partial x_1}$$

$$+ \frac{1}{2} x_1^T(t)(T_1 + \delta I)x_1(t) - \frac{\partial J_1^{(i,j)^T}}{\partial x_1} S_1(x_1(t)) \frac{\partial J_1^{(i,j-1)}}{\partial x_1} = 0 \qquad (11.53)$$

Update the disturbance:

$$\tilde{\omega}_1^{(i,j+1)}(t) = \gamma^{-2} H_1^T \frac{\partial J_1^{(i,j)}}{\partial x_1}$$

Set $j = j + 1$.
    End j loop.
    Update the control law:

$$\tilde{u}_1^{(i+1)}(t) = -\tilde{B}_1(x_1(t))^T \frac{\partial J_1^{(i,j)}}{\partial x_1}$$

Set $i = i + 1$.
    End i loop.

Since the GHJI equation (Equation 11.53) is linear partial differential equation, it is still difficult to solve. In this section, we seek an approximate solution of this equation using Galerkin's projection method. A detailed description of the SGA method can be found in Beard (1995) and Kim et al. (2003a,b).

Given an initial control $\tilde{u}_1^{(0)}(t)$, we compute an approximation to its cost $J_{1N_1}^{(0,0)} = c_{1N_1}^{T(0,0)} \Phi_{1N_1}$ where $c_{1N_1}^{(0,0)}$ is the solution of Galerkin approximation of Equation 11.52, i.e.

$$a_1^{(0,0)} c_{1N_1}^{(0,0)} + b_1^{(0,0)} = 0 \qquad (11.54)$$

where

$$a_1^{(0,0)} = \left\langle \nabla \Phi_{1N_1} A_1 x_1(t), \Phi_{1N_1} \right\rangle_{\Omega_1} + \left\langle \nabla \Phi_{1N_1} \tilde{B}_1(x_1(t)) \tilde{u}_1^{(0)}(t), \Phi_{1N_1} \right\rangle_{\Omega_1}$$

$$b_1^{(0,0)} = \frac{1}{2} \left\langle x_1^T(t)(T_1 + \delta I)x_1(t), \Phi_{1N_1} \right\rangle_{\Omega_1} + \frac{1}{2} \left\langle \tilde{u}_1^{T(0)}(t) \tilde{u}_1^{(0)}(t), \Phi_{1N_1} \right\rangle_{\Omega_1}$$

After duel iterative steps, we can obtain the approximation to its cost $J_{1N_1}^{(i,j)} = c_{1N_1}^{T(i,j)} \Phi_{1N_1}$ where $c_{1N_1}^{(i,h)}$ is the solution of Galerkin approximation of GHJB equation (Equation 11.53), i.e.

$$a_1^{(i,j)} c_{1N_1}^{(i,j)} + b_1^{(i,j)} = 0 \qquad (11.55)$$

where

$$
\begin{aligned}
a_1^{(i,j)} &= \left\langle \nabla\Phi_{1N_1} A_1 x_1(t),\ \Phi_{1N_1} \right\rangle_{\Omega_1} \\
&\quad - \left\langle \nabla_1\Phi_{1N_1} S_1(x_1(t))\nabla_1\Phi_{1N_1}^{\mathrm{T}} \mathbf{c}_{1N_1}^{(i-1,j-1)},\ \Phi_{1N_1} \right\rangle_{\Omega_1} \\
b_1^{(i,j)} &= \frac{1}{2}\left\langle x_1^{\mathrm{T}}(t)(T_1 + \delta I)x_1(t),\ \Phi_{1N_1} \right\rangle_{\Omega_1} \\
&\quad + \frac{1}{2}\left\langle \mathbf{c}_{1N_1}^{\mathrm{T}(i-1,j-1)} \nabla_1\Phi_{1N_1} S_1(x_1(t))\nabla_1\Phi_{1N_1}^{\mathrm{T}} \mathbf{c}_{1N_1}^{(i-1,j-1)},\ \Phi_{1N_1} \right\rangle_{\Omega_1}
\end{aligned}
$$

Moreover, we can obtain the updated disturbance that is based on the approximated solution, $J_{1N_1}^{(i,j)}$:

$$
\tilde{\omega}_{1N_1}^{(i,j+1)}(t) = \gamma^{-2} H_1^{\mathrm{T}} \frac{\partial J_{1N_1}^{(i,j)}}{\partial x_1} = \gamma^{-2} H_1^{\mathrm{T}} \nabla_1\Phi_{1N_1}^{\mathrm{T}} \mathbf{c}_{1N_1}^{(i,j)} \tag{11.56}
$$

and the updated control law:

$$
\tilde{u}_{1N_1}^{(i+1)}(t) = -\tilde{B}_1(x_1(t))^{\mathrm{T}} \frac{\partial J_{1N_1}^{(i,j)}}{\partial x_1} = -\tilde{B}_1(x_1(t))^{\mathrm{T}} \nabla_1\Phi_{1N_1}^{\mathrm{T}} \mathbf{c}_{1N_1}^{(i,j)} \tag{11.57}
$$

Similarly, given an initial control $u_2^{(0,0)}(t)$, we can compute an approximation to its cost $J_{2N_2}^{(i,j)} = \mathbf{c}_{2N_2}^{\mathrm{T}(i,j)} \Phi_{2N_2}$.

The following theorem shows the existence of a unique solution of SGA.

THEOREM 11.3

Suppose that $\{\phi_k\}_1^N$ is linearly independent and $\partial\phi_k/\partial x \neq 0$, then there exists an unique solution, $\mathbf{c}_N$.

Proof Suppose that $\{\phi_k\}_1^N$ is linearly independent, then Φ_N is linearly independent. Suppose $\partial\phi_k/\partial x \neq 0$, such that $\nabla\Phi_N \neq 0$, then linearly independent Φ_N implies that $\nabla\Phi_N$ is linearly independent. This implies that $\langle \nabla\Phi_N A x(t),\ \Phi_N \rangle_{\Omega} - \langle \nabla\Phi_N S(x(t))\nabla\Phi_N^{\mathrm{T}}\mathbf{c}_N,\ \Phi_N \rangle_{\Omega}$ is invertible. This implies that $a_1^{(i,j)}$ is invertible in Equation 11.55 for all i and j. Therefore, there exists an unique solution to a linear equation (Equation 11.55). ∎

From the solutions of Galerkin approximations of Equations 11.49 and 11.50, P_1 and P_3 can be determined. Then, we can obtain the approximate solution of Equation 11.51.

Defining $J_{3N_3} = \mathbf{c}_{3N_3}^{\mathrm{T}} \Phi_{N_3}$, we denote that $\partial J_3/\partial x_1 = \nabla_1\Phi_{3N_3}^{\mathrm{T}} \mathbf{c}_{3N_3}$ and $\partial J_3/\partial x_2 = \nabla_2\Phi_{3N_3}^{\mathrm{T}} \mathbf{c}_{3N_3}$. Using these notations, we can derive the Galerkin approximation of Equation 11.51 as follows

$$
a_3 \mathbf{c}_{3N_3} + b_3 = 0 \tag{11.58}
$$

where

$$a_3 = \left\langle \nabla_1 \Phi_{3N}\{A_1 - S_1(x_1(t))P_1\}x_1(t), \ \Phi_{3N} \right\rangle_{\Omega_3}$$
$$+ \left\langle \nabla_2 \Phi_{3N}\{A_4 - S_3(x_2(t))P_3\}x_2(t), \ \Phi_{3N} \right\rangle_{\Omega_3}$$
$$b_3 = \left\langle x_1^T(t)\{P_1A_2 + A_3^T P_3 - P_1S_2(x(t))P_3 + T_2\}x_2(t), \ \Phi_{3N} \right\rangle_{\Omega_3}$$

In this case, $\Omega_3 = \Omega_1 \cup \Omega_2$, and P_3 can be determined without an iterative step.

Hence, we propose a new algorithm which is used to design an H_∞ control law with two independent reduced-order HJB equations (Equations 11.49 and 11.50) and Equation 11.51 using the SGA method for weakly coupled bilinear systems with time-varying parameter uncertainties and exogenous disturbance.

ALGORITHM 11.3:   Duel Successive Galerkin Approximation

Let an initial control law, $\tilde{u}_1^{(0)}(t): R^{m_1} \times \Omega_1 \to R$, be stabilizing for the system $\dot{x}_1(t) = A_1x_1(t) + \tilde{B}(x_1(t))\tilde{u}_1(x_1(t))$ with no uncertainty and no disturbance (i.e., $\Delta A_1 = 0, \omega_1^{(0,0)}(t) = 0$).

Compute

$$a_1^{(0,0)} = \left\langle \nabla_1 \Phi_{1N_1} A_1 x_1(t), \ \Phi_{1N_1} \right\rangle_{\Omega_1} + \left\langle \nabla_1 \Phi_{1N_1} \tilde{B}_1(x_1(t))\tilde{u}_1^{(0)}(t), \ \Phi_{1N_1} \right\rangle_{\Omega_1}$$
$$b_1^{(0,0)} = \frac{1}{2}\left\langle x_1^T(t)C_1x_1(t), \ \Phi_{1N_1} \right\rangle_{\Omega_1} + \frac{1}{2}\left\langle \tilde{u}_1^{T(0)}(t)R_1\tilde{u}_1^{(0)}(t), \ \Phi_{1N_1} \right\rangle_{\Omega_1}$$

and

$$a_2^{(0,0)} = \left\langle \nabla_2 \Phi_{2N_2} A_4 x_2(t), \ \Phi_{2N_2} \right\rangle_{\Omega_2} + \left\langle \nabla_2 \Phi_{2N_2} \tilde{B}_3(x_2(t))\tilde{u}_2^{(0)}(t), \ \Phi_{2N_2} \right\rangle_{\Omega_2}$$
$$b_2^{(0,0)} = \frac{1}{2}\left\langle x_2^T(t)C_3x_2(t), \ \Phi_{2N_2} \right\rangle_{\Omega_2} + \frac{1}{2}\left\langle \tilde{u}_2^{T(0)}(t)R_2\tilde{u}_2^{(0)}(t), \ \Phi_{2N_2} \right\rangle_{\Omega_2}$$

Find $c_{1N_1}^{(0,0)}$ and $c_{2N_2}^{(0,0)}$ satisfying the following linear equations:

$$a_1^{(0,0)} c_{1N_1}^{(0,0)} + b_1^{(0,0)} = 0$$
$$a_2^{(0,0)} c_{2N_2}^{(0,0)} + b_2^{(0,0)} = 0$$

Routine for P_1
While $\| c_{1N_1}^{(i,j)} - c_{1N_1}^{(i-1,j)} \| > \alpha$

Set $j = 0$ and $\omega_1^{(i,0)}(t) = 0$.
While $\| c_{1N_1}^{(i,j)} - c_{1N_1}^{i,j-1)} \| > \alpha$
Compute

$$a_1^{(i,j)} = \left\langle \nabla \Phi_{1N_1} A_1 x_1(t), \ \Phi_{1N_1} \right\rangle_{\Omega_1}$$
$$- \left\langle \nabla_1 \Phi_{1N_1} S_1(x_1(t)) \nabla_1 \Phi_{1N_1}^T c_{1N_1}^{(i,j-1)}, \ \Phi_{1N_1} \right\rangle_{\Omega_1}$$

$$b_1^{(i,j)} = \frac{1}{2} \left\langle x_1^T(t)(T_1 + \delta I)x_1(t), \, \Phi_{1N_1} \right\rangle_{\Omega_1}$$
$$+ \frac{1}{2} \left\langle c_{1N_1}^{T(i,j-1)} \nabla_1 \Phi_{1N_1} S_1(x_1(t)) \nabla_1 \Phi_{1N_1}^T c_{1N_1}^{(i,j-1)}, \, \Phi_{1N_1} \right\rangle_{\Omega_1}$$

Find $c_{1N_1}^{(i,j)}$ satisfying the following linear equation:

$$a_1^{(i,j)} c_{1N_1}^{(i,j)} + b_1^{(i,j)} = 0$$

Update the disturbance:

$$\tilde{\omega}_{1N_1}^{(i,j+1)}(t) = \gamma^{-2} H_1^T \nabla_1 \Phi_{1N_1}^T c_{1N_1}^{(i,j)}$$

Set $j = j + 1$.
    End j loop.
    Update the control law:

$$\tilde{u}_{1N_1}^{(i+1)}(t) = -\tilde{B}_1^T(x_1(t)) \nabla_1 \Phi_{1N_1}^T c_{1N_1}^{(i,j)}$$

Set $i = i + 1$.
    End i loop.
    Determine P_1.

Routine for P_3
While $\| c_{2N_2}^{(i,j)} - c_{2N_2}^{(i-1,j)} \| > \alpha$
    Set $j = 0$ and $\omega_2^{(i,0)}(t) = 0$.
    While $\| c_{1N_1}^{(i,j)} - c_{1N_1}^{(i,j-1)} \| > \alpha$
    Compute

$$a_2^{(i,j)} = \left\langle \nabla_2 \Phi_{2N_2} A_4 x_2(t), \, \Phi_{2N_2} \right\rangle_{\Omega_2}$$
$$- \left\langle \nabla_2 \Phi_{2N_2} S_3(x_2(t)) \nabla_2 \Phi_{2N_2}^T c_{2N_2}^{(i,j-1)}, \, \Phi_{2N_2} \right\rangle_{\Omega_2}$$
$$b_2^{(i,j)} = \frac{1}{2} \left\langle x_2^T(t)(T_1 + \delta I)x_2(t), \, \Phi_{2N_2} \right\rangle_{\Omega_2}$$
$$+ \frac{1}{2} \left\langle c_{2N_2}^{T(i,j-1)} \nabla_2 \Phi_{2N_2} S_3(x_2(t)) \nabla_2 \Phi_{2N_2}^T c_{2N_2}^{(i,j-1)}, \, \Phi_{2N_2} \right\rangle_{\Omega_2}$$

Find $c_{2N_2}^{(i,j)}$ satisfying the following linear equation:

$$a_2^{(i,j)} c_{2N_2}^{(i,j)} + b_2^{(i,j)} = 0$$

Update the disturbance:

$$\tilde{\omega}_{2N_2}^{(i,j+1)} = \gamma^{-2} H_4^T \nabla_2 \Phi_{2N_2}^T c_{2N_2}^{(i,j)}$$

Set $j = j + 1$.
$\quad$ End j loop.
$\quad$ Update the control law:

$$\tilde{u}_{2N_2}^{(i+1)}(t) = -\tilde{B}_4^{\mathrm{T}}(x_2(t)) \nabla_2 \Phi_{2N_2}^{\mathrm{T}} \mathbf{c}_{2N_2}^{(i,j)}$$

Set $i = i + 1$.
$\quad$ End i loop.
$\quad$ Determine P_3.

Routine for P_2
Compute

$$a_3 = \left\langle \nabla_1 \Phi_{3N} \{ A_1 - S_1(x_1(t)) P_1 \} x_1(t), \, \Phi_{3N} \right\rangle_{\Omega_3}$$
$$+ \left\langle \nabla_2 \Phi_{3N} \{ A_4 - S_3(x_2(t)) P_3 \} x_2(t), \, \Phi_{3N} \right\rangle_{\Omega_3}$$
$$b_3 = \left\langle x_1^{\mathrm{T}}(t) \{ P_1 A_2 + A_3^{\mathrm{T}} P_3 - P_1 S_2(x(t)) P_3 + T_2 \} x_2(t), \, \Phi_{3N} \right\rangle_{\Omega_3}$$

Find $\mathbf{c}_3 N_3$ satisfying the following linear equation:

$$a_3 \mathbf{c}_3 N_3 + b_3 = 0$$

Determine P_2, then the approximate parallel H_∞ control law is given by

$$u_{pN}(t) = -\tilde{B}^{\mathrm{T}} \begin{bmatrix} P_1 & \varepsilon P_2 \\ \varepsilon P_2^{\mathrm{T}} & P_3 \end{bmatrix} x(t)$$

The following theorem shows that the approximate parallel H_∞ control law, $u_{pN}(t)$, designed by the proposed algorithm converges to the H_∞ optimal control law, $u^*(t)$.

THEOREM 11.4

For any small positive constant β, we can choose N for a sufficiently large i to satisfy that

$$\| u^*(t) - u_{pN}^{(i)}(t) \| < \beta \tag{11.59}$$

Proof It was proved that $u^*(t)$ converges to $u_N(t)$ pointwise on Ω for finite N in Beard (1995), where $u_N(t)$ is a control law designed using the SGA. It implies that for a sufficiently large i, we can choose N satisfying $\| u_p(t) - u_{pN}^{(i)}(t) \| < \tilde{\beta}$, where $u_p(t)$ is the parallel H_∞ control law obtained by the reduced order scheme for weakly coupled bilinear systems and $\tilde{\beta}$ is a small positive constant. By the help of weakly coupling theory, $u_p(t) = u^*(t) + O(\varepsilon^2)$. This implies that for any small positive constant β, we can choose N for a sufficiently large i satisfying Equation 11.59. ∎

11.3.3 CASE STUDY: A PAPER MAKING MACHINE

In order to demonstrate the efficiency of the proposed method for the parallel H_∞ control for weakly coupled bilinear systems with time-varying parameter uncertainties and exogenous disturbance using the SGA, we have run a fourth-order real example, a paper making machine control problem reported in Ying et al. (1992).

The problem matrices have the following values:

$$A = \begin{bmatrix} -1.93 & 0 & 0 & 0 \\ 0.394 & -0.426 & 0 & 0 \\ 0 & 0 & -0.63 & 0 \\ 0.095 & -0.103 & 0.413 & -0.426 \end{bmatrix}$$

$$B = \begin{bmatrix} 1.274 & 1.274 \\ 0 & 0 \\ 1.34 & -0.65 \\ 0 & 0 \end{bmatrix}, \quad M_1 = \begin{bmatrix} 0 & 0 \\ 0 & 0 \\ 0.755 & 0.366 \\ 0 & 0 \end{bmatrix}$$

$$M_2 = M_4 = \begin{bmatrix} 0 & 0 \\ 0 & 0 \\ 0 & 0 \\ 0 & 0 \end{bmatrix}, \quad M_3 = \begin{bmatrix} 0 & 0 \\ 0 & 0 \\ -0.718 & -0.718 \\ 0 & 0 \end{bmatrix}$$

$$C^T C = \begin{bmatrix} 1 & 0 & 0.13 & 0 \\ 0 & 1 & 0 & 0.09 \\ 0.13 & 0 & 0.1 & 0 \\ 0 & 0.09 & 0 & 0.2 \end{bmatrix}, \quad H = \begin{bmatrix} 1 & 0 \\ 0 & 0 \\ 0 & 1 \\ 0 & 0 \end{bmatrix}$$

Initial states are chosen as $x(t_0) = [3.7 \ 3.2 \ 4 \ 2.8]^T$, time-varying parameter uncertainties are chosen as $1.2 \sin(0.5\pi t)I$, and exogenous disturbance is chosen as $[0.4 \sin(\pi t) - 0.7 \cos(\pi t) \ 0.8 \cos(\pi t) - 0.6 \sin(\pi t)]^T$. The simulation results are presented in Figures 11.6 through 11.10, where the dashed lines are the trajectories that are obtained from full-order SGA method and the solid lines are the trajectories that are obtained from the proposed algorithm. Figure 11.10 shows that the performance criterion trajectory of the proposed algorithm is better than that of the full-order SGA method, because errors of the full-order SGA method are bigger than those of the proposed algorithm. In the full-order SGA method, eight-dimensional basis are used and four-dimensional integrals of $8 \times (1 + 8 + 64) = 584$ times are performed. But, in the proposed algorithm, we can use only three-dimensional basis and compute two-dimensional integrals of $3 \times (1 + 3 + 9) = 39$ times for each reduced-order problem in parallel, and compute four-dimensional integrals of $8 \times (1 + 8) = 72$

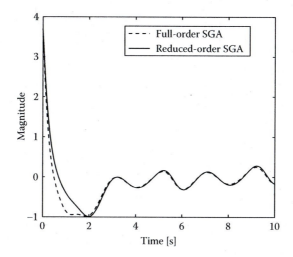

FIGURE 11.6 Trajectories of $x_1(t)$.

times based on eight-dimensional basis for the problem according to Equation 11.16. Therefore, the computational complexity is greatly reduced.

11.4 CONCLUSION

We have presented the closed-loop optimal control scheme for weakly coupled bilinear systems using the successive Galerkin approximation. In addition, we have studied the closed-loop H_∞ control scheme for weakly coupled bilinear systems with

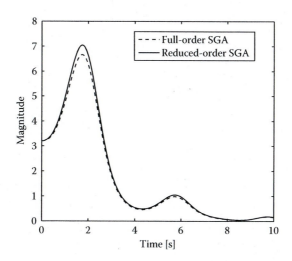

FIGURE 11.7 Trajectories of $x_2(t)$.

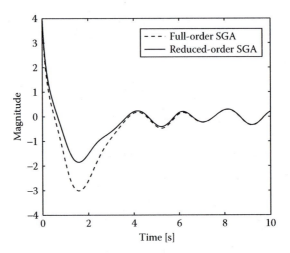

FIGURE 11.8   Trajectories of $x_3(t)$.

time-varying parameter uncertainties and exogenous disturbance and developed a new algorithm using the duel successive Galerkin approximation for the scheme. The difficulty of the SGA method is a computational complexity, but in the proposed algorithms, it can be greatly reduced. The presented simulation results for a fourth-order real example, a paper making machine control problem, show that the performance trajectories of the proposed algorithms are better than those of the full-order SGA method. Furthermore, it should be noted that the proposed algorithms are more effective than the full-order SGA method.

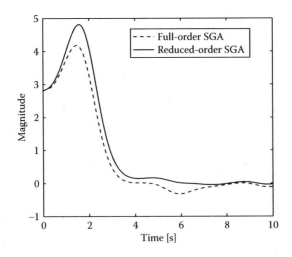

FIGURE 11.9   Trajectories of $x_4(t)$.

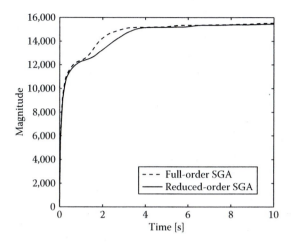

FIGURE 11.10 Values of the performance criterion.

REFERENCES

Z. Aganović and Z. Gajić, Optimal control of weakly coupled bilinear systems, *Automatica*, 29, 1591–1593, 1993.

Z. Aganović and Z. Gajić, *Linear Optimal Control of Bilinear Systems: With Applications to Singular Perturbations and Weak Coupling*, Springer, London, UK, 1995.

R. Beard, *Improving the Closed-Loop Performance of Nonlinear Systems*, PhD dissertation, Rensselaer Polytechnic Institute, Troy, New York, 1995.

R. Beard, G. Saridis, and J. Wen, Galerkin approximation of the generalized Hamilton–Jacobi–Bellman equation, *Automatica*, AC-33, 2159–2177, 1996.

R. Beard and T. McLain, Successive Galerkin approximation algorithms for nonlinear optimal and robust control, *International Journal of Control*, 71, 717–743, 1998.

W. Cebuhar and V. Costanza, Approximation procedures for the optimal control for bilinear and nonlinear systems, *Journal of Optimization Theory and Applications*, 43, 615–627, 1984.

J.C. Doyle, K. Glover, P.P. Khargonekar, and B.A. Francis, State space solution to standard H_2 and H_∞ control problems, *IEEE Transactions on Automatic Control*, AC-34, 831–846, 1989.

G. Figalli, M. Cava, and L. Tomasi, An optimal feedback control for a bilinear model of induction motor drives, *International Journal of Control*, 39, 1007–1016, 1984.

Z. Gajić and X. Shen, Decoupling transformation for weakly coupled linear systems, *International Journal of Control*, 50, 1515–1521, 1989.

Z. Gajić and X. Shen, *Parallel Algorithms for Optimal Control of Large Scale Linear Systems*, Springer, London, UK, 1992.

E. Hoffer and B. Tibken, An iterative method for the finite-time bilinear quadratic control problem, *Journal of Optimization Theory and Applications*, 57, 411–427, 1988.

Y.J. Kim, B.S. Kim, and M.T. Lim, Composite control for singularly perturbed nonlinear systems via successive Galerkin approximation, *Dynamics of Continuous, Discrete, and Impulsive Systems, Series B: Applications and Algorithms*, 10, 247–258, 2003a.

Y.J. Kim, B.S. Kim, and M.T. Lim, Composite control for singularly perturbed bilinear systems via successive Galerkin approximation, *Proceedings of IEE, Part D, Control Theory and Application*, 150, 483–488, 2003b.

Y.J. Kim and M.T. Lim, Parallel robust H_∞ control for weakly coupled bilinear systems with parameter uncertainties using successive Galerkin approximation, *International Journal of Control, Automation, and Systems*, 4, 689–696, 2006.

Y.J. Kim and M.T. Lim, Parallel optimal control for weakly coupled bilinear systems using successive Galerkin approximation, *IET Control Theory and Applications*, 1, 909–914, 2007.

D. Kirk, *Optimal Control Theory*, Prentice-Hall, Englewood Cliffs, NJ, 1970.

P. Kokotović, W. Perkins, J. Cruz, and G. D'Ans, ε-coupling for near-optimum design of large scale linear systems, *Proceedings of IEE, Part D, Control Theory and Application*, 116, 889–892, 1969.

R. Mohler, *Nonlinear systems—Applications to Bilinear Control*, Prentice-Hall, Englewood Cliffs, NJ, 1991.

A. van der Schaft, L_2-gain analysis of nonlinear systems and nonlinear state-feedback H_∞ control, *IEEE Transactions on Automatic Control*, AC-37, 770–784, 1992.

L. Xie and C.E. Desouza, Robust H_∞ control for class of uncertain linear time-invariant systems, *Proceedings of IEE, Part D, Control Theory and Application*, 138, 479–483, 1991.

L. Xie and C.E. Desouza, Robust H_∞ control for linear systems with norm-bounded time-varying uncertainty, *IEEE Transactions on Automatic Control*, AC-37, 1188–1191, 1992.

Y. Ying, M. Rao, and X. Shen, Bilinear decoupling control and its industrial application, *Proceedings of American Control Conference*, Chicago, 1163–1167, 1992.

K. Zhou, J. Doyle, and K. Glover, *Robust and Optimal Control*, Prentice-Hall, New York, 1996.

Index